X-rays in atomic and nuclear physics

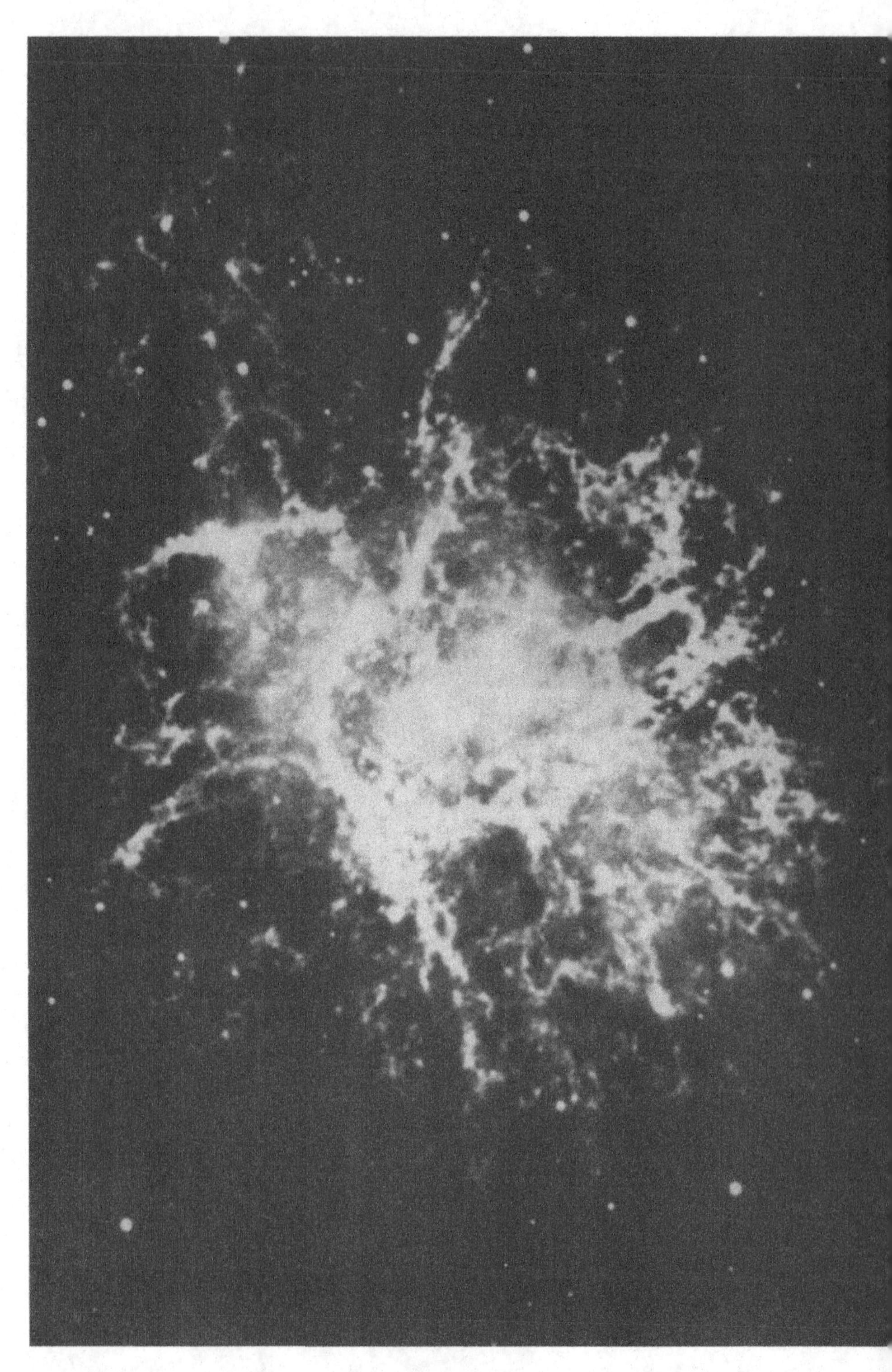

The frontispiece is a photograph of the Crab nebula. This consists of the debris followin
supernova explosion in AD 1054, and contains sources of radio- and X-ray emission as wel
visible light. The characteristics of its X-radiation are referred to in section 8.5 of this bo

X-RAYS

in atomic and nuclear physics

Second edition

N.A. DYSON

Senior Lecturer,
School of Physics and Space Research,
University of Birmingham

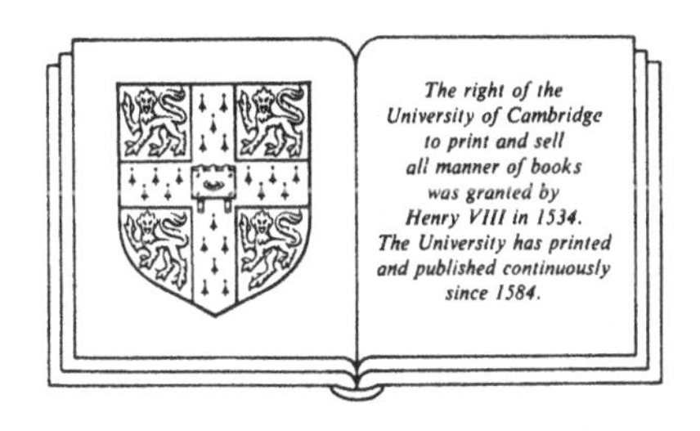

CAMBRIDGE UNIVERSITY PRESS

Cambridge

New York Port Chester Melbourne Sydney

CAMBRIDGE UNIVERSITY PRESS
Cambridge, New York, Melbourne, Madrid, Cape Town, Singapore, São Paulo

Cambridge University Press
The Edinburgh Building, Cambridge CB2 2RU, UK

Published in the United States of America by Cambridge University Press, New York

www.cambridge.org
Information on this title: www.cambridge.org/9780521262804

First published by Longman Group Limited 1973
Second edition published by Cambridge University Press 1990
This digitally printed first paperback version 2005

A catalogue record for this publication is available from the British Library

Library of Congress Cataloguing in Publication data

Dyson, N.A. (Norman Allen), 1929–
X-rays in atomic and nuclear physics/N.A. Dyson. – 2nd ed.
p. cm.
Bibliography: p.
Includes index.
ISBN 0 521 26280 1
1. X-rays. I. Title.
QC481.D95 1990
539.7′222 – dc 19 88-38330 CIP

ISBN-13 978-0-521-26280-4 hardback
ISBN-10 0-521-26280-1 hardback

ISBN-13 978-0-521-01722-0 paperback
ISBN-10 0-521-01722-X paperback

To Elizabeth

Contents

Preface to the second edition

In the second edition, new material on characteristic X-ray production has been added in Chapter 3. During the intervening period the study of X-ray production by proton bombardment has been actively pursued, and this has merited the inclusion of a new chapter (Chapter 6). Chapter 8 has been extended and up-dated.

The S.I. units are now in very widespread use, and this has justified the conversion of the whole book into this system of units. Occasionally, in topics where the old system of units is particularly entrenched, older units have been retained.

Although many aspects of X-ray physics have expanded very considerably since the first edition of this book appeared, I have attempted to keep the book within reasonable proportions, and hope that not too much of the work of recent years has been omitted.

I am indebted to Mrs Erica Gaize, Mrs Pauline Goddard and Mrs Eileen Shinn for re-typing the whole work, and to many friends and colleagues over the years for discussion of X-ray processes. I am especially indebted to Dr R. Sokhi, Dr R.G. Harris, and Dr M. Church for reading Chapters 6, 7, and parts of Chapter 8 respectively, and for their valuable comments. My thanks are due also to Longman Group UK Limited, publishers of the first edition of this work, for permission to reproduce here a substantial number of line diagrams which appeared originally in that edition.

The need for up-to-date information on X-ray processes is now very wide indeed, and I hope that this book will go some way towards meeting this need.

Birmingham, January 1988 N.A. Dyson

Preface to the first edition

(ABRIDGED)

The main purpose of this book is to give a concise account of the production and properties of X-rays. This may at first sight seem a rather restricted aim, but, as the title of the book indicates, its coverage extends beyond the 'conventional' aspects of the subject to include other branches of physics in which X-rays play an important rôle. I have brought together several of the areas of physics in which X-rays are encountered, because it has become very evident in recent years that much of the great body of X-ray knowledge acquired several decades ago is highly relevant to other fields such as the recent developments in radioactivity, plasma physics and astrophysics.

In a book of moderate size it is not possible to give a comprehensive treatment, in depth, of the whole of X-ray physics. But I have included a reasonably full account of the continuous X-ray spectrum at low and medium energies, and have also described the production of characteristic X-rays (by electron bombardment) in sufficient detail to enable the research workers to make useful predictions about what is happening or is likely to happen in a wide variety of circumstances when electrons impinge on matter.

X-ray research necessarily involves the use of radiation detectors, and an account of some of these techniques is given. But X-ray physics is inseparably linked, through the photoelectric effect and the Compton effect, with electron physics, and the whole subject of β-ray spectrometry would therefore demand treatment in any text purporting to be comprehensive. A line has to be drawn somewhere, and I have excluded this subject, and also the subject of photographic emulsions, from my terms of reference. Furthermore, no attempt has been made to cover the many important uses to which X-rays are put the worker with X-rays is already well-furnished with books on X-ray crystallography, industrial radiography, and the physics of diagnostic radiology and radiotherapy. The present work is designed rather to meet the needs of the X-ray worker who wishes to study the basic physics of X-rays in more detail than is usually found in books dealing with applications thereof. It is hoped that the book will be of interest also to those whose work lies in the field of radioactivity, and to graduate students who wish simply to know more about X-rays than they can conveniently find elsewhere. However, sufficient reference is made to the actual applications of X-rays to give the book an essentially practical outlook, and I have incorporated a good deal of experimental data throughout the whole book.

The many bibliographic references have been arranged by chapters rather than in one mammoth alphabetical list. In this way it becomes possible to see at a glance the general area of X-ray physics to which a reference belongs, and renders the bibliography rather more useful than might otherwise be the case.

A word about Chapter 8 no observer of the current X-ray scene can fail to be impressed and stimulated by the pace at which certain areas are being studied. To provide a full bibliography for X-ray astronomy, for example, would be a large task in itself. I have felt that this subject should be included if only to convey in a general way something of the excitement of current X-ray research in this area. The subject of X-ray microscopy and micro-analysis also merits at least a brief account, if only to lead the reader into the comprehensive specialised texts on the subject.

I should like to thank Professor P.B. Moon, FRS, and Professor W.E. Burcham, FRS for their interest and encouragement throughout this project. I am indebted to Professor A.J.C. Wilson, FRS and Dr* P.J. Black, and also again to Professor Moon and Professor Burcham for reading and criticising substantial sections of the text. I wish to thank Dr* G.R. Isaak and Miss R. Barany for translations of certain papers in the German literature, and owe a special word of gratitude to Miss (Dr) P. Aspden for checking the tabulated data in Appendix 3 and for preparing the decay schemes illustrated there.

Finally, I should like to express my thanks to Dr V.E. Cosslett, FRS, of the Cavendish Laboratory, Cambridge for his advice and interest. He introduced me to the study of X-ray physics in 1952, and provided me with a basis from which at least some of my subsequent interests have stemmed.

Birmingham, March 1972 N.A. Dyson

* Now Professor

Acknowledgements

We are indebted to the following publishers and societies for permission to reproduce certain illustrations and tables from their copyright publications: Longman Group UK Limited (figures and tables from the first edition); The Royal Society (figs. 3.23, 4.24); The Institute of Physics (figs. 3.13, 3.21, 3.26, 3.27, 3.28, 4.5, 4.15, 4.28, 5.12, 8.8; tables 3.9 and 3.10); The North-Holland Publishing Company, Amsterdam (fig. 4.17). Other acknowledgements appear in the text and in the captions to diagrams and tables.

1

Historical introduction

When planning a historically-orientated introduction to a book on X-rays it becomes clear at an early stage that there is no shortage of material, so well documented are the facts relating to the discovery and early investigation of X-radiation. Many of the facts are well-known, but the very richness of the subject presents interesting opportunities for discussion, not only because of the large amount of material published during the twelve months or so following the discovery (it has been stated that almost 1000 communications were published during that one year), but also because of the great increase in X-ray studies in recent years, which naturally stimulates an interest in the origins and early days of the subject.

The discovery of X-rays, which took place on 8 November 1895 in Würzburg, Bavaria, constituted an event the importance of which was immediately obvious to the discoverer, and which became apparent to the world at large within a very few weeks of Röntgen's announcement. The subject of X-rays has, of course, always enjoyed the twin privilege of being important as a branch of physics in its own right and of having applications in medical practice the importance of which is known to all. Therein lies a clue to the great speed with which the early researches were put in hand.

Röntgen reported his discovery on 28 December, and dispatched some X-ray photographs and a few copies of his paper to friends on New Year's Day, 1896. The first press report appeared in Vienna on 5 January, by 9 January it had been reported in an American newspaper and by 12 January a radiograph had been obtained in at least one American laboratory (at Davidson College). Experiments were already in progress in several other laboratories in Europe and America. On 16 January a short note appeared in *Nature*, and from 23 January onwards that journal carried advertisements offering for sale slides of X-ray photographs. On 30 January and henceforward, advertisements of apparatus for 'the new photography' were displayed.

The rapidity with which X-ray demonstrations and research programmes were set in motion was helped greatly by the fact that the necessary equipment was already to hand – the Crookes tube for studying cathode rays was available and could readily be connected to an induction coil or electrostatic machine. The tube used by Crookes for early X-ray photographs had in fact been made several years earlier, in 1879, and one can be reasonably certain that X-rays were emitted from these devices long before they were knowingly observed. In fact, in some of Lenard's experiments (using evacuated tubes with a thin aluminium window) X-rays may have contributed to certain observed external effects.

A translation (by A. Stanton) of Röntgen's paper appeared in *Nature* during January 1896 in which the early observations are recorded. The radiation was found to penetrate paper, packs of cards, thick blocks of wood, glass (but to a much reduced extent in lead glass), 15 mm aluminium, and thin sheets of other metals. Certain substances were found to fluoresce, and photographic plates were found to be sensitive. No deviation of the rays was observed in prisms of water, carbon disulphide and mica. (Certain other substances showed possible small deviations with the photographic plate, but not when viewed with the fluorescent screen). Lenses were found not to focus the beam, and the X-rays remained undeviated by a magnet, in sharp distinction to the known behaviour of cathode rays. The inverse square law was found to be obeyed, from which it was inferred that if X-rays were absorbed in air, it was to a much less extent than was the case with cathode rays. Finally, Röntgen proposed that the X-rays might be the long sought longitudinal 'ether waves' which, it was supposed, ought to exist in nature, but which had so far escaped detection. This last proposition was, of course, erroneous, though aesthetically pleasing.

During 1896 many publications appeared in the pages of the *Proceedings of the Royal Society*, *Nature*, and the *Philosophical Magazine*. A.A.C. Swinton (1896) repeated many of Röntgen's original experiments with success. J. J. Thomson (1896) and others established that the higher the vacuum in the Crookes tube, the more penetrating the X-rays, due to the fact that higher EMFs could be sustained across the higher impedance offered by the tube under these conditions. C.T.R. Wilson (1896) observed that the presence of X-rays greatly increased the number of drops formed in an expansion chamber. Other investigators examined in detail the properties of different absorbers, and the non-homogeneous nature of the continuous spectrum was established in the course of these investigations. In the realm of X-ray technology, metal and oil-immersion tubes were constructed, the application of X-rays to non-medical radiography was instituted (by, for example, the detection of foreign matter, such as dust or sand, in foodstuffs). Of course diagnostic radiology flourished and the first X-ray journal *Archives of Clinical Skiagraphy** was published, in London, in the Spring of 1896.

All this was achieved using instrumentation of the simplest kind, coupled with the most careful and systematic interpretation of results. In the years following the discovery of X-rays, absorption was examined in detail, characteristic radiation was discovered, comprising the K and the L series (Barkla, 1909) all by the use of absorption curves which was at that time the only method of investigating the spectral distribution of X-rays. This method has never in fact fallen into complete disuse – it has been found useful in the context of medical physics (e.g. Jones, 1940; Greening, 1950) and has been employed in certain investigations in plasma physics (e.g. Jahoda *et al.*, 1960) where all other methods would have been impossible for technical reasons.

However, although X-ray physics flourished from its earliest days, it took an enormous step forward with the appearance of the first X-ray spectrum obtained by crystal diffraction (Bragg & Bragg, 1913). The classical work of Moseley (see chapter 3) then became possible. What with this and the invention of the hot-cathode (or Coolidge) tube in the same year, both the physics and the technology of X-rays were placed on the sure footing which they hold to this day.

The importance of having some means of recording the quantity of radiation delivered during medical procedures, particularly radiotherapy, was realised by many early users of X-rays. Colour changes in various materials were used in 1902 (by Holzknecht)† for standardising exposure to the radiation, and the (reversible) changes of colour observed in celluloid were used in conjunction with colour matching against a standard and by selecting an appropriate position for the celluloid midway between the tube and the patient the sensitivity of the method could be increased. In 1903 the use of photographic film was proposed. The ionizing effect of X-rays was observed within a few weeks of the discovery of X-rays, and in 1905 an electrostatic dosimeter, in which the position of a suspended aluminium fibre in an electrostatic field depended upon discharge caused by exposure to X-rays, was demonstrated.

In 1908, Villard proposed a definition of dose which, after some lapse of time, became known as the Röntgen, and was adopted for use in Germany in 1923, although it did not become universal until 1929. The importance of

* Later *Archives of the Roentgen Ray*

† The article by Quimby (1945) is of great interest in connection with the historical aspects of dosimetry, and is hereby acknowledged.

knowing something about the energy absorbed by the tissue was pointed out by Christie in 1914, and is of course the basis of modern dosimetry.

At the time of the discovery of X-rays, the experiments of Becquerel were in progress and were yielding interesting results. Becquerel had been exposing certain fluorescent salts of uranium to sunlight and found that a radiation was emitted which could penetrate aluminium or cardboard. This radiation was also highly persistent and clearly did not resemble the normal phosphorescent processes. Shortly afterwards, in 1901, it was realised that the radiation was spontaneous, and did not require previous exposure to visible or ultraviolet radiation. Radioactivity was thus discovered. The study of X-rays and radioactivity therefore started at about the same time as each other and have developed hand-in-hand. They have called for similar techniques of instrumentation, and have both been associated with medical applications. More recently they have both been found valuable in industrial radiography, and, in the realm of pure physics, the study of radioactivity has involved detailed examination of the X-rays emitted by radioactive isotopes. The availability of radioactive materials in a large variety and quantity has opened up a rich field of study in connection with their X- and γ-radiation, and the theme of this book is that much of X-ray physics is relevant to the study of radioactivity and other branches of nuclear physics. In general, the intensity from radioactive sources is much less than that available from X-ray tubes by several orders of magnitude – with the exception of the very large sources of caesium-137 and cobalt-60 in use in medical 'teletherapy' units – thereby precluding the observation of the weaker lines of the characteristic spectrum. The X-ray spectroscopy of radionuclides is therefore less well-developed than spectroscopy using characteristic radiation from X-ray tubes, or fluorescent radiation. However, much work has been done in this field, and, as shown in chapter 7, this has greatly added to our understanding of radioactive decay processes, and of the detailed behaviour of atoms in which nuclear transformations have occurred.

The impact of nuclear physics upon X-ray work has been felt very strongly in the field of instrumentation, and the great advances made during the 1940s and 1950s in proportional and scintillation counting made considerable impact on the course of X-ray research. The study of the continuous spectrum provides a good example of this. The early studies of Duane and Hunt, and Ulrey, led to the work of Kuhlenkampff discussed in detail in chapter 2, in which the instrumentation consisted mainly of ionization chambers and point-discharge counters. The work continued until about 1941, during which time it was extended to higher energies, but very little work was carried out using targets which were thin enough to eliminate the effects of electron scattering, and which would enable the basic process of Bremsstrahlung production to be examined closely. The limitation stemmed from the very low intensities available from thin targets, and the subject lay in abeyance for many years. However, in 1955 the first of a new series of papers appeared, also by Kuhlenkampff and co-workers, in which the subject was taken up again, using proportional counters and associated electronics. This 'nuclear' instrumentation was an essential factor in determining the course of this work. The use of coincidence techniques (Nakel, 1966) has further extended our knowledge in this field.

A feature of recent X-ray research has been the revival of studies of phenomena which were discovered during the development of 'atomic' physics, and which have recently acquired a new interest on account of their relevance to another field. Two examples may be given – one relating to applications in solid-state physics and one to nuclear physics. Our first illustration is the Compton effect, known originally as X-ray scattering with change of wavelength, which received its interpretation in terms of the momentum of incident and scattered photons in or about 1922.

The detailed evaluation of this effect, including calculations of differential cross-sections, angular distribution of scattered radiation, energy dependence, etc., took place during the next ten years or so, and this phenomenon became part of the basic subject-matter of all discussion of 'modern' physics. However, the consequences of the thermal motion of the scattering electrons, although appreciated at the time, attracted very little attention until relatively recently, when, in 1965, consideration of this was resumed in connection with studies of electron momentum distribution in solids. The study of the Compton profile appears to be becoming an important tool in solid-state research. An interesting discussion of this technique, including an illuminating historical survey, is given by Cooper (1971), and further reviews by the same author have appeared more recently, and are referenced in chapter 8.

Our second example is in the field of characteristic X-radiation. X-ray spectroscopy was pursued during the 1920s, associated with the classical work of M. Siegbahn, and a basic body of detailed information had been accumulated by the end of that decade. The subject then ceased to stand in the fore-front of physics research, owing perhaps to the development of nuclear physics during the following years, but in 1952 with the discovery of mesonic X-rays, characteristic X-rays became, once again, an important tool for the elucidation of some quite fundamental processes resulting from the capture of the μ-meson by an atom. Interest was enhanced when it became clear that the energy of muonic X-rays from heavy elements is very sensitive to nuclear radius, thereby enabling nuclear radii to be studied with an accuracy which exceeded that which was previously obtainable. The study of pionic and kaonic X-rays has brought further additions to our understanding of meson physics.

In view of the foregoing one might almost be tempted to divide any historical view of X-rays into three main periods. The period of discovery would cover the years 1895 1922, from the discovery of X-rays to the elucidation of the Compton effect. During these years all the main phenomena of X-ray physics were discovered and investigated, and all the foundations were laid. The second period (the period of Consolidation perhaps) might extend from 1922 to 1945. These were the years when the detailed study of X-ray spectroscopy, and the devising of sensitive instruments for this purpose, was pursued. The subject of X-ray crystallography and structure analysis was developed into a major scientific discipline, and tremendous strides were made in the theoretical interpretation of atomic energy levels. The technology of X-ray tubes also underwent great development. The quest for higher voltages and energies is discussed, and several machines are described by Kaye (1936). Our third period, of revival and new discovery, would extend from 1945 to the present day. The fruits of nuclear research became available in the form of new instrumentation, and an abundance of radioactive isotopes. During this period, mesonic X-rays have been discovered, and X-ray physics has entered new and rather unexpected fields*, such as solar and stellar physics. New phenomena have been revealed such as the intense beams of soft X-radiation associated with electrons in the electron synchrotron, and the 'isotope effect' in X-ray emission. Much work has been carried out on the production of X-rays by proton and heavy-ion bombardment, and these phenomena have been applied to the elemental analysis of a wide variety of materials. X-ray physics has re-emerged as an advancing subject of considerable fundamental interest, and will continue to claim the interest of growing numbers of physicists and others for many years to come.

* But see Lea (1896) and Cajori (1896) for comments on this.

2

The continuous X-ray spectrum

2.1 Introduction

Among the earliest established facts regarding the X-ray spectrum which is produced when an electron beam impinges upon a target were the observations that the radiation is not monoenergetic, and that its intensity is a function of the atomic number of the target material. As examples of early studies of the shape of the continuous spectrum we may cite the work of Duane and Hunt (1915), and Ulrey (1918), who examined the radiation from a target of tungsten, for several values of the electron energy. Ulrey's results are reproduced in fig. 2.1. The most prominent features of these measurements are the occurrence of a maximum in all the curves, and the existence of a short-wave limit advancing in the direction of shorter wavelengths as the applied voltage increases. The short-wave limit (known as the Duane–Hunt limit) is seen to be inversely proportional to applied voltage, and is a direct consequence of the quantum nature of electromagnetic radiation – no photon can be emitted with an energy greater than that of the bombarding electrons. The exact position of the high energy limit has been the subject of very careful investigation and is discussed in section 2.10, but the shape of any continuous spectrum depends upon the way in which it is plotted and the familiar shape of the continuous X-ray spectrum is simply a consequence of the use of the intensity per unit interval of wavelength, a usage which stems from the widespread employment of Bragg spectrometers for studies of X-ray spectra. From the Bragg law

$$2a \sin\theta = n\lambda$$

it follows that

$$2a \cos\theta \, d\theta = n \, d\lambda$$

The use of a constant value for $d\theta$ (i.e. a constant slit-width), followed by a simple $\cos\theta$ correction, leads immediately to the intensity per unit wavelength. Methods developed more recently for the recording of continuous X-ray spectra give the intensity per unit frequency or energy interval, I_ν or I_E. They are related to each other and to the intensity per unit wavelength interval by the expressions

$$I_E|dE| = I_\nu|d\nu| = I_\lambda|d\lambda|$$

from which are obtained the relationships

$$I_E = \frac{1}{h} I_\nu = \frac{\lambda^2}{ch} I_\lambda \qquad (2.1)$$

The theoretical treatment of the subject is also expressed more naturally

Fig. 2.1. The continuous spectrum from a tungsten target, plotted as intensity per unit wavelength interval (Ulrey, 1918).

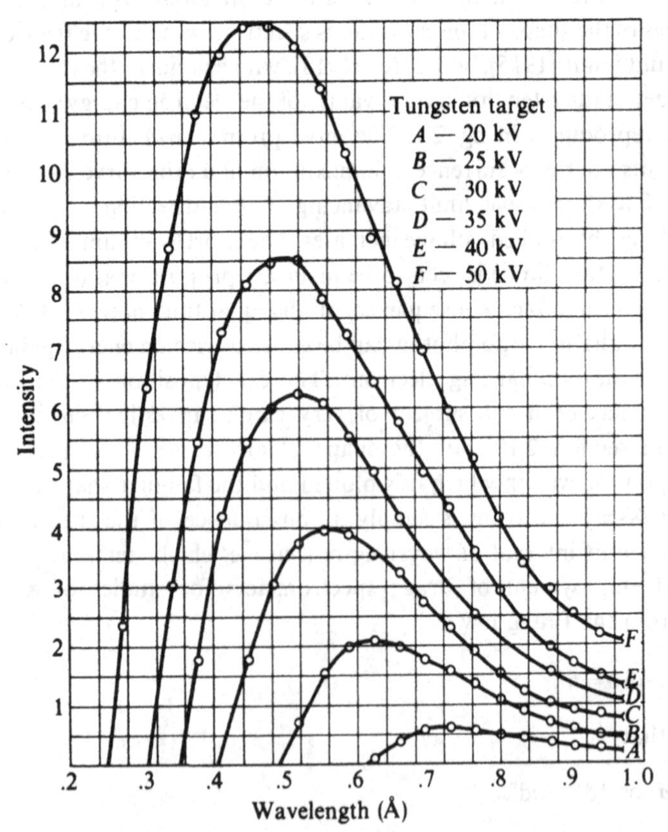

in this form. The number of photons per unit frequency or energy interval is often used, and may be obtained by writing

$$n_\nu = \frac{I_\nu}{h\nu}$$

Investigations of the continuous spectrum may be divided into two groups. In the one group (which is the first in historical order but not in simplicity of interpretation) the targets are sufficiently thick to stop *all* the incident electrons. The targets may be massive, as in the anticathodes of conventional X-ray tubes, or they may be foils which are thin enough to transmit a substantial proportion of the X-rays, but nevertheless sufficiently thick to arrest all the electrons. These electron-opaque targets are used in several types of tube for specialised practical applications (see chapter 4 and chapter 8). The other group of investigations consists of experiments using extremely thin foils in order to allow the electrons to pass through with only a very small number of interactions, and with as little loss of energy as possible. This type of target is used in studies of the elementary process of X-ray production.

2.2 Energy distribution of the continuous spectrum from thin targets

The first studies of this kind appear to have been made by Duane (1927, 1928) and Nicholas (1929). Duane, using targets of mercury vapour and an accelerating voltage in the region of 12 kV, examined the radiation at 0° and 90° to the forward direction and established its lack of homogeneity by the use of absorption methods. Nicholas examined the radiation from thin aluminium (0.7 μm) and gold (0.09 μm) bombarded by electrons accelerated to 45 kV. Measurements were made at angles of 40°, 90°, and 140° to the forward direction, using a crystal spectrometer and photographic recording, and the results were displayed in the form of intensity per unit frequency interval, primarily for comparison with the early theory of Kramers, which is referred to in section 2.7.

Nicholas showed that I_ν was approximately constant up to the high energy limit, and zero thereafter, although there was some evidence to suggest that the shape of the spectrum depends to some extent on the angle of observation. This is discussed further in section 2.3. No further detailed measurements of the spectral distribution from thin targets at these energies appears to have been carried out until those of Amrehn and Kuhlenkampff (1955), who made a series of measurements at 90° to the forward direction, using thin targets of aluminium (25.5 nm in thickness), nickel (7 nm) and tin (~4 nm)*.

* The electron beam was incident at an angle of 45° to the target, so the effective thicknesses exceeded these values by a factor of $\sqrt{2}$

The intensity of radiation from such thin targets is, of course, very low, but by using a proportional counter and pulse-height analyser it was possible to obtain data of good statistical accuracy in reasonably short times (30 seconds for each observation). Their data compare well with the results expected from theoretical considerations. The distribution from aluminium for a bombardment energy of 34 keV is shown in fig. 2.2(*a*) together with a theoretical curve (normalised for best fit) from the Sommerfeld theory (section 2.7). In this work the accelerating voltage was varied in the range 20 34 kilovolts, and the spectral distribution from an aluminium target for three values of accelerating voltage is shown in fig. 2.2(*b*). The constancy of I_ν observed originally by Nicholas is seen to be well verified, and in addition the intensity per unit energy interval was found to be inversely proportional to accelerating voltage, for photon energies in the range $0.4\nu_0$–ν_0, where $h\nu_0$ is the high energy limit of the spectrum.

Amrehn and Kuhlenkampff compared the spectral distribution curves for the three elements under investigation, and fig. 2.3 illustrates that the variation of shape with atomic number is only slight, for radiation measured at 90° to the incident electrons. The total intensity per atom per cm ^{2} was found to be approximately proportional to Z^2.

Amrehn (1956) measured the shape of the spectrum for a wider range of target elements (Z=6 to 79), and confirmed these observations, but the

Fig. 2.2. The continuous spectrum from a thin target of aluminium; (*a*) bombardment energy 34 keV (curve calculated from the Sommerfeld theory; (*b*) bombardment energies 25, 34, and 40 keV (Curves are best fit) (Amrehn and Kuhlenkampff, 1955).

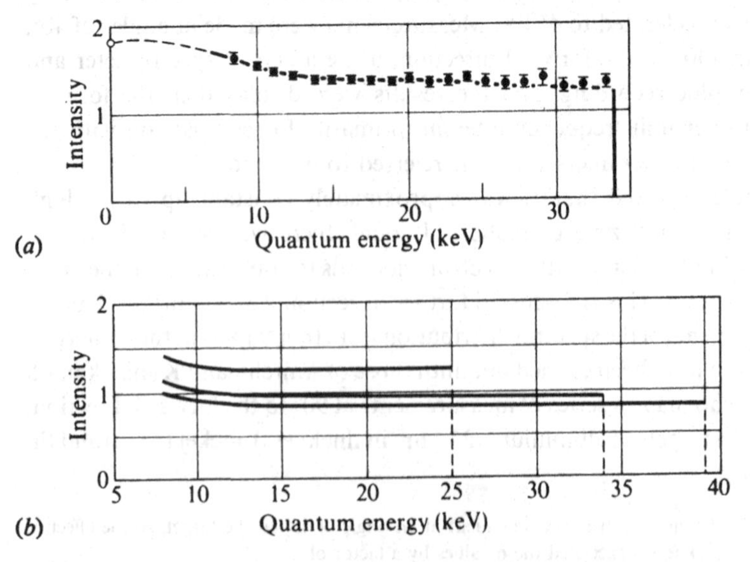

proportionality with Z^2 was found to be only very approximate, falling well below this at small values of Z.

Studies of the spectral distribution from thin targets at higher energies are described in section 2.6.

2.3 Angular distribution of the continuous spectrum from thin targets

This has been subjected to more intensive investigation than the energy distribution – useful results were obtainable even at a time when only a limited degree of energy discrimination was possible. Moreover, the angular distribution curves show certain marked features which can readily be compared with theory. Although the theoretical aspects will be discussed below, we may note here that the simple polar diagram associated with dipole radiation from an accelerated (or decelerated) electron is strongly modified by relativistic considerations, even at bombarding energies as low as 10 keV, and that the maximum which is predicted by classical (non-relativistic) theory at right-angles to the forward direction is in fact displaced towards the forward direction. This has been studied at energies up to 40 keV, by Kuhlenkampff (1928), Böhm (1937,8), Honerjäger (1940), Kerscher and Kuhlenkampff (1955) and Doffin and Kuhlenkampff (1957).

The main features of the angular distribution curves of Kuhlenkampff are the presence of a minimum of intensity in the forward and backward directions and a gradual movement of the maximum of intensity towards smaller angles as the accelerating voltage is increased. For example at a bombarding energy of 16.4 keV, the maximum occurs at 63°, falling to 51° at 37.8 keV, for X-radiation in the region of the high energy limit. For lower energy radiation (at the same bombarding energy) the maximum was found

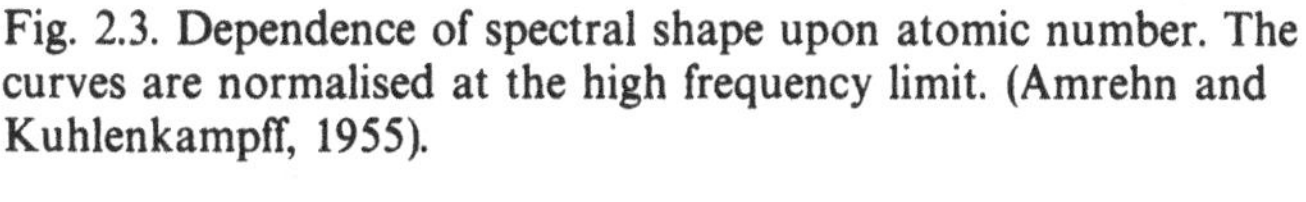

Fig. 2.3. Dependence of spectral shape upon atomic number. The curves are normalised at the high frequency limit. (Amrehn and Kuhlenkampff, 1955).

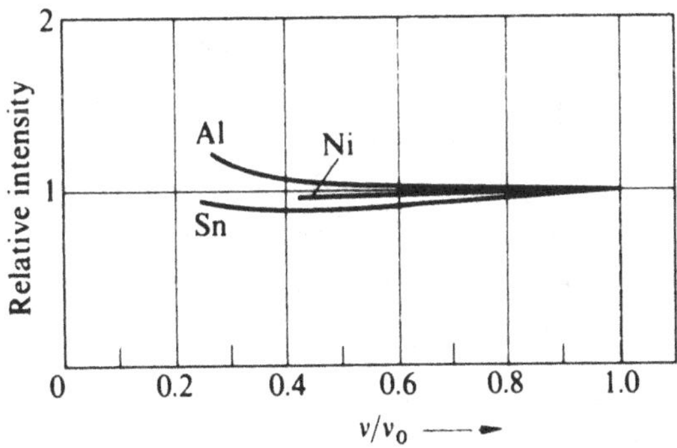

to occur more forward (i.e. at lower angles) than that of the high energy limit, and the minimum in the forward direction was less marked.

The target used in this work was aluminium, with a thickness of 600 nm (1.6×10^{-4} g cm^{-2}). This did not represent an adequate approach to ideal thinness, and subsequent work was planned to approach this condition more closely. Böhm used thin targets of magnesium, which was more convenient for the preparation of evaporated foils than aluminium on a thin backing of celluloid; his thinnest foils were less than 100 nm in thickness, which is less than 2% of the electron range at 30 kV. Böhm's measurements included the dependence of angular distribution on voltage for the high energy limit of the spectrum, and the dependence of angular distribution on position in the spectrum, at several different accelerating voltages. Fig. 2.4,

Fig. 2.4. Angular distributions from thin targets of magnesium (Böhm, 1938) (*a*): at the high energy limit (bombardment energies) (1): 20 keV; (2) 31 keV; (3); 40 keV. (*b*) dependence of angular distribution on photon energy for a bombardment energy of 20 keV (1): 19 keV; (2) 8 keV.

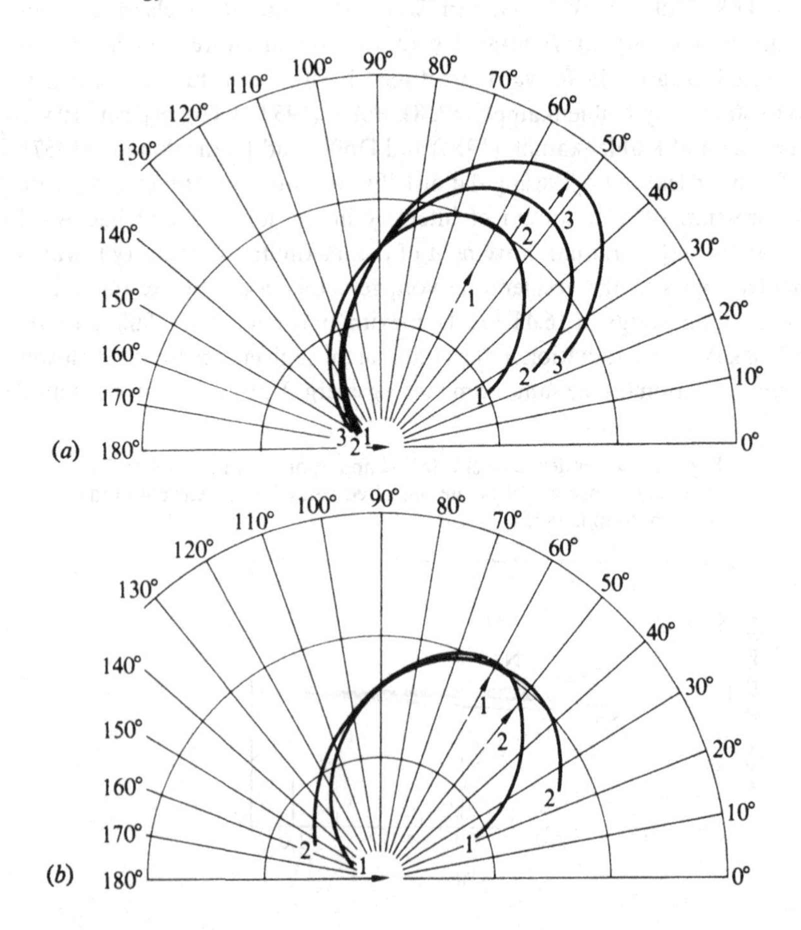

taken from his 1938 paper, illustrates some of these results, confirming the general correctness of the earlier work. It was also shown that a gradual reduction in thickness caused the curves to become less broad and the forward minimum to become more prominent. The effect of using thinner targets was most pronounced near the high energy limit of the spectrum.

Theoretical considerations at about that time (Scherzer, 1932) suggested that the intensity in the forward direction should fall to zero, at least at the high energy limit of the spectrum, and it was thought that only the finite thickness of the target foils prevented this from being so in practice. The matter was tested by Honerjäger (1940) who worked down to target thicknesses in the region of 10 nm representing a real approach to ideal thinness. Fig. 2.5, taken from Honerjäger's paper, cast some doubt on the theoretical prediction of a zero in the forward direction, but the point was not finally clarified until the work of Kerscher and Kuhlenkampff in 1955, who established conclusively that there is a finite emission in the forward direction, using a target of aluminium 25 nm in thickness, and a bombarding energy of 34 keV.

Detailed considerations of the criteria for ideal thinness (section 2.7) show that for targets of 25 nm or less the amount of electron scattering is in fact so small that it would not be expected to affect the 'ideal' distribution to any appreciable extent.

When a proportional counter and single channel pulse-height analyser are used to select photons in a particular energy interval, the boundaries of the interval are somewhat ill-defined because of the limited energy resolution of the system. To select an energy interval with better-defined boundaries (though not necessarily a narrower band) Doffin and Kuhlenkampff (1957) used the method of 'balanced filters' (chapter 4). Fig. 2.6 illustrates their results on aluminium at 34 kV.

From the foregoing one may infer that the spectral distribution at angles

Fig. 2.5. Angular distribution for targets of several thicknesses (at the high energy limit) and a comparison with the theory of Scherzer. Accelerating voltage 34 kV ((Honerjäger, 1940).

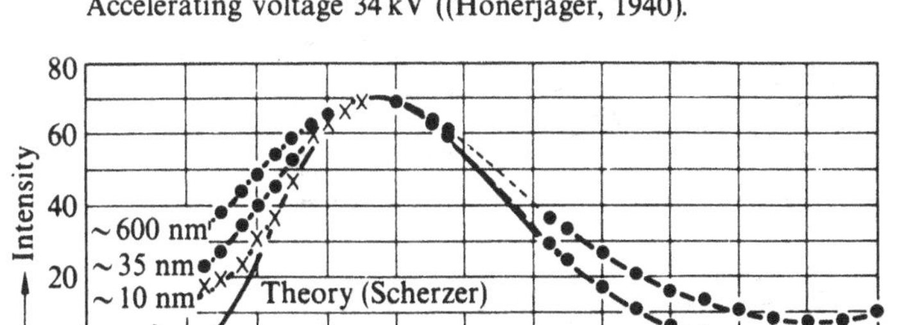

Fig. 2.6. Angular distribution for several energies in the spectrum (thin aluminium target, 34 kV) (*a*): $\nu=\nu_o$ (*b*) $\nu=0.7\ \nu_o$ (*c*) $\nu=0.4\nu_o$ (*d*) $\nu=0.26\ \nu_o$ (Doffin and Kuhlenkampff, 1957).

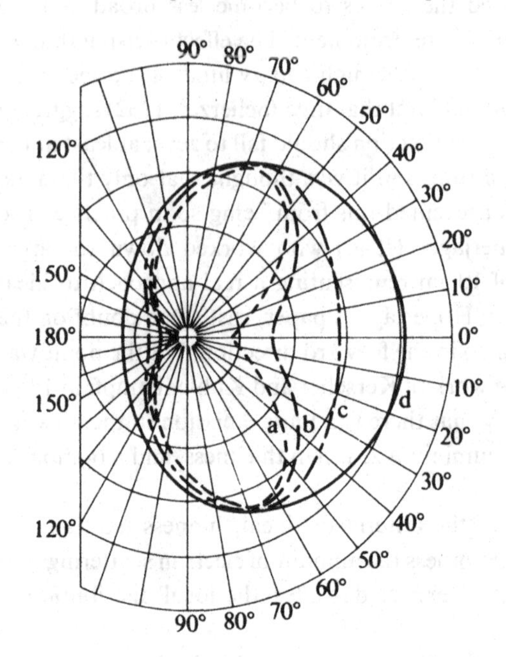

Fig. 2.7. Spectral distribution for various angles of emission (thin aluminium target 34 kV) (Kerscher and Kuhlenkampff, 1955).

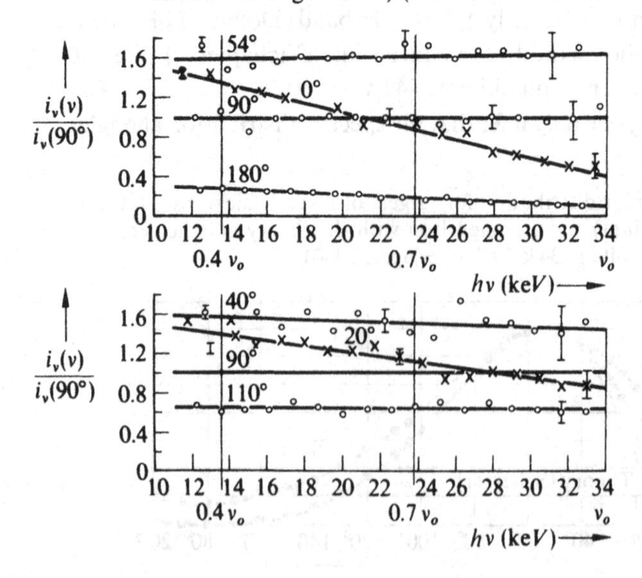

in the region of maximum emission will be relatively stronger in photons near the high energy limit. Spectra at such angles (in practice between about 40° and 110° at this electron energy) are in fact relatively flat, whereas outside this angular range the intensity per unit energy interval falls steadily as the high energy limit is approached. Fig. 2.7 illustrates this. For further details the papers by Kuhlenkampff (1938), Kerscher and Kuhlenkampff, and Kuhlenkampff and Ross (1961) may be referred to.

A recent study of the angular distribution of Bremsstrahlung from aluminium and gold at 50 and 200 keV has been described by Rester, Edmonson and Peasley (1970). They made use of a lithium-drifted germanium spectrometer which, because of its relatively good energy resolution, could be used very selectively, even at photon energies very close to the high energy limit.

At a beam energy of 200 keV, the radiation with photon energy centred on 196 keV (i.e. very near to the high energy limit) from an aluminium target showed a maximum at 30° to the forward direction, which is consistent with the Sommerfeld theory as depicted in fig. 2.15.

Studies of angular distribution at higher energies are described in section 2.6.

2.4 Energy distribution from electron-opaque targets

After the early work of Ulrey detailed measurements were undertaken by Kuhlenkampff (1922) using massive anticathodes of platinum, tin, silver, copper and aluminium, and a range of accelerating voltages from 7 to 11.98 kV.

The data were presented as intensities per unit frequency interval, and the very simple form of these graphs is illustrated in fig. 2.8(*a*). These measurements were carried out at an angle of approximately 90° to the forward direction. It is clear that a substantially linear relation obtains, although a small term independent of accelerating voltage and quantum energy is needed to accommodate the observed deviation from linearity near the high frequency limit. Kuhlenkampff proposed the expression

$$I_\nu = A_c Z(\nu_0 - \nu) + B_c Z^2 \tag{2.2}$$

A_c and B_c are constants independent of applied voltage or atomic number. The ratio B_c/A_c is found to be in the region of 0.0025. The term in Z^2 is thus small except near the high frequency limit.

Kuhlenkampff and Schmidt (1943) extended this work to somewhat higher accelerating voltages, and found the linear relationship to be closely obeyed, without the need for an additional $B_c Z^2$ term. The total intensity, was proportional to V^2 to within about 10% (Fig. 2.8(*b*)). At lower voltages

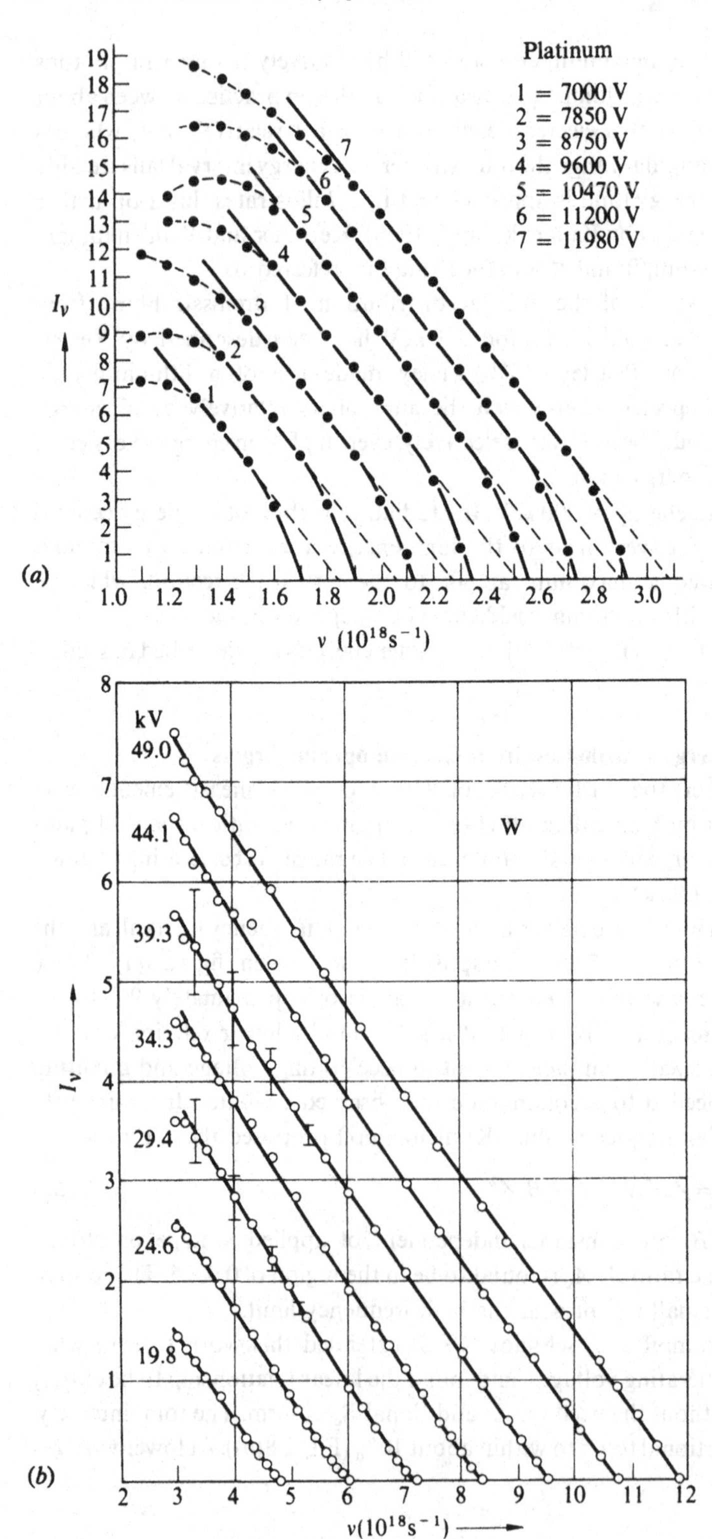
Platinum
1 = 7000 V
2 = 7850 V
3 = 8750 V
4 = 9600 V
5 = 10470 V
6 = 11200 V
7 = 11980 V
I_ν
(a)
ν (10^{18}s^{-1})
kV
49.0
44.1
39.3
34.3
29.4
24.6
19.8
W
I_ν
(b)
ν(10^{18}s^{-1})

Fig. 2.8. Spectral distribution for thick targets (*a*) Kuhlenkampff (1922) (*b*) Kuhlenkampff and Schmidt (1943). (Data obtained at 90 degrees to the electron beam).

experimental difficulties become greater because of low output and the increasing absorption to which the softer radiations are subject. Neff (1951) obtained data using electron accelerating potentials from 2 kV down to 900 volts, using targets of platinum, silver and nickel, and a grating vacuum spectrometer. His data for platinum are shown in fig. 2.9. The total intensity is still proportional to V^2, as shown by the approximate parallelism of these spectra, but the total intensity increases less rapidly in proportion to the atomic number than would be expected from (2.2). Absorption losses in the target are much greater at high atomic numbers, and this may help to explain this result. Insufficient data points were obtained to establish whether an additional term of the form $B_c Z^2$ is needed at these low accelerating potentials. Stephenson and Mason (1949) investigated the continuous spectrum from tungsten, copper and aluminium, using Bragg reflection from mica, for a smaller range of electron energies. When their data are plotted in the same manner as that of Neff a straight-line relationship is not obtained, and the spectrum is found to be very deficient in the softer components. Their data show a less than linear dependence on Z, though their inclusion of a correction for target absorption causes the departure from linearity with Z to be not so great as that noticed by Neff.

Peterson and Tomboulian (1962) have examined the spectrum from magnesium, aluminium, manganese, copper, germanium and silver, in the extreme long-wavelength part of the spectrum. Accelerating voltages of 600–3000 volts were used, and the region of wavelength investigated was 8–18 nm, using a grating spectrometer. Equation 2.2 led to the prediction of an inverse square dependence of intensity per unit wavelength interval for points remote from the short-wave limit. The results of this investigation were expressed in the form $I_\lambda = 1/\lambda^\alpha$ where $1.8 < \alpha < 2.7$. The effect of self-absorption in the target is shown to be negligible under the conditions of this experiment.

All the measurements cited so far have provided data at angles in the region of 90° to the forward direction. Measurements in the forward direction using targets consisting of electron-opaque foils have been described by Cosslett and Dyson (1957) and by Dyson (1959). In this work targets of beryllium, aluminium, copper and gold were used, the last two being in the form of thin layers evaporated on to distrene. Four values of accelerating voltage were used, approximately 6, 8.4, 10 and 12 kilovolts. The X-radiation was detected using a proportional counter and single-channel pulse-height analyser, and corrections were applied for the absorption in the target and the air path between target and counter, and for variation in counter efficiency with photon energy. Data for aluminium and gold are shown in fig. 2.10. The data resemble very closely those obtained by Kuhlenkampff at right-angles to the incident beam, suggesting that the anisotropy of the radiation is rather small on account of electron scattering and diffusion within the target, except perhaps near the high energy limit. (Angular distributions were measured in a separate series of measurements using the same apparatus, and are referred to in section 2.5).

The non-linearities near the high energy limit are present, and are more pronounced in the heavier elements. This is in accordance with the earlier work; but in the case of the lighter elements, the forward radiation near the

Fig. 2.9. Spectral distributions at lower energies (Neff, 1951).

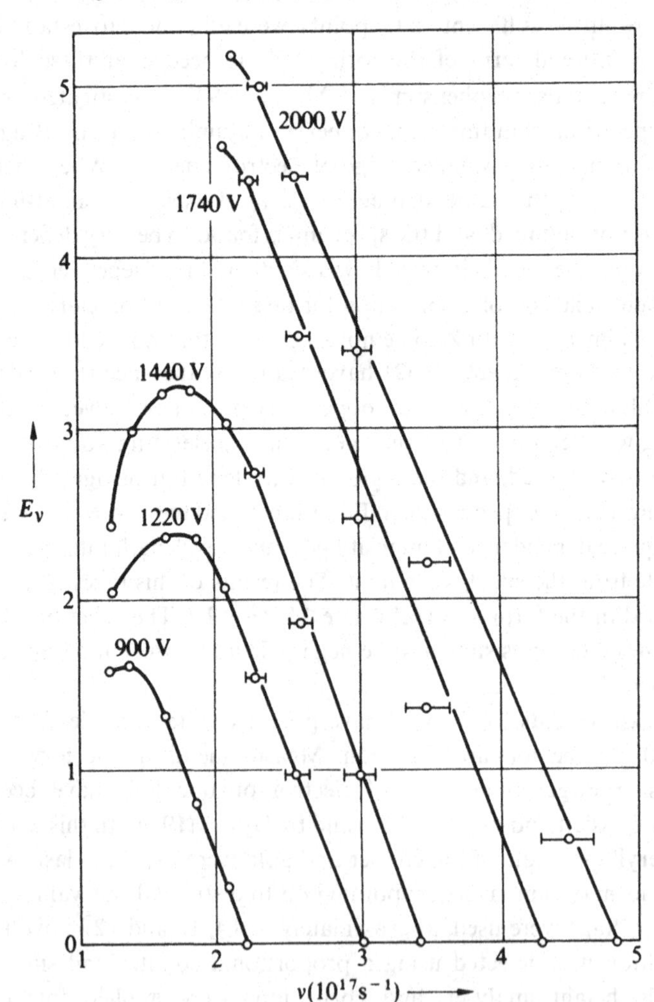

Fig. 2.10. Spectral distributions in the forward direction (*a*) aluminium (*b*) gold (Dyson, 1959).

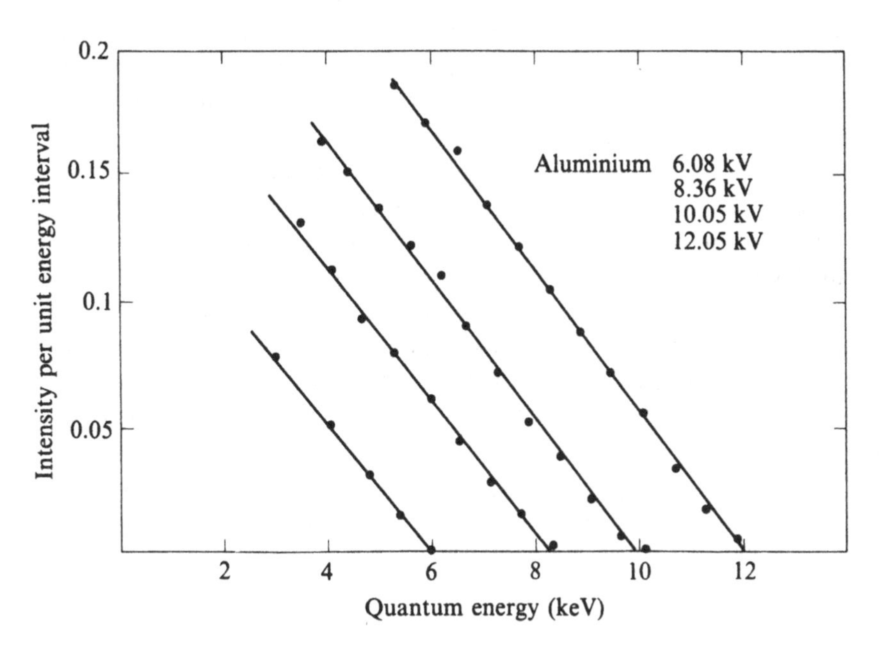

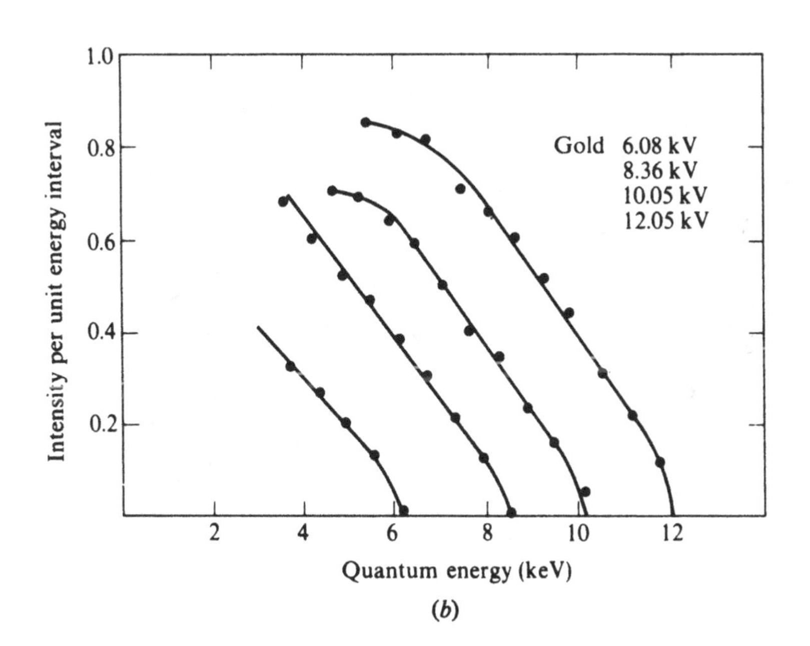

(*b*)

high energy limit is relatively weaker than that observed by Kuhlenkampff at 90°, suggesting that some trace of the thin target anisotropy remains even when electron-opaque targets are used.

The spectral distribution from thick targets at 1.0, 1.5, and 2.0 MeV has been examined in detail by Edelsack *et al.* (1960) using a van de Graaff accelerator, and a sodium iodide scintillation counter. An example of their data is given in fig. 2.11, and it is seen that the spectra for all the elements studied are similar in shape, except in the low energy region of the spectrum. Here the heavy elements show a marked deficiency in low energy photons (except for the appearance of characteristic peaks) compared with the spectra from polythene and aluminium, and this difference may be attributable to screening effects in the heavier elements. A detailed comparison with theoretical expressions is not given, but if the data are plotted in the manner of Kuhlenkampff and co-workers, marked departures from linearity are observed in the heavier elements which are additional to the effects at the low-energy end of the spectrum just referred

Fig. 2.11. Spectral distributions at 2 MeV (Edelsack *et al.*, 1960). Each curve has been multiplied by 10^{α} for clarity of presentation. The value for α is given alongside each curve.

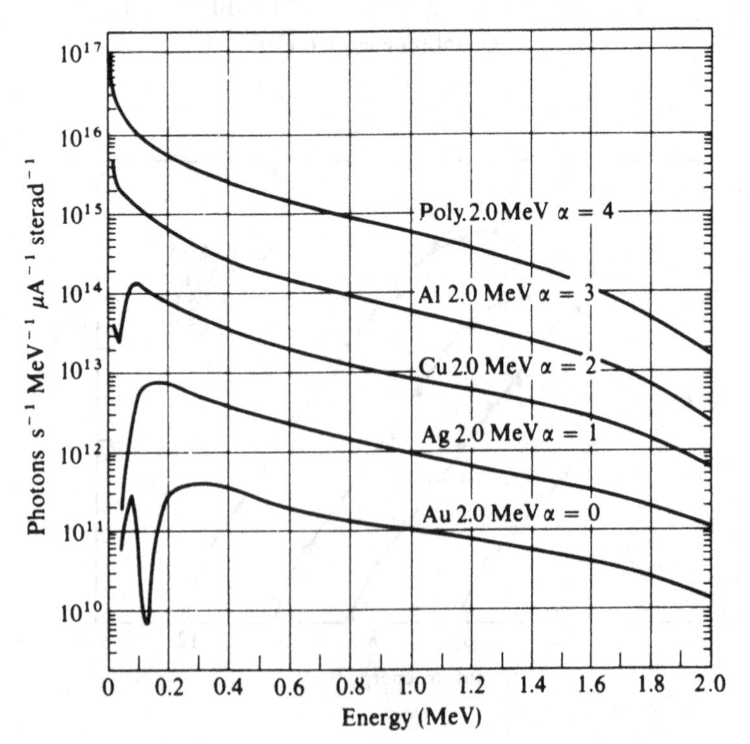

to. It is also clear that proportionality with Z no longer obtains, and that the intensity rises much less than linearly with Z. This is to be compared with the work of Buechner *et al.* (1948) in the same energy region, in which the total intensity *integrated over all directions* was found to be accurately proportional to Z.

2.5 Angular distribution from electron-opaque targets

Data obtained in electron-opaque conditions are of less fundamental interest than when very thin targets are used, but are of considerable practical importance. The angular distribution from conventional X-ray tubes with massive anticathodes has naturally been studied for practical purposes; Bouwers and Diepenhorst (1933) measured the distribution of the total radiation from a tube with a tungsten anticathode using a voltage in the region of 100 kV. The surface of the anode made an angle of about 45° with the electron beam, and the radiation was found to be isotropic except for a falling off at angles of less than 5° to the anticathode surface, which was attributed to absorption effects. The minimum expected at 180° to the beam is thus completely obscured by the scattering and diffusion of the electrons. At lower energies, Oosterkamp and Proper (reported by Botden *et al.* 1952) have measured the angular distribution in the forward hemisphere from a small tube designed for radiotherapy in conditions where the dose rate is required to fall off very rapidly with distance. The target consisted of a beryllium sheet 1.5 mm thick (forming the end window of the tube) on which was deposited an electron-opaque layer of gold. Their results (fig. 2.12) show that at 25 kV the effects of electron scattering and diffusion do not entirely obliterate the forward minimum but that at 10 kV it is in fact

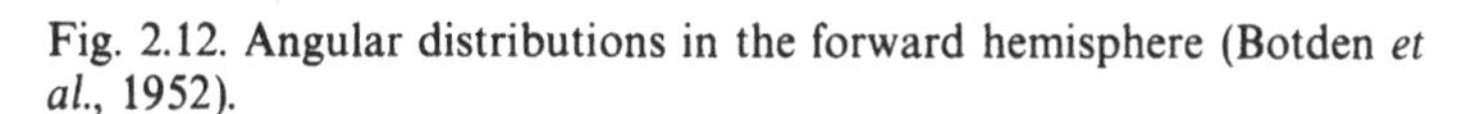
Fig. 2.12. Angular distributions in the forward hemisphere (Botden *et al.*, 1952).

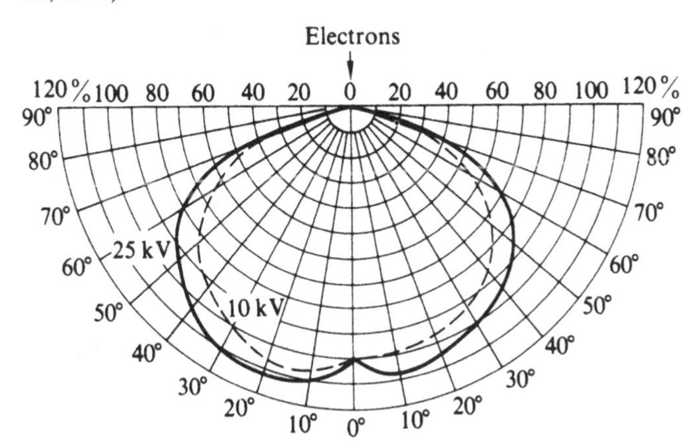

absent. These measurements were made using an ionization chamber, which would give weight preferentially to the lower quantum energies in the spectrum. Further, the graphs shown are not corrected for target absorption but show the distribution actually obtained under practical conditions. Both these factors contribute to the absence of a forward minimum at 10 kV.

Measurements made in connection with X-ray tubes for projection microscopy have been reported by the writer (Dyson, 1959). The targets and kilovoltages were the same as those used for the spectral distributions referred to above. Data obtained for gold and aluminium using accelerating voltages of 12.05 kV are shown in fig. 2.13. Data at approximately 6 and 8 kV for these elements are given in the paper just referred to, and data at 10 keV are quoted by Cosslett and Dyson (1957).

Fig. 2.13. Angular distributions in the forward hemisphere (*a*) Aluminium (*b*) Gold (Cosslett and Dyson, 1957).

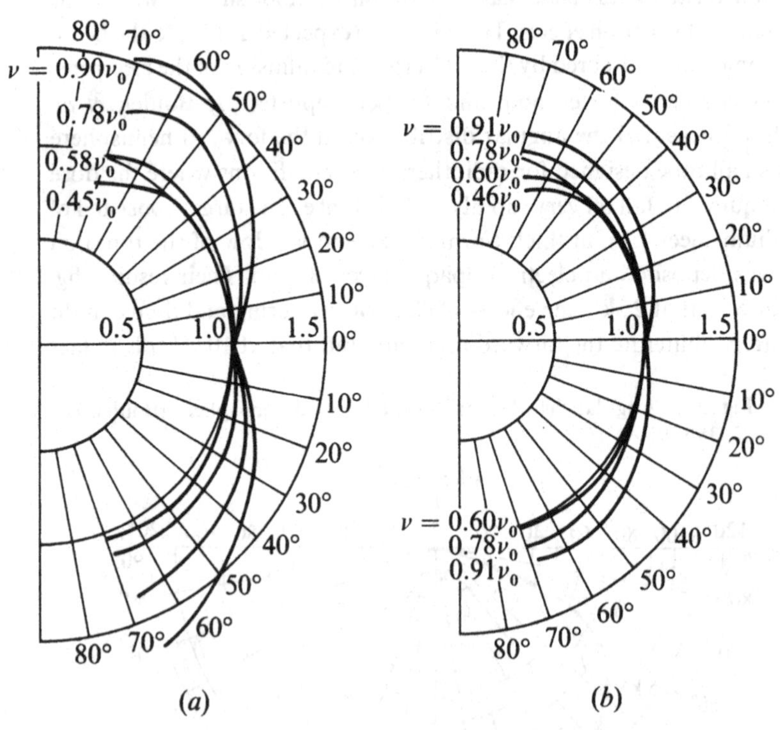

Lower quadrant:
corrected for absorption in the target
Upper quadrant:
uncorrected

Data for beryllium at 8.36 kV, and distrene at 10.05 kV are illustrated in fig. 2.14.

The general trend is towards increasing anisotropy as the high energy limit is approached, and as the atomic number is decreased. This pattern of variation is determined by the fundamental processes of Bremsstrahlung production, modified by the effects of electron scattering and diffusion, and is discussed below.

Work at higher energies has been reported by Thordarson (1939). Anticathode materials of aluminium and tungsten were used, and the accelerating voltage was in the region 60–170 kV. The variation of angle of maximum intensity with kilovoltage was measured, and found to agree with the theory of Sommerfeld up to about 110 kV; above this voltage, the theoretically calculated angles were appreciably larger than the observations. Further work on a massive anticathode was reported by Determann

Fig. 2.14. Angular distributions in the forward hemisphere (*a*) Beryllium at 8.36 kV (*b*) Distrene at 10.05 kV (Dyson, 1959).

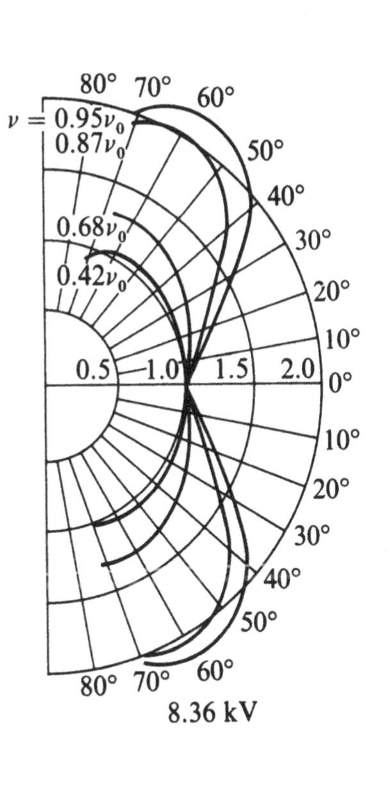

(*a*)

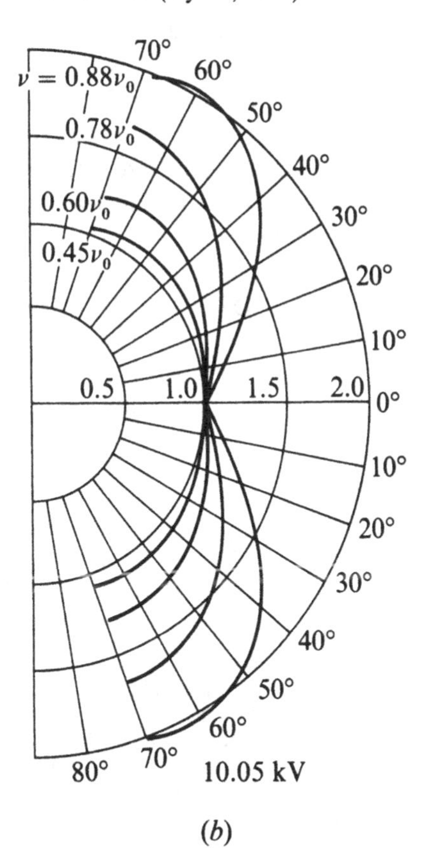

(*b*)

(1937), at similar kilovoltages. Determann verified that the position of the intensity maximum does not depend on the anticathode material (carbon, beryllium, aluminium). The anisotropy of the radiation was observed to increase with increasing tube voltage and decreasing atomic number. This latter is in accord with what is observed at much lower voltages also (see above). Determann observed that the angle of maximum intensity was less than that obtained from the Sommerfeld theory, the discrepancy amounting to 7° at 140 kV, and that filtering the radiation to separate out the softer components made little difference to the position of the intensity maximum. Further work of a similar nature has been reported by Sesemann (1941) and Hinz (1941), and the subject has been reviewed by Stephenson (1957).

Data reported by several authors are compared with the Sommerfeld theory in fig. 2.15.

2.6 Bremsstrahlung at higher energies

Studies of the spectral distribution of Bremsstrahlung at energies above about 1 MeV have been directed mainly towards comparison with the theory of Bethe and Heitler (1934) who considered the problem in detail for energies up to about 50 MeV. Bourgoignie *et al.* (1965) have studied the radiation emitted from a 4.3 MeV linac, using electron kinetic energies in the range 2.5–4.0 MeV, from targets of Be, C, Al, Ti, Cu, Zr, Ag, and Au. Relative photon yields were determined above a 1.69 MeV threshold using the $^{9}Be\ (\gamma,n)^{8}Be$ photonuclear reaction. The yield was found to be proportional to Z^2.

Absolute thick target spectra have been recorded for 5.3–20.9 MeV electrons by O'Dell *et al.* (1968). They used a target of tungsten–gold with a

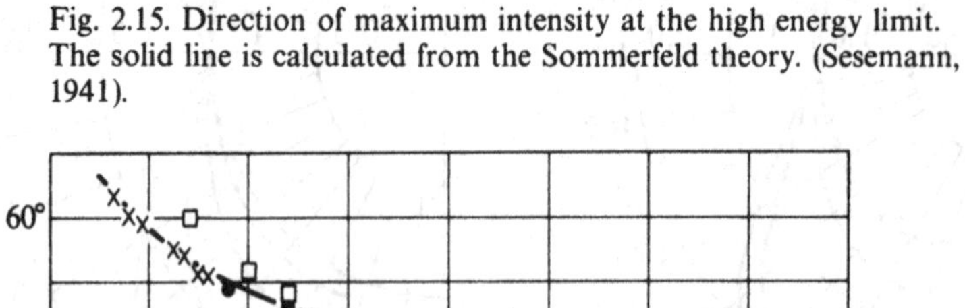

Fig. 2.15. Direction of maximum intensity at the high energy limit. The solid line is calculated from the Sommerfeld theory. (Sesemann, 1941).

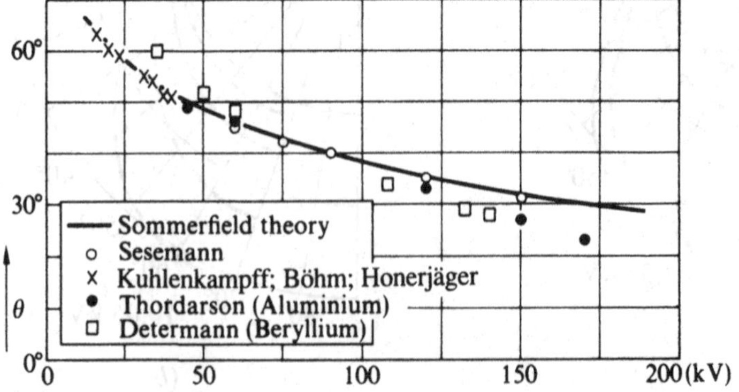

thickness of 0.2 radiation lengths, and detected the Bremsstrahlung photon using the d(γ,n)p reaction (photodisintegration of the deuteron). The system therefore had a detection threshold for γ-rays at 2.23 MeV. Measurements were made in the forward direction. This, in effect, averages the photon spectra over all emitted directions, because the radiation is concentrated strongly within a cone of semi angle $\sim \frac{m_0 c^2}{E_0}$ and appreciable electron scattering through angles of this order will already have occurred in the target before photon emission takes place. Koch and Carter (1950) studied the radiation from a thin (0.005″) target of platinum in a betatron. The electron energy was 19.5 MeV, and the Bremsstrahlung was observed in the forward direction. The experimental intensities are stated to be somewhat higher in the region 8–12 MeV than expected from theory. Felbinger *et al.* (1960) have obtained spectral distributions from platinum at 29 MeV at 0° and 2.6° to the forward direction (fig. 2.16); Studies at 1 GeV have been reported by Diambrini *et al.* (1961) and spectral distribution curves are given by them; data at higher energies are reported by Bem *et al.* (1966) at 2.4 GeV and Lohrmann (1961) for energies greater than 100 GeV, using electron–positron pairs created in nuclear emulsions exposed to cosmic radiation.

As the electron energy is increased, the angle of maximum intensity of emission falls gradually. At 1.5 MeV, for example, it has fallen to approximately 6° (Buechner *et al.*, 1948).

At higher energies it is clear that even in targets much thinner than the

Fig. 2.16. Spectral distribution from a platinum target at 29 MeV. (Felbinger *et al.*, 1960).

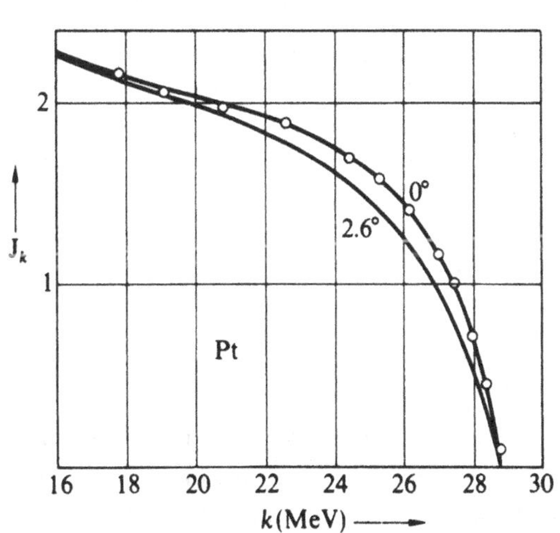

electron range the observed distributions will often be strongly influenced by electron scattering. In these circumstances the minimum at 0° is difficult to observe, and is often absent (Lanzl and Hanson, 1951), but if adequately thin targets are used the minimum can still be seen. Zdarko *et al.* (1964) have examined the angular distribution near the high energy limit of the Bremsstrahlung of photon energy 320 MeV produced by electrons of 375 and 550 MeV. For the electrons of the lower of these energies the photon energy was 0.86 of the high energy limit. The intensity rose by about 7% from the forward direction to the angle of maximum emission which was of the order of m_0c^2/E_0 (only a few minutes of arc at this energy), in accordance with theoretical prediction for extreme relativistic energies.

Studies of Bremsstrahlung at 8.5 MeV have been carried out by Kimura *et al.* (1959). A detailed review of Bremsstrahlung cross-section formulae has been compiled by Koch and Motz (1959), and for details of experimental work carried out at high energies the monograph by Bogdankevich and Nikolaev (1966) may be consulted.

2.7 Theories of the continuous spectrum

(a) *Angular distribution*

In the development of electromagnetic theory it is established that an accelerated charge loses its energy progressively in the form of electromagnetic radiation. In vector notation and rationalised MKS units we may write, for the radiation field associated with a charge with displacement f from an origin

$$\mathbf{H}=\frac{e}{4\pi cr^2}\ddot{\mathbf{f}}\times\mathbf{r};\ \mathbf{E}=\frac{e}{4\pi\varepsilon c^2r^3}(\mathbf{r}\times(\mathbf{r}\times\ddot{\mathbf{f}})) \tag{2.3a}$$

where the point at which the fields are measured has radius vector **r** from the origin, (e.g. Panofsky and Philips, 1962, p. 358)

For radiation emitted in the xy plane of fig. 2.17,

$$H_z=\frac{\ddot{f}e}{4\pi cr}\sin\theta \qquad (H_r=0, H_\theta=0) \tag{2.3b}$$

and

$$E_\theta=\frac{\ddot{f}e}{4\pi\varepsilon_0c^2r}\sin\theta \quad (E_r=0, E_z=0) \tag{2.3c}$$

If the electron (e.g. in the form of a beam) is accelerated (or decelerated) in the direction parallel to its motion, θ is then the direction at which the fields are measured relative to the direction of the electron beam.

The outward flow of energy associated with the acceleration is given by the Poynting vector **N**

$$\mathbf{N}=\mathbf{E}\times\mathbf{H} \tag{2.4a}$$

For the flow of energy in fig. 2.17, we may write

$$N=E_\theta H_z=\frac{1}{4\pi\varepsilon_0}\frac{c}{4\pi}\left(\frac{\ddot{f}e}{rc^2}\right)^2\sin^2\theta \tag{2.4b}$$

The $\sin^2\theta$ term gives rise to the familiar polar diagram for dipole radiation.

When this is integrated over all angles, we obtain for the total radiated power I,

$$I=\int_0^\pi N 2\pi r^2 \sin\theta \mathrm{d}\theta=\frac{1}{4\pi\varepsilon_0}\frac{2e^2\ddot{f}^2}{3c^3}$$

In the theory of X-ray production the radiation is associated with the slowing down of electrons in the target material. Most of the interactions during the slowing down process are essentially with orbital electrons, giving rise to ionization and energy-loss. But interactions also take place with the coulomb fields of nuclei, and it is supposed in the classical theory that those are associated with large decelerations and consequent radiation. In the theory outlined so far, the assumption was made that the slowing down is rectilinear but we may note here that, according to more detailed treatment, the quantity $\ddot{f}$ is not the same direction as the electron path, but is determined by the Coulomb force exerted on the electron by the nucleus,

$$\ddot{f}=\frac{1}{4\pi\varepsilon_0}\cdot\frac{Ze^2}{ma^2}$$

where a is the time-varying distance between the moving electron and the nucleus. The screening effect of the orbital electrons is neglected. Combining this with (2.4), we obtain the important result that the radiated power is

Fig. 2.17.

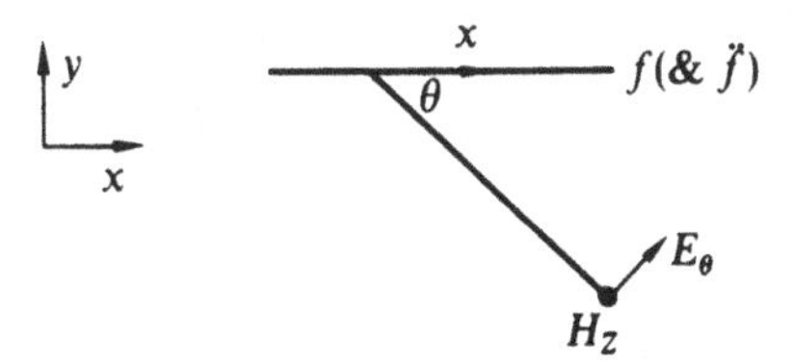

proportional to Z^2, a result which is found to be approximately true in practice.

The assumption has also been made that the electron velocity is small compared with the velocity of light. For electrons with energies greater than a few keV, relativistic treatment is necessary, giving rise to the expression

$$I=\frac{1}{4\pi\varepsilon_0}\frac{c}{4\pi}\left(\frac{\ddot{f}e}{rc^2}\right)^2\frac{\sin^2\theta}{(1-\beta\cos\theta)^6} \tag{2.5}$$

for the instantaneous value of the intensity. I is the energy flux per unit area per unit time, and $\beta=v/c$ (e.g. Compton and Allison, appendix 1; Panofsky & Philips, p. 359).

The maximum in the radiated energy flux thus moves forward from 90° to smaller angles, the magnitude of the shift depending on the velocity of the electron during radiation. To carry the calculation through, however, it is necessary to integrate the radiated intensity over the time occupied by the slowing down. Energy considerations alone require that the electron be brought completely to rest, if radiation at the high energy limit is being considered.

So we can write $S=\int I\,\mathrm{d}t$

where the integral is taken over the whole time occupied by the slowing down.

Radiation will arrive at the point of observation at a time t which is *later* than the time of emission. Let $\mathrm{d}t'$ and $\mathrm{d}t$ denote time increments in the emission from O and arrival at P respectively. Referring to fig. 2.18 we can write (e.g. Compton and Allison, p. 98; Panofsky and Philips, p. 356)

$$\mathrm{d}t'=\mathrm{d}t-\frac{\mathrm{d}r}{c},\ \text{or}\ \mathrm{d}t=\mathrm{d}t'-\frac{v\,\mathrm{d}t'\cos\theta}{c}$$

Fig. 2.18.

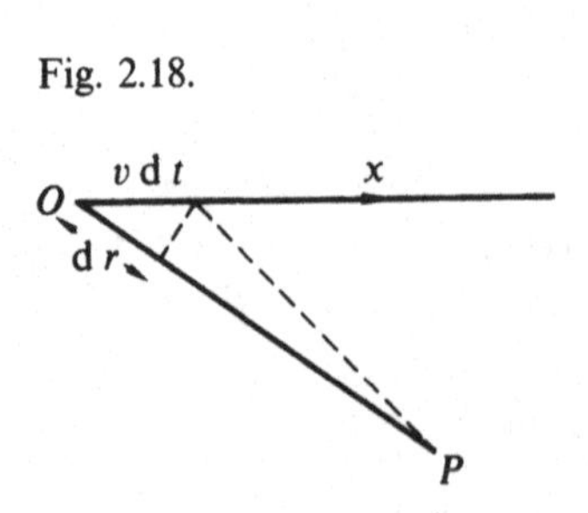

i.e.

$$\mathrm{d}t = \mathrm{d}t'(1-\beta\cos\theta)$$

$$= (1-\beta\cos\theta)\frac{\mathrm{d}t'}{\mathrm{d}\beta}\mathrm{d}\beta$$

$$= (1-\beta\cos\theta)\frac{c}{\ddot{f}}\mathrm{d}\beta$$

So the integral $\int I\mathrm{d}t$ becomes

$$\frac{1}{4\pi\varepsilon_0}\frac{c}{4\pi}\left(\frac{\ddot{f}e}{rc^2}\right)^2\int_0^\beta \frac{\sin^2\theta}{(1-\beta\cos\theta)^5}\frac{c}{\ddot{f}}\mathrm{d}\beta$$

$$= \frac{1}{4\pi\varepsilon_0}\frac{c^2}{16\pi}\ddot{f}\left(\frac{e}{rc^2}\right)^2\left[\frac{1}{(1-\beta\cos\theta)^4}-1\right]\frac{\sin^2\theta}{\cos\theta} \tag{2.6}$$

Equation (2.6) must be multiplied by $r^2\mathrm{d}\Omega$, to convert from energy per unit area to energy radiated into an element of solid angle, $\mathrm{d}\Omega$, in order to achieve conformity with the result of Panofsky and Philips, p. 360. The electric vector of the radiation lies in the plane formed by the radius vector and the direction of deceleration.

It must be remembered that this expression will hold only for photons emitted at the high energy limit. For other parts of the spectrum, Sommerfeld (1929) referred back to (2.5) and used a *mean* value (instead of the initial value) for β. For the long-wavelength parts of the spectrum (which are produced by collisions which do *not* greatly reduce the speed of the incident electron) β would be greater, and the position of the maximum *more forward* than for photons nearer the high energy limit. This treatment of Sommerfeld is thus in qualitative agreement with the measurements of Kuhlenkampff referred to earlier.

The theory of rectilinear stopping implied that the radiation in the forward direction would be zero at all parts of the spectrum. Clearly only an interaction yielding a photon near the high energy limit can be approximated to a 'head-on' collision (properly a hyperbolic path of small impact parameter); less energetic interactions will give rise to changes of velocity with a component perpendicular to the direction of the incident beam, and which will emit some radiation in the forward direction (fig. 2.19). This is observed to be the case in practice (e.g. the data of Amrehn and Kuhlenkampff referred to earlier); but even at the high energy limit the forward radiation is finite, showing that the change of electric dipole moment has a perpendicular component in cases where classically this should not occur.

More modern treatments of the continuous spectrum at low and medium energies treat the electron in the form of spherical waves, as it approaches and recedes from the nucleus, and sum the outgoing electron over all possible directions. The effective moment of the electronic charge can then be used to compute the radiation intensity. If x is the direction of the incident electron, we can write $M_{x,y,z}$ as the components of the associated electric dipole, and $I_{x,y,z}$ (proportional to the square of these moments) as the corresponding 'radiation components'.

A non-relativistic expression for the radiation intensity emitted at an angle θ to the forward direction in the xy plane would then be

$$I = I_x \sin^2\theta + I_y \cos^2\theta + I_z \tag{2.7}$$

For radiations at the high energy limit, M_x is the predominant component, but it is a feature of the theory and all subsequent treatments of it that M_y and M_z are finite, even at the high energy limit, yielding a finite intensity of radiation in the forward direction. This was experimentally confirmed by Kerscher and Kuhlenkampff, although it had been observed in many of the earlier studies of angular distribution, and was at that time thought to be due to the targets being insufficiently thin.

Sommerfeld (1931) found that the intensity from a thin target, as a function of frequency and angle, would, by the use of certain approximations, be given by,

$$i(\nu,0) = \text{const.} \times \left\{ M_x^2 \frac{\sin^2\theta}{(1-\beta\cos\theta)^4} + M_y^2 \frac{\cos^2\theta}{(1-\beta\cos\theta)^4} + M_z^2 \right\} \tag{2.8}$$

Near the high energy limit, M_x is the dominant component of the electric dipole, and so, to obtain the direction of maximum emitted intensity, we may write

$$\frac{\mathrm{d}}{\mathrm{d}\theta} \frac{\sin^2\theta}{(1-\beta\cos\theta)^4} = 0$$

Fig. 2.19. Change in linear momentum of bombarding electron (*a*) Small impact parameter (*b*) Large impact parameter.

or

$$\cos\theta_m = \frac{1}{2\beta}[(1+8\beta^2)^{\frac{1}{2}} - 1] \tag{2.9}$$

which agrees well with the measurements available at that time.

It may be noted that (2.9) reduces to

$$\cos\theta = 2\beta \tag{2.10}$$

when β is small.

A very useful formulation of the Sommerfeld theory has been given by Scheer and Zeitler (1955). For an unpolarized electron beam, $M_y = M_z$, and so equation (2.8) may be written

$$i(v,\theta) = \text{const.} \times \left\{ M_x^2 \frac{\sin^2\theta}{(1-\beta\cos\theta)^4} + M_y^2\left(1 + \frac{\cos^2\theta}{(1-\beta\cos\theta)^4}\right)\right\} \tag{2.11}$$

At right-angles to the forward direction this becomes

$$i(v,90) = \text{const.} \times (M_x^2 + M_y^2) \tag{2.12}$$

Scheer and Zeitler then use the 'degree of polarization' p_s, where

$$p_s = \frac{M_x^2 - M_y^2}{M_x^2 + M_y^2} \tag{2.13}$$

and make the further step of expressing the intensity in the form of the ratio to that at 90 degrees. Thus they obtain

$$Q(v,\theta) = \frac{i(v,\theta)}{i(v,90)} = F(\theta) + (1 - p_s(v))G(\theta) \tag{2.14}$$

where

$$F(\theta) = \sin^2\theta/(1-\beta\cos\theta)^4 \tag{2.15a}$$

and, with the approximation that $(\beta\cos\theta)^2 \ll 1$,

$$G(\theta) = \frac{1}{(1-\beta\cos\theta)^2}\left\{1 - \frac{\sin^2\theta}{(1-\beta\cos\theta)^2}\right\} \tag{2.15b}$$

$F(\theta)$ and $G(\theta)$ can readily be evaluated in terms of accelerating voltage. The moments $M_{x,y,z}$ and the parameter p_s are available from the detailed numerical calculations of Kirkpatrick and Wiedmann (1945), and will be discussed in connection with the spectral distribution and the polarization effects (section 2.9). Two further points should be made regarding expression (2.12). First, it is assumed that the incident electron beam is itself unpolarized. Hence there is circular symmetry about the x axis and M_y equals M_z. M_z has thus been eliminated from (2.11). Secondly, in situations where the transverse accelerations, and hence M_y and M_z, are small, (i.e. for photons emitted near the high energy limit), (2.9) would give the same values for the angle of maximum intensity as (2.11).

Although the equations of Scheer and Zeitler are convenient when considering the shape of the angular distribution, the position of the intensity maximum is best calculated by reverting to an earlier formulation, in which the 'Depolarization Coefficient', D, is introduced. This is defined as

$$D=\frac{M_z^2}{M_x^2}=\frac{M_y^2}{M_x^2}$$

and enables (2.11) to be written in the form

$$i(\nu,\theta)=\text{const.}\times\left\{\frac{D^{-1}\sin^2\theta}{(1-\beta\cos\theta)^4}+\frac{\cos^2\theta}{(1-\beta\cos\theta)^4}+1\right\}M_z^2 \tag{2.16}$$

This may now be differentiated and set equal to 0 giving

$$\cos\theta_{\text{m}}=\frac{1}{2\beta}\left[\left(1+\frac{8\beta^2}{1-D}\right)^{\frac{1}{2}}-1\right] \tag{2.17}$$

Equation 2.17 reduces to

$$\cos\theta_{\text{m}}=2\beta/(1-D) \tag{2.18}$$

Furthermore, (2.17) and (2.18) reduce to (2.9) and (2.10) when $D=0$, i.e. when polarization is complete.

Sesemann (1941) collected values of the angle of maximum emission from several published observations and compared them with values calculated from (2.17) (fig. 2.15). Evidently the approximations in the Sommerfeld theory begin to break down at energies above 100 keV.

Discussions of Sommerfeld's paper are given by Weinstock (1942, 1943), Kirkpatrick and Wiedmann (1945) and Kuhlenkampff, Scheer and Zeitler (1959). It is instructive to consider the treatment of Kuhlenkampff (1959). Here, the electron is regarded as a moving system emitting in accordance with (2.7) and the observations are made in the stationary (laboratory) system. If the angle in the moving system is θ', Kuhlenkampff writes

$$N(\theta')=N_x\sin^2\theta'+N_y\cos^2\theta'+N_z \tag{2.19}$$

where the quantities N_x, N_y, N_z correspond to the squares of matrix elements or similar quantities.

We now introduce the relativistic transformations of angles

$$\sin\theta' = \frac{(1-\beta^2)^{\frac{1}{2}}\sin\theta}{1-\beta\cos\theta}; \; \cos\theta' = \frac{\cos\theta - \beta}{1-\beta\cos\theta} \tag{2.20}$$

and the transformation of solid angle given by

$$N(\theta')\sin\theta'\mathrm{d}\theta' = N(\theta)\sin\theta\,\mathrm{d}\theta$$

Using (2.20), we obtain

$$N(\theta) = N(\theta')\frac{\mathrm{d}(\cos\theta')}{\mathrm{d}(\cos\theta)} = N(\theta')\frac{(1-\beta^2)}{(1-\beta\cos\theta)^2} \tag{2.21}$$

Inserting (2.20) and (2.21) into (2.19) we obtain

$$N(\theta) = \frac{1-\beta^2}{(1-\beta\cos\theta)^2}\left\{N_x\frac{(1-\beta^2)\sin^2\theta}{(1-\beta\cos\theta)^2} + N_y\frac{(\cos\theta-\beta)^2}{(1-\beta\cos\theta)^2} + N_z\right\} \tag{2.22}$$

which is identical in form with the equation given by Kuhlenkampff, Scheer and Zeitler, which in turn is a more exact form of the treatment of Sommerfeld.

The treatment just outlined shows that the problem is formally similar to the aberration of light, which is familiar from astronomical observations, but with the radiation proceeding in the opposite direction through the telescope.

(b) *Energy distribution*

The energy distribution of the continuous spectrum has been considered by Kramers (1923), who assumed a parabolic path for the incident electrons, and applied the laws of classical electrodynamics to obtain expressions for the radiation intensity. The motion of the incident electron in the Coulomb field of the nucleus gives rise to a pulse of X-radiation which Kramers then expressed in the form of its Fourier components. The high-energy limit was put in at the point where $h\nu = eV$ in accordance with the requirements of quantum theory.

Kramers obtained the result

$$i_\nu \mathrm{d}\nu \mathrm{d}x = \frac{32\pi^2}{3\sqrt{3}}\frac{Z^2e^6n\mathrm{d}x\mathrm{d}\nu}{c^3m^2v^2} \tag{2.23a}$$

for the intensity in a frequency interval $\mathrm{d}\nu$, emitted from a single electron with velocity v traversing a thin target consisting of n atoms per unit volume. At low energies, putting the electron energy eV equal to $\frac{1}{2}mv^2$, this may be written*

$$i_\nu\,d\nu dx = \frac{16\pi^2}{3\sqrt{3}}\frac{Z^2e^5 n d\nu dx}{c^3 mV}\ (\text{when } \nu < \nu_0) \tag{2.23b}$$

$$i_\nu = 0\ (\text{when } \nu > \nu_0)$$

The essential features of this expression are the constancy of i_ν over the spectrum, the proportionality with Z^2 and the inverse dependence on V. The latter implies that the integral over the whole spectrum is given by

$$i\,dx = \frac{16\pi^2}{3\sqrt{3}}\frac{Z^2 ne^6 dx}{c^3 mh} \tag{2.23c}$$

which is independent of accelerating voltage. It is assumed that the target is sufficiently thin that the electrons traverse it with negligible energy loss. The experimental work of Nicholas, and of Amrehn and Kuhlenkampff, referred to earlier, indicated that the general shape of the spectrum is given approximately correctly by this theory.

To obtain detailed information regarding the spectral distribution we return to the Sommerfeld theory, as interpreted by Kirkpatrick and Wiedmann.

In the Sommerfeld treatment, wave functions for the incoming and outgoing electron are used to obtain expressions of the form

$$e\int \psi_f^{\,*}\, r\psi_i\, d\tau \tag{2.24}$$

in which the integration is carried out over all space. These matrix elements are then summed over all directions of the emergent electron to give the 'effective electric moments' M_x, M_y and M_z.

The 'radiation components' are then given by

$$I_{x,y,z} = \frac{2\pi\nu^3 k^2}{\hbar c^3} M^2_{x,y,z} \tag{2.25}$$

where ν is the frequency of the radiation and k the momentum of the electron after radiating. Kirkpatrick and Wiedmann then evaluated the radiation components for a wide range of the variables, and found that over most of the spectrum the accelerating voltage and atomic number appear

* When discussing the work of Kramers, which has a historical flavour, it has not been thought necessary to introduce additional factors of the type needed to achieve consistency with the S.I. system of units. Elsewhere in this book, appropriate factors are introduced whenever necessary.

only in the combination Z^2/V, so they calculate the radiation components in terms of this parameter. At the low frequency end of the spectrum, the calculations become more difficult because of the effects of screening of the nucleus by the orbital electrons, and the intensity is found to depend on Z and V separately to some extent.

Kirkpatrick and Wiedmann show that the component I_x is in general the dominant one, and that it does not vary greatly for different frequencies within the spectrum. The Kramers expression thus has some degree of validity. I_y does however vary rather strongly with photon energy, being small at the high energy limit, rising for the longer-wavelength parts of the spectrum, and becoming equal to I_x at about $0.2\,\nu_0$. I_y at the high energy limit is least important when V/Z^2 is large, and becomes progressively more important as this parameter is reduced. The forward radiation (which depends solely on I_y and not at all on I_x) is thus most pronounced for high atomic numbers and low accelerating voltages. The physical reason for this is that under these conditions the Coulomb field of the nucleus has a greater effect, and tends to change the direction of the incident electron *before* the photon is emitted.

Kirkpatrick and Wiedmann obtain the total intensity, in the non-relativistic approximation, using the expression

$$I_x \sin^2\theta + I_y(1+\cos^2\theta) \tag{2.26}$$

The intensity in the forward direction and at right-angles to the forward direction is thus given simply by

$$I_0 = 2I_y;\; I_{90} = I_x + I_y \tag{2.27}$$

Fig. 2.20 drawn from their data illustrates I_0 and I_{90} for an accelerating voltage of 20 kV.

Equation (2.26) may be integrated to give the energy emitted over all directions:

$$\begin{aligned} I &= \int_0^\pi [I_x \sin^2\theta + I_y(1+\cos^2\theta)]2\pi \sin\theta \mathrm{d}\theta \\ &= \tfrac{8}{3}\pi(I_x + 2I_y) \end{aligned} \tag{2.28}$$

Fig. 2.21 illustrates this for various values of V/Z^2. The Kramers approximation (giving a flat spectrum) is seen to be most useful for high atomic numbers and low accelerating voltages. The total intensity at the high energy limit is shown in fig. 2.22. It is approximately proportional to Z^2/V, conforming to Kramers' expression.

Fig. 2.20. Theoretical spectral distributions at 0° and 90° calculated from the data of Kirkpatrick and Wiedmann for an accelerating voltage of 20 kV. (*a*) $Z=9$ (*b*) $Z=65$. The unit of emitted energy is joule sterad^{-1} Hz^{-1} cm^2 atom^{-1} electron^{-1}.

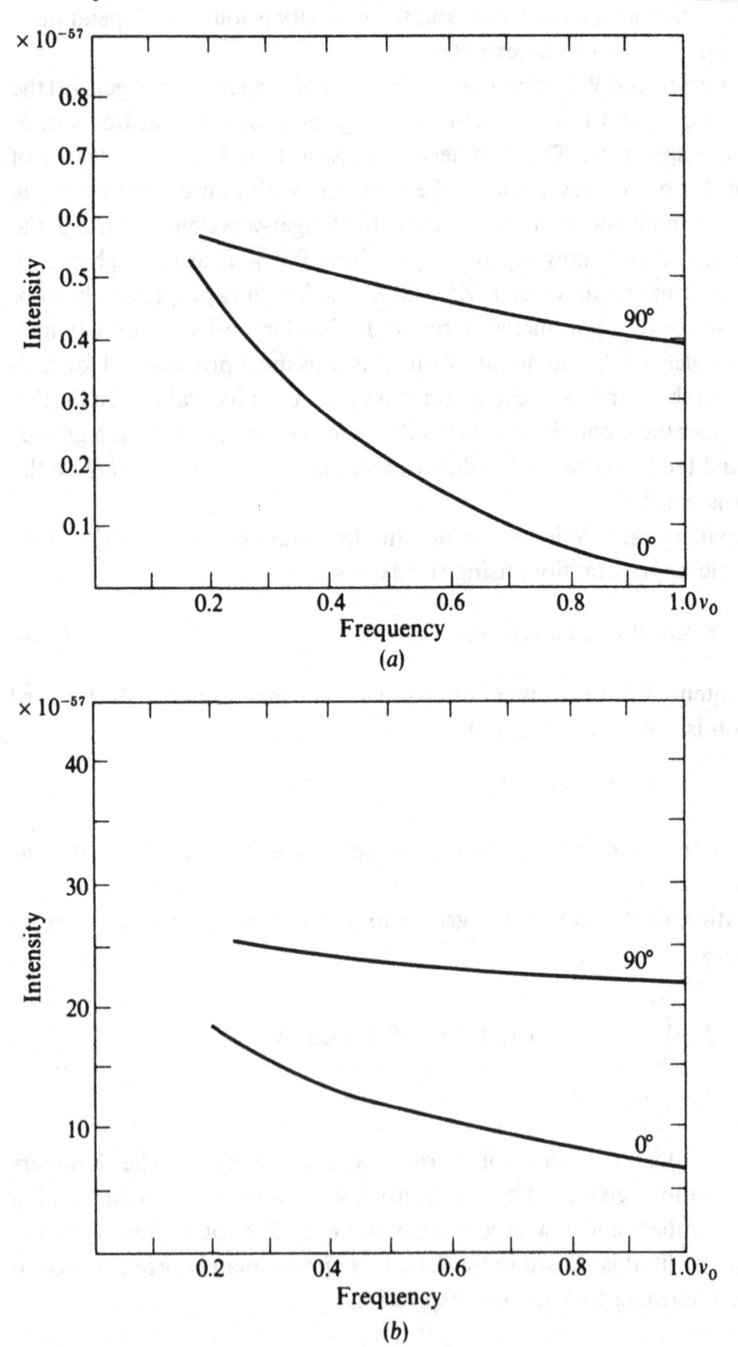

Fig. 2.21. Variation of emitted energy (summed over all directions) with quantum energy, for various values of V/Z^2 (V in volts). The unit of emitted energy is joule $\mathrm{Hz}^{-1}\,\mathrm{cm}^2\,\mathrm{atom}^{-1}\,\mathrm{electron}^{-1}$ (after Kirkpatrick and Wiedmann, 1945).

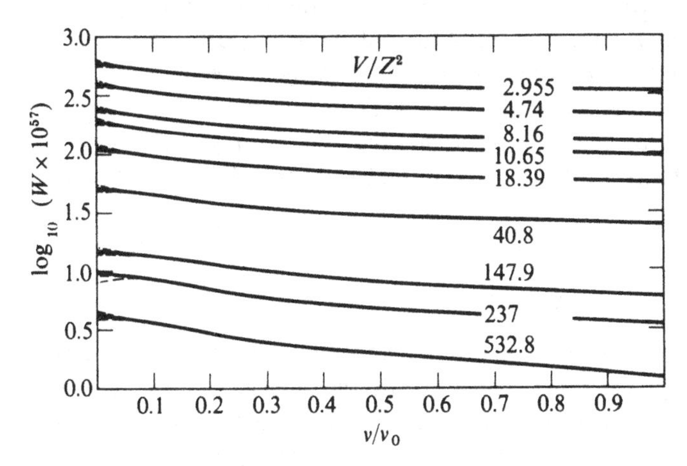

Fig. 2.22. Emitted radiation at the high energy limit as a function of Z^2/V (V in volts). (Units as in fig. 2.21.)

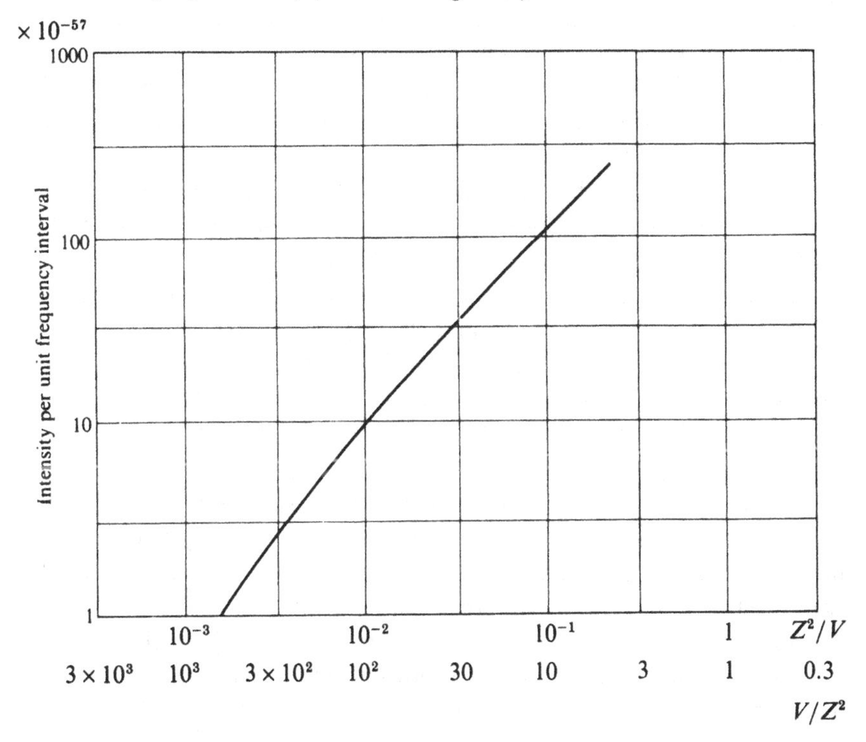

Kirkpatrick and Wiedmann's use of the non-relativistic expression (2.26) means that their work cannot be used immediately for the calculation of angular distributions. But when combined with the expressions of Scheer and Zeitler referred to earlier, detailed calculations become possible, from which the angular distribution of the radiation can be evaluated for any value of Z and accelerating voltage, for all parts of the spectrum.

Approximations in the Sommerfeld expressions make them unsuitable when the electron velocity approaches the velocity of light. The theory of Bethe and Heitler (1934) is valid under the conditions of the Born Approximation.

$$2\pi\frac{Ze^2}{hv_0} \ll 1;\ 2\pi\frac{Ze^2}{hv_e} \ll 1$$

where v_0 and v_e are the velocities of the incoming and outgoing electron. The spectral distribution is found to have the same general characteristics as at low energies, and the cross-section (integrated over all directions) is approximately proportional to Z^2. The probability of an electron emitting more than one photon before coming to rest becomes appreciable, and Bethe and Heitler give expressions for the average number of protons emitted as a function of electron energy.

The radiation becomes very sharply peaked in or near the forward direction; and even with targets much thinner than the electron range the effects of small-angle multiple scattering are appreciable compared with the intrinsic effects of the Bremsstrahlung process. Some authors (e.g. Lawson, 1950) have preferred to neglect the intrinsic effect, and comparisons with experiment are then to be regarded as tests of electron scattering theory rather than of the Bremsstrahlung process (e.g. Muirhead, Spicer and Lichtblau, 1952). It has however been established (e.g. Bethe, 1934) that a transverse momentum of the order mc is expected in the emitted photon, so in the case where all the momentum of the (extreme relativistic) electron is transferred to the photon it would be expected to be emitted at an angle of the order of m_0c^2/E_0. The measurements of Zdarko *et al.* (1964) referred to in section 2.6 are interesting in this connection.

(c) *Thick targets and the effect of scattering and energy-loss*

The range of electrons in matter has been the subject of many investigations. The data between 10 keV and 10 MeV has been reviewed by Katz and Penfold (1952) and that below 10 keV by Kanter (1961). Theoretical treatment is usually expressed in the form of the Bethe–Bloch formula for electron energy-loss. This formulation is discussed in section 3.5a. However, many workers have expressed their results in the form of a power law, operative over a restricted range of electron energy. If we write $R = \text{const.}\ V^\alpha$, α is found to be about 1.7 for energies in the region of 20 keV, and falls progressively until it becomes almost a linear law ($\alpha = 1$) with a small constant correction term, above about 1 MeV. Whiddington (1912, 1914) found that $\alpha = 2$ is serviceable in the range 8–30 keV, and the so-called Thomson–Whiddington law has been used frequently for approximate calculations in this field. Its use is to some extent justified by the work of Lane and Zaffarano (1954), who found $\alpha = 1.8$ for aluminium between 2 and 20 keV. For a more detailed account of range–energy relations see appendix 1.

Writing Kramers' expression in the form

$$\begin{aligned} i_\nu \mathrm{d}\nu \mathrm{d}x &= \frac{16\pi^2}{3\sqrt{3}} \frac{Z^2 e^5 n}{c^3 m V} \mathrm{d}x \mathrm{d}\nu \quad (\nu < \nu_0) \\ i_\nu &= 0 \qquad\qquad (\nu > \nu_0) \end{aligned} \tag{2.29}$$

substitution of an expression for rate of energy-loss followed by integration over electron-energy leads to an expression for the spectral distribution from a thick target.

The Thomson–Whiddington Law for energy-loss may be written in the form suitable for use with (2.29)

$$V^2 = 4\pi e^2 n Z \ell x \tag{2.30}$$

This makes use of Bohr's (1913) value for the constant of proportionality as simplified by Kramers, in which ℓ is a numerical factor in the region of 6,

We get

$$\frac{\mathrm{d}V}{\mathrm{d}x} = -4\pi e^2 n Z \ell \frac{1}{2V} \tag{2.31}$$

Hence

$$\begin{aligned} i_\nu \mathrm{d}\nu \mathrm{d}x &= \frac{16\pi^2}{3\sqrt{3}} \frac{Z^2 e^5 n}{c^3 m} \mathrm{d}\nu \frac{1}{2\pi e^2 n \ell Z} \mathrm{d}V \\ &= \frac{8\pi}{3\sqrt{3}\ell} \frac{Z e^3}{c^3 m} \mathrm{d}\nu \mathrm{d}V \end{aligned} \tag{2.32}$$

and this is to be integrated with respect to electron energy from eV_0 the incident energy, to eV_ν the lowest energy at which the beam is able to produce quanta of energy $h\nu$.

Hence

$$I_\nu \mathrm{d}\nu = \frac{8\pi}{3\sqrt{3}\ell} \frac{h e^2 Z}{c^3 m} (\nu_0 - \nu) \mathrm{d}\nu \tag{2.33}$$

or

$$I_\nu = A_c Z(\nu_0 - \nu) \tag{2.33a}$$

where A_c is the constant which appears in the first term of Kuhlenkampff's expression (2.2).

This result shows a useful degree of agreement with the work of Kuhlenkampff and others, in which the non-linearity near the high energy limit is often quite small. It is, however, limited in its applicability by three important considerations.

(1) The thin target spectrum is not flat, but tends to diminish towards the high-energy limit (see for example fig. 2.7), especially for directions of emission which differ markedly from 90°.
(2) The simple (Thomson–Whiddington) form of the electron energy-loss expression is valid over only a restricted range of energies.
(3) Self-absorption of the X-radiation emerging from the target will attenuate the output, particularly at low photon energies.

Limitations (1) and (3) tend to cancel each other, thereby causing (2.33a) to be a more useful approximation than would otherwise be the case. This aspect of the continuous X-ray spectrum has been examined in detail by Soole (1972, 1977).

This treatment readily lends itself to extension:

If we write the range R in the form V^{q+1}, where q is somewhat less than unity, it can readily be shown that the intensity now assumes the form

$$\begin{aligned} I_\nu &= a_1 Z(\nu_0{}^q - \nu^q) \quad (\nu < \nu_0) \\ I_\nu &= 0 \qquad\qquad\qquad\quad (\nu > \nu_0) \end{aligned} \tag{2.34}$$

The spectra might thus be expected to be of the form shown in fig. 2.23, in which q is taken to be typically 0.6. In practice, however, all X-ray spectra from thick targets are considerably modified by the effects of backscattering of electrons from the target. Any electrons backscattered with reduced energy will clearly cause the spectrum to be deficient in photons of lower energy, and this effect will offset the curvature apparent in fig. 2.23. For heavier elements the backscattering is in fact sufficient to produce a slight curvature of opposite sense.

(d) *Semi-thick targets*

It has been assumed that the thickness of the target is sufficient to reduce the electrons to rest, but it is interesting to consider the situation when the targets is of intermediate thickness. If the incident and emergent energies are eV_0 and eV_e respectively, we can write,

$$i_\nu \mathrm{d}\nu \mathrm{d}x = \frac{k_1 Z^2 n \mathrm{d}\nu \mathrm{d}x}{V} \qquad \left(k_1 = \frac{16\pi^2 e^5}{3\sqrt{3c^3 m}}\right)$$

$$\text{for } h\nu < eV_0 \tag{2.35}$$

$$i_\nu \mathrm{d}\nu \mathrm{d}x = 0 \qquad \text{for } h\nu > eV_0$$

For the energy loss, following the previous argument

$$\frac{\mathrm{d}V}{\mathrm{d}x} = -\frac{1}{V} k_2 n Z, \text{ where } k_2 = 2\pi e^2 \ell \tag{2.36}$$

The intensity as a function of quantum energy has to be considered in three parts:

(a) $h\nu < eV_e$: integrating from V_0 to V_e

$$I_\nu \mathrm{d}\nu = \frac{k_1}{k_2} Z \mathrm{d}\nu \int_{V_e}^{V_0} \mathrm{d}V = \frac{k_1}{k_2} Z(V_0 - V_e) \mathrm{d}\nu \tag{2.37}$$

$$= \frac{k_1}{k_2} \frac{h}{e} Z(\nu_0 - \nu_e) \,\mathrm{d}\nu \text{ where } h\nu_e = eV_e \tag{2.38a}$$

Fig. 2.23. Thick target spectral distributions calculated on the assumption of a range-energy relation $x = V^{1+q}$ when $q = 0.6$ (see text).

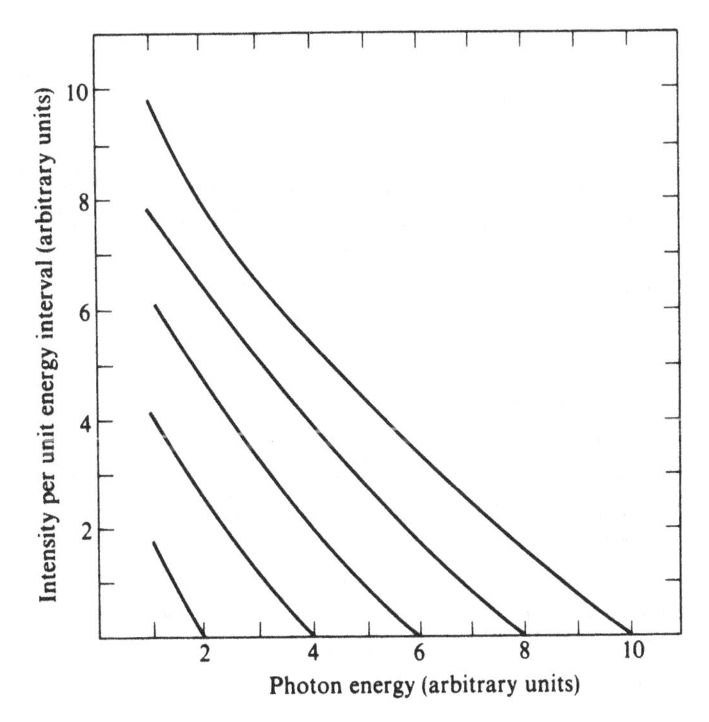

(b) $eV_e < h\nu < eV_0$: Integrating from V_0 to $h\nu/e$

$$I_\nu \mathrm{d}\nu = \frac{k_1}{k_2}\frac{h}{e}Z(\nu_0 - \nu)\,\mathrm{d}\nu$$

(c) $h\nu > eV_0$

$$I_\nu \mathrm{d}\nu = 0$$

If $V_e = 0$ we have of course

$$I_\nu \mathrm{d}\nu = \frac{k_1}{k_2}\frac{h}{e}Z(\nu_0 - \nu)\mathrm{d}\nu \qquad \text{for all } \nu \text{ up to } \nu_0,$$

as in (2.33).

The intensity per unit frequency interval for a semi-thin target is thus the same as for an electron-opaque target for that part of the spectrum from the high energy limit down to the photon energy corresponding to the emergent energy of the electrons, and is constant for all smaller photon energies.

From (2.37) we see that *layers causing equal electron energy-loss give equal intensities per unit frequency interval.* This applies to layers of finite thickness as well as to infinitesimal layers.

By integrating (2.35) over all photon energies, to give the total energy for a thickness $\mathrm{d}x$, we get

$$I\,\mathrm{d}x = k_1 Z^2 n \frac{e}{h}\mathrm{d}x \qquad (2.39)$$

Integration over finite x then shows that the *total intensity from a semi-thin target is independent of electron energy* and that *equal thicknesses yield equal total intensities.* This was deduced without making any assumptions regarding the form of the energy-loss relation, and requires only that the electrons emerge from the target with finite velocity. These rules are illustrated in fig. 2.24(*a*)–(*c*).

(e) *Electron scattering effects*

In practice the shapes of spectra from semi-thin targets are considerably affected by the straggling of the electrons, and by departures from Kramers' simple expression. Few experimental data are available, but fig. 2.25 illustrates spectra obtained in the forward direction from semi-thin gold targets evaporated on to a distrene backing, together with a spectrum from an electron-opaque target for comparison. The spectral shapes show some degree of conformity with the above predictions, and the 'equal thickness' rule is approximately obeyed.

Fig. 2.24. Spectra from semi-thin targets: (*a*) layers causing equal energy-loss give equal intensities per unit frequency interval (*b*): The four semi-thin target spectra produced from a single layer of target material bombarded at four different energies have equal areas, and may be added together to give the thick target spectrum shown in (*c*).

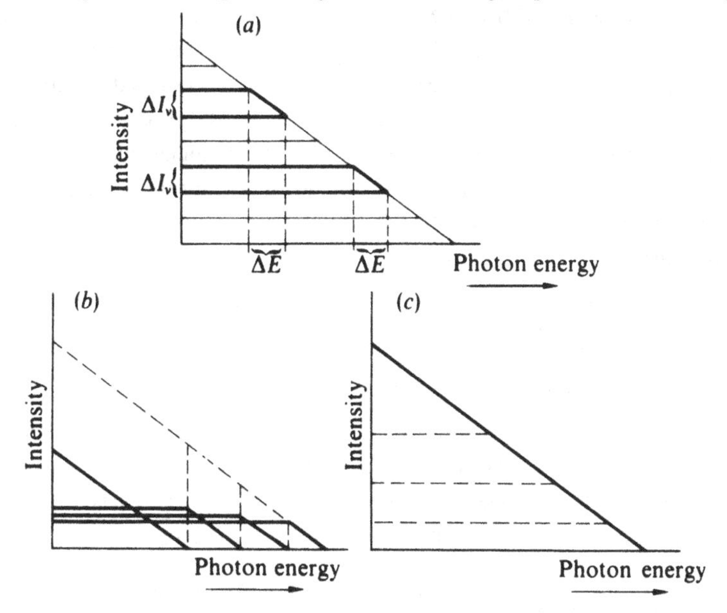

Fig. 2.25. Semi-thin target spectra for gold, with an electron-opaque target for comparison. Curves *B* are corrected for radiation produced in the backing foil. The uncorrected data are shown as curves *A*. (Dyson, unpublished).

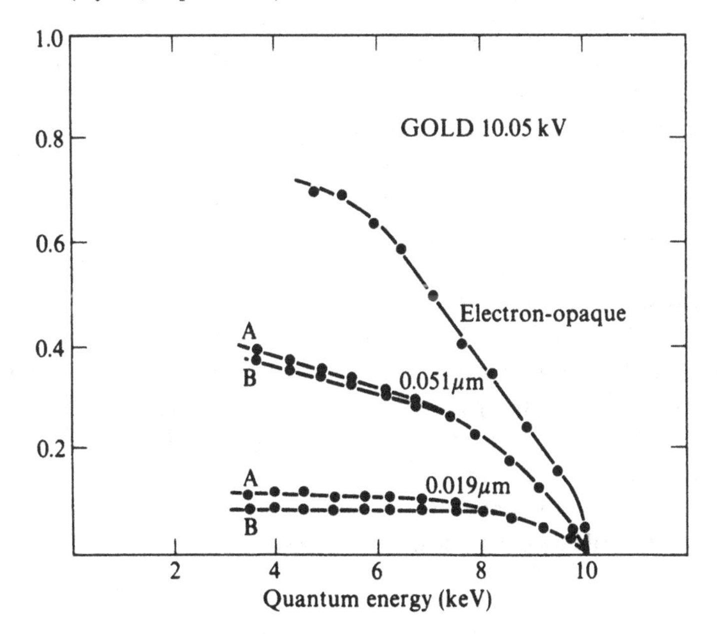

In discussing the spectral distributions from thick and semi-thin targets we have been concerned primarily with the energy losses which the incident electrons experience as a result of inelastic scattering events with other electrons in the target. To consider the angular distribution, we must examine nuclear scattering, because it is these collisions which are mainly responsible for the change of direction of the incident electrons. The scattering of electrons is discussed in appendix 1. It is pointed out there that the elastic scattering can be treated by regarding the electron as being subjected to Rutherford scattering by the atomic nuclei, but modified by screening effects. From the work of Lenz (1954), elaborated by Cosslett and Thomas (1964a, b), it is possible to express the elastic scattering in the simple form

$$\sigma_{el} = \frac{Z^{4/3}h^2}{2\pi mE}, \qquad (2.40)$$

where E is the electron energy.

From this expression the number of scattering events in a given thickness of target can be calculated. For example the thickness in which an average of one elastic scattering occurs is given by $x = (\sigma n)^{-1}$ (where n is the number of atoms per unit volume), or, when expressed in terms of mass per unit area, by $\rho x = 2\pi \frac{m}{h^2} \frac{A}{N} \frac{E}{Z^{4/3}}$. This thickness is of importance in the theory of the electron microscope, as it establishes a criterion for the optimum thickness of the specimen. We shall call this the 'single elastic scattering thickness'. It is seen not to vary greatly from element to element, and it is proportional to electron energy. For example at 10 keV the single elastic scattering thickness is approximately 2×10^{-6} and 1×10^{-6} g cm^{-2} for aluminium and gold respectively. Combining these data with the expressions for energy-loss, it is easy to determine how many elastic scattering events occur during a given amount of slowing down in a target. The distance in which a 10 keV electron loses 10% of its energy in these elements is approximately 40 and 80×10^{-6} g cm^{-2} respectively, so it is immediately clear that a large number of elastic scattering events (20 and 80 respectively in our example) take place before much energy is lost by the inelastic processes; even radiation near the high energy limit will be produced by electrons which have been scattered many times, so the 'thin target' angular distributions will be grossly modified when thicker targets are used. The degree of 'smearing out' will be greater for elements of high atomic number. The data of fig. 2.13 substantiate this.

Turning back to the data on the angular distribution from very thin targets, we can calculate that the data of Kerscher and Kuhlenkampff, and that of Honerjäger, represent a real approach to the condition of ideal thinness. For example, the 25 nm nickel target used by the former workers is only four times the single elastic scattering thickness at 34 keV, and Honerjäger's magnesium target (10 nm in thickness) is only one half the single elastic scattering thickness at this electron energy. In these experiments, electron scattering would thus not affect the angular distribution to any real extent.

Further details on the scattering and energy-loss of electrons will be found in appendix 1.

2.8 The efficiency of production of continuous X-radiation

Kirkpatrick and Wiedmann compared the total intensity given by (2.28) with the energy lost by the electrons and found the efficiency to be

$$\eta = 2.8 \times 10^{-6} ZV \qquad (2.41)$$

where V is the accelerating potential expressed in kilovolts, although their calculations showed that it is not strictly proportional to the product ZV, but depends on these variables separately to some extent.

Absolute measurements on thin targets have been reported by several authors, although only the more recent measurements have yielded reasonably good agreement with the Sommerfeld theory. Smick and Kirkpatrick (1941) made an absolute measurement of the radiation from a thin target of nickel bombarded with 15 keV electrons, using a band of the X-ray spectrum centred at about 8 keV. They obtained a value of 2.2×10^{-57} joule sterad^{-1} Hz^{-1} cm^2 atom^{-1} electron^{-1} for the energy radiated at 88° to the forward direction, which is to be compared with a value of 6×10^{-57} (in the same units) from the calculations of Kirkpatrick and Wiedmann. Clark and Kelly (1941) made a measurement at 60° to the forward direction, using a thin target of aluminium (accelerating voltage 31.7 kV, quantum energy 26.2 keV) and found the energy radiated to be $6.2 \times 10^{-58} (\pm 33\%)$ in the above units, compared with 4.23×10^{-58} obtained by calculation. The calculated values are those given by Massey and Burhop (1952) rather than those to be found in the original papers.

More recent work has been reported by Amrehn and Kuhlenkampff (1955) who, in the course of their studies of the energy distribution from thin targets, state that good absolute agreement between theory and experiment was noted in the case of aluminium at 34 keV. Amrehn (1956) has compared his data for carbon, aluminium, nickel, silver and gold with theory. Agreement is often better than 10% and usually better than 20%. This represents a useful verification of the theory at this electron energy.

Motz and Placious (1958) have obtained photon spectra on an absolute basis at various angles to the forward direction using very thin targets of aluminium and gold and an accelerating voltage of 50 kV. They find that the Kirkpatrick and Wiedmann cross-sections are most accurate at angles in the region of 90°, (where the denominators $(1-\beta\cos\theta)$ are near to unity). Correction using a factor $(1-\beta\cos\theta)^{-2}$ brings reasonable absolute agreement at other angles also especially in the case of gold, but their angular distribution data indicate that a correction of $(1-\beta\cos\theta)^{-4}$ gives a closer resemblance between theory and experiment.

This calculation may be extended to the case of electron-opaque targets. Using the data of Williams (1931) for the rate of energy-loss, Kirkpatrick and Wiedmann obtained an expression for the thick target efficiency

$$\eta = 1.3\times10^{-6} ZV \text{ (}V\text{ in kilovolts)} \tag{2.42a}$$

This is to be compared with similar expressions obtained by Kramers (by integration of (2.33)) and Compton and Allison (1935, who reviewed the experimental data available at that time) in which the constants of proportionality are 0.92×10^{-6} and 1.1×10^{-6} respectively. This equation is often written in the form

$$P = KZV^2 \tag{2.42b}$$

where P is the power radiated per unit current.

The absolute intensity of production in the forward hemisphere was measured for electron-opaque targets of aluminium, copper, and gold, by the present writer (1959). The extrapolation of the low energy end of a continuous X-ray spectrum is attended by some uncertainty, because the exact shape is not known in this region; but the efficiency was found to be proportional to applied potential quite accurately, and to atomic number approximately, with a constant of proportionality of 1.3×10^{-6} (V in kilovolts). Values of the quantity K have been determined by Green and Cosslett (1968) for applied potentials up to 50 kV. The value of K depends to some extent on Z. It passes through a minimum of $0.75\times10^{-6}\,\text{kV}^{-1}$ at $Z=60$, rising smoothly on either side to 1.5×10^{-6} for $Z=13$ and 0.85×10^{-6} for $Z=79$. These data are summarised in fig. 2.26.

Referring back to (2.2) and neglecting the small term in Z^2, we note that if I_ν is treated as the power radiated per unit frequency interval per unit current, integration over the range $0\rightarrow\nu\rightarrow\nu_0$ yields the result

$$P = A_c Z\frac{\nu_0{}^2}{2} \tag{2.43a}$$

Substituting for photons at the high energy limit $h\nu_0 = Ve$, we get

$$P = \frac{A_c}{2}\left(\frac{e}{h}\right)^2 ZV^2 \tag{2.43b}$$

where P is the *total radiated power per unit current.*

The production efficiency (i.e. radiated power divided by electrical input power) then becomes

$$\frac{P}{V} = \eta = \frac{A_c}{2}\left(\frac{e}{h}\right)^2 ZV \tag{2.44}$$

and the quantity

$$\frac{A_c}{2}\left(\frac{e}{h}\right)^2$$

may therefore be identified with K in (2.42b). If V is in kilovolts, the dimensions of K are (kV^{-1}.

Fig. 2.26. Efficiency of continuous X-ray production. (Green and Cosslett, 1968).

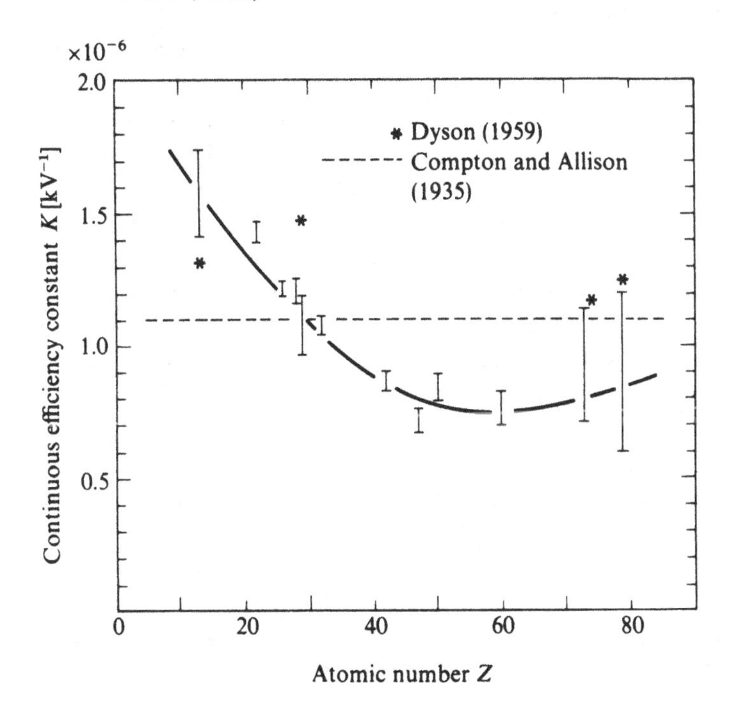

We may note at this point that the terms in the efficiency expression (2.44) may be regrouped to give

$$\eta = \frac{A_c}{2}\frac{e}{h^2}Z(Ve) = \frac{A_c}{2}\frac{e}{h^2}ZE \qquad (2.45)$$

where E is the electron energy. $\frac{A_c}{2}\frac{e}{h^2}$ (which we may temporarily denote as K') is now in reciprocal energy units but if E is expressed in kiloelectron-volts, K' is numerically unchanged, although it now has the dimensions of keV^{-1}. The same efficiency expressions (2.42a and 2.42b) may be used to give the *fraction of radiated energy per incident electron*, with similar numerical values for K. We shall have occasion to use this formulation later in this book.

2.9 Polarization of the continuous radiation

Electromagnetic radiation from an oscillating electric dipole is emitted with the electric vector parallel to the plane containing the axis of the dipole and the direction of emission, and perpendicular to the direction of propagation. One may therefore expect to find the continuous X-radiation to be polarized to some extent. The old theory of rectilinear stopping of the electrons implied complete linear polarization of all parts of the spectrum, but this was soon found not to be the case. A more detailed consideration in terms of the hyperbolic orbits of the incident electrons suggests that this may well be so for electrons giving up most or all their energy in a close encounter, because the deceleration is then substantially in line with the direction of the incident electrons, leading to the prediction that radiation near the high energy limit will be completely polarized with the electric vector parallel to the emission plane.* In practice this is modified, even with ideally thin targets, by the possibility of an electron being deviated in the nuclear field before emitting the Bremsstrahlung photon; and there is an additional requirement to the effect that radiation of all energies in the forward direction must show a complete absence of linear polarization because of symmetry requirements.

Photons with an energy which is far from the high energy limit are radiated by electrons which experience little change in velocity, and which are deflected through relatively small angles in the nuclear field. The deceleration is thus essentially transverse and may occur in any direction

* Defined as the plane containing the direction of the incident electrons and the direction of observation of the emitted photons.

relative to the emission, giving rise to some degree of polarization perpendicular to this plane.

A reversal of polarization is thus expected to occur at some point within the spectrum, and this is observed in practice. These remarks apply to electrons of low or moderate energy only. Consideration of extreme 'relativistic' electrons is deferred until later.

The linear polarization of a beam of emitted X-rays may be defined by the expression

$$p=\frac{I_{\parallel}-I_{\perp}}{I_{\parallel}+I_{\perp}} \tag{2.46}$$

where $I_{\parallel}$ and $I_{\perp}$ refer respectively to the radiation components with polarization parallel and perpendicular to the plane of emission (fig. 2.27). Complete polarization with the electric vector parallel to the emission plane thus corresponds to $p=+1$. (This definition follows the practice adopted during the development of the subject, although more recent work has tended to use a definition which is opposite in sign to this.)

For emission in the xy plane at an angle θ to the forward direction, (using the non-relativistic approximation), we have

$$I_{\parallel}=I_x \sin^2\theta+I_y \cos^2\theta$$
$$I_{\perp}=I_z$$

Fig. 2.27. See text.

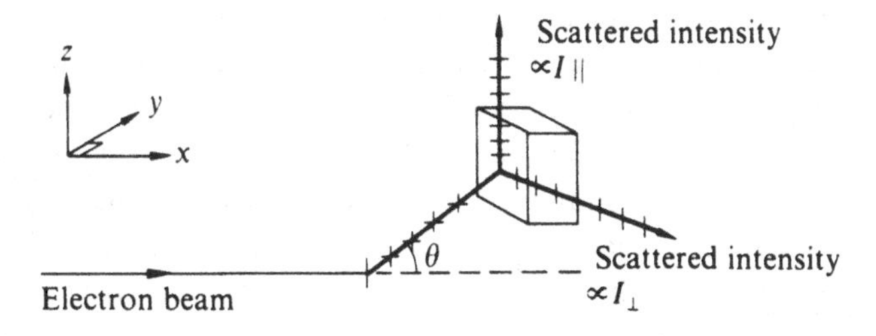

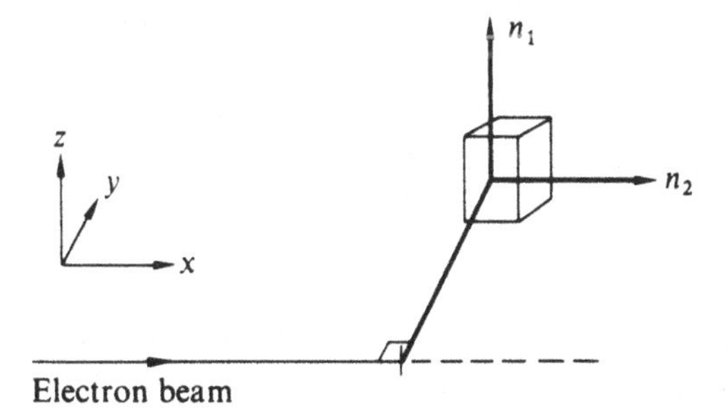

Hence,

$$p = \frac{I_x \sin^2\theta + I_y \cos^2\theta - I_z}{I_x \sin^2\theta + I_y \cos^2\theta + I_z} \tag{2.47}$$

Putting $I_y = I_z$ we get

$$p = \frac{I_x \sin^2\theta - I_y \sin^2\theta}{I_x \sin^2\theta + I_y(1+\cos^2\theta)}$$

or

$$p = \frac{1-(I_y/I_x)}{1+(I_y/I_x)(2\,\mathrm{cosec}^2\,\theta - 1)} \tag{2.48}$$

For the case of $\theta = 90°$, p becomes $(I_x - I_z)/(I_x + I_z)$ $(= p_{90})$. p_{90} is the same as the 'degree of polarization', p_s used by Scheer and Zeitler, and others (section 2.7).

From (2.48) we see that $p = 1$ for all angles of emission if $I_y/I_x = 0$, except for the limiting case of $\theta = 0$. I_y/I_x is always finite in practice; hence in the forward direction the polarization is always equal to zero.

The incomplete nature of the polarization seems to have been clearly established by Kirkpatrick (1923), and by Duane (1929). Duane, using a mercury vapour target and a voltage of about 12 kV, obtained $p_{90} = 0.474$ for the average degree of polarization over the whole spectrum.

Piston (1936) used the method of balanced filters (chapter 4) to look at a portion of the spectrum between the K edges of tantalum and tungsten, using targets of aluminium and silver. The excitation voltage was varied from the appropriate quantum limit up to 120 kV. Piston considered that the polarization should be very nearly complete for *low* atomic numbers and high accelerating voltages. This generalization has been confirmed by subsequent work, and is also in accordance with theoretical considerations.

Kirkpatrick and Wiedmann have calculated the parameter p_{90} (as a function of ν within the continuous spectrum) for various values of V/Z^2. Their curves are reproduced in fig. 2.28, from which it is seen that the general trends observed by earlier workers are confirmed. The degree of polarization is seen to depend mainly upon position within the spectrum, but is influenced to some extent by atomic number and accelerating voltage, particularly near the high energy limit.

Polarization as a function of angle has been given in (2.48). At small angles to the forward direction this reduces to

$$\frac{(1-D)\sin^2\theta}{2D} \tag{2.49}$$

which may be either positive or negative, according as to whether I_y is less than or greater than I_x.

Kuhlenkampff and Zinn (1961) have made measurements using targets of aluminium oxide and gold for 35 and 45 kV, at 90° to the forward direction, as a function of photon energy. They find the null point to occur at about 15% of the high energy limit. For gold the polarization at the high energy limit is less, and falls off less steeply with decreasing quantum energies than for aluminium. They also found that the polarization was reduced as the target thickness was increased, due to increased electron scattering under these conditions.

Equation (2.48) predicts a maximum polarization for radiation emitted at 90° to the forward direction, but for higher electron energies, the expression for p has to be modified by the inclusion of the denominators $(1-\beta\cos\theta)^4$ for the radiation components I_x and I_y. The maximum polarization then occurs at angles smaller than 90°.

In addition, at higher energies the effect of electron spin becomes more important – the emission of Bremsstrahlung can take place as a result of a 'spin-flip' (a change of 'spin current') and this process supplements the normal (change of 'orbital current') mechanism. The spin-flip process will produce circularly polarized Bremsstrahlung, and the interference between this and the linearly polarized radiation will in general produce elliptically polarized radiation. However, the plane of maximum linear polarization (i.e. the major axis of the polarization ellipse) is not necessarily that of the

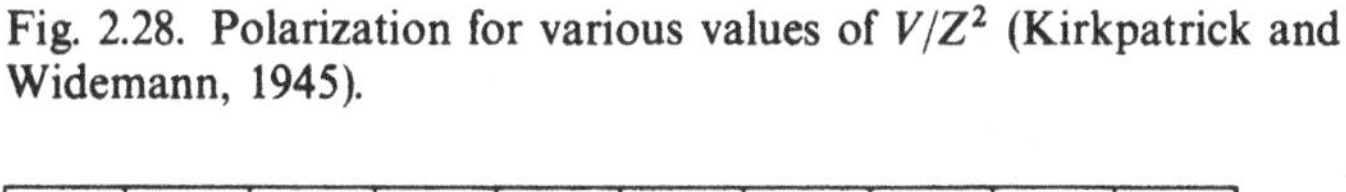
Fig. 2.28. Polarization for various values of V/Z^2 (Kirkpatrick and Widemann, 1945).

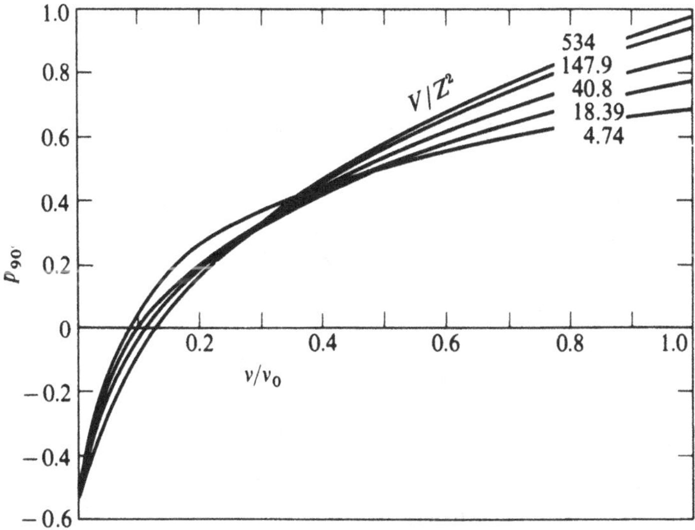

orbital current contribution and in circumstances where the interference is destructive it may be perpendicular to the emission plane.

For these reasons the maximum polarization is displaced to angles of less than 90° to the forward direction, and if the perpendicularly polarized component dominates, the polarization becomes negative. This reversal occurs at progressively smaller angles as the incident electron energy is increased. It is of course quite distinct from the reversal which occurs at fixed angles as one goes from the high energy limit of the spectrum to low photon energies.

The polarization at these high energies has been discussed by Fano *et al.* (1959) and by Motz and Placious (1960). Fig. 2.29 (from their paper) is based on the calculations of Gluckstern and Hull (1953), for radiation at 0.9 of the high energy limit in each case. A study at 50 and 180 keV is reported by Scheer *et al.* (1968).

The polarization at low photon energies within the spectrum is not influenced markedly by spin effects, and is in any case negative; no reversal occurs with angle, but the denominators $(1-\beta\cos\theta)^4$ are still effective in reducing the angle of maximum (negative) polarization to angles considerably below 90°.

Fig. 2.29. Polarization as a function of angle of observation. (Motz and Placious, 1960; their definition of p is opposite in sign to ours).

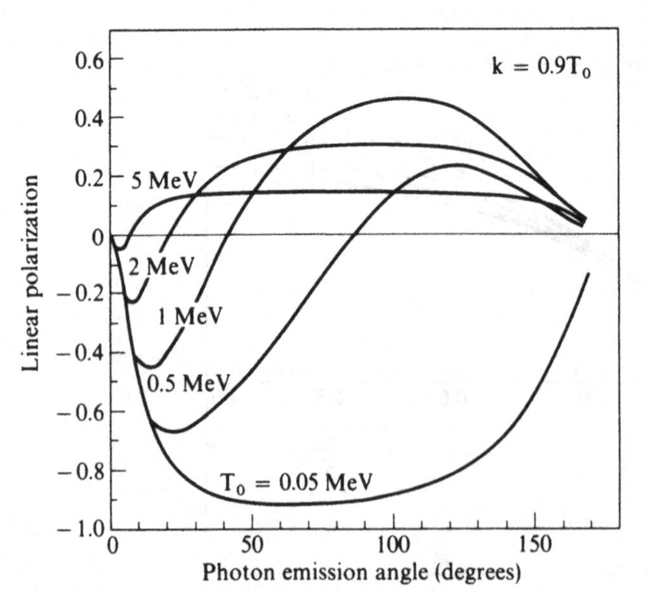

More recent experimental data has been presented by Kuckuck and Ebert (1973), for thin targets of aluminium, copper, silver and gold, for incident electron energies of 50, 75 and 100keV. A sample of their data is illustrated in fig. 2.30, again with theoretical data of Gluckstern & Hull (1953) for comparison. We see that the polarization is less than theory would indicate, and that the discrepancy becomes greater with increasing atomic number. This is known as the Coulomb effect, and is caused by the electron being deflected before the emission of the photon takes place. This effect would of course be expected to increase with increasing nuclear charge. The calculations of Gluckstern and Hull do not take account of it, although the influence of Z is in fact apparent in the curves of Kirkpatrick and Wiedmann reproduced in fig. 2.28.

Fig. 2.30. Polarization as a function of angle of observation (Kuckuck and Ebert, 1973). (The definition of p is as for fig. 2.29).

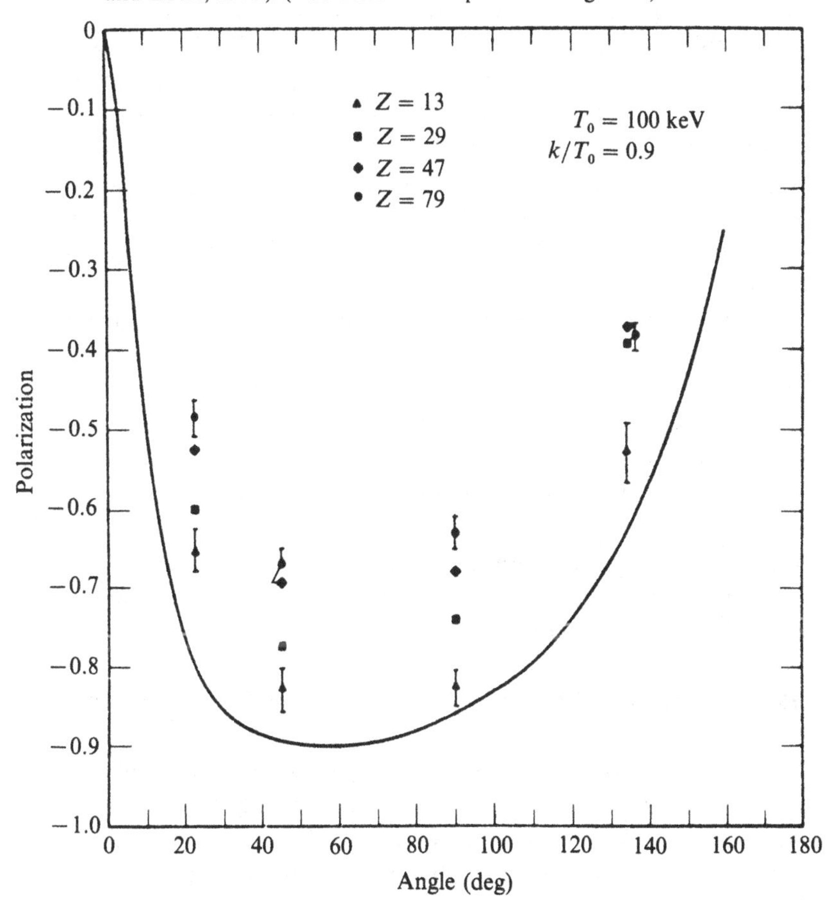

A discussion of the effects of electronic screening is to be found in the papers of Gluckstern and Hull and of Kirkpatrick and Wiedmann.

Discussion of polarization in the region 1–1000 MeV is to be found in Motz and Placious' paper. An account of measurements made at a photon energy of 15.1 MeV in the Bremsstrahlung from 25 MeV electrons, is given by Jamnik and Axel (1960).

In all the above work it has been assumed that the incident electron beam is unpolarized. If this is not so, additional effects become possible. If the beam is *transversely* polarized, the cross-section for Bremsstrahlung production may exhibit azimuthal variations, as a function of the angle between the emission plane of the photons and the plane of polarization of the incident electrons (Johnson and Rozics, 1962). It is found however that the degree of linear polarization of the photons does not depend on the state of polarization of the electrons, at least under the conditions of the Born Approximation. If the electrons are *longitudinally* polarized, the Bremsstrahlung is in general *circularly* polarized, to an extent which may be considerable (Goldhaber *et al.*, 1957; McVoy, 1958; Olsen and Maximon, 1958; Fronsdal and Uberall, 1958; Bisi and Zappa, 1959). The photons emitted in the Bremsstrahlung process have spin direction $(+1)$ which are the same as the spins of the incident electrons $(+\frac{1}{2})$, the angular momentum being balanced by the spin of the recoil electrons $(-\frac{1}{2})$. The degree of circular polarization is approximately proportional to the longitudinal component of the spin of the incident electron. Bisi, Fasana and Zappa (1963) have observed that the degree of circular polarization falls with increasing atomic number of the emitting target. Galster (1964) attributes this to the effect of electron diffusion in the target.

Neumcke (1966) has reported the production of circularly polarized Bremsstrahlung at 1 GeV, (together with a linearly polarized component) using longitudinally polarized electrons.

If the nuclei of the target are allowed to be oriented, the production of circularly polarized Bremsstrahlung even with unpolarized electrons, and without taking into consideration the 'spin-flip' process described above, becomes possible. This has been discussed theoretically by Sarkar (1963) and by Sokolov and Kerimov (1966).

2.10 Isochromats: The determination of h/e, and the structure near the high energy limit

If the voltage applied to an X-ray tube is varied and the intensity is examined in a fixed narrow waveband, the resulting curve is called an isochromat. The form of such curves may easily be inferred for thin and thick targets from expressions (2.29) and (2.33) respectively.

For the thin target

$$i_\nu \mathrm{d}\nu \mathrm{d}x \propto \frac{Z^2}{V} \mathrm{d}\nu \mathrm{d}x \text{ (for } eV > h\nu;\ i_\nu = 0 \text{ otherwise)} \tag{2.50}$$

and for the thick target

$$I_\nu \mathrm{d}\nu \propto A_c Z\left(V - \frac{h\nu}{e}\right) \text{(for } eV > h\nu;\ I_\nu = 0 \text{ otherwise)} \tag{2.51}$$

The latter result is particularly simple, and data supporting this linear expression were obtained by Wagner (1918), Webster and Hennings (1923), and Nicholas (1927). The radiation from thin targets of nickel has been measured by Harworth and Kirkpatrick (1942) in two wavebands (defined by Ross filters) centred on 0.0486 and 0.1012 nm, with an applied voltage in the range 12 to 180 kV. Their data are shown in fig. 2.31 and the similarity with curves plotted on the basis of the Sommerfeld theory is seen to be quite close.

The approximate relation $I \propto V^{-1}$ is approached most closely when the electron bombarding energy and the photon energy are both high. It should be noted that because the isochromats are observations within a fixed waveband, no corrections are needed for variations of sensitivity with wavelength. Particular interest is attached to the exact value of the voltage at which emission in the chosen band begins, and also the details of the isochromat in the immediate environment of this threshold. It was early realized that the threshold voltage is given by the relation $eV = h\nu$, and that determinations of the ratio h/e could be made from accurate measurements

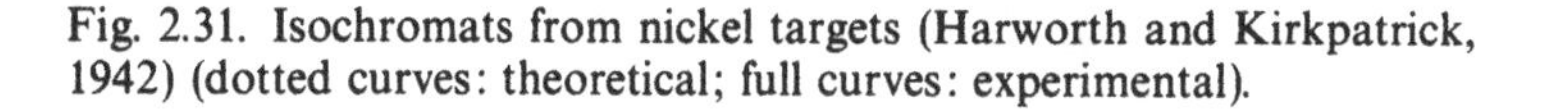

Fig. 2.31. Isochromats from nickel targets (Harworth and Kirkpatrick, 1942) (dotted curves: theoretical; full curves: experimental).

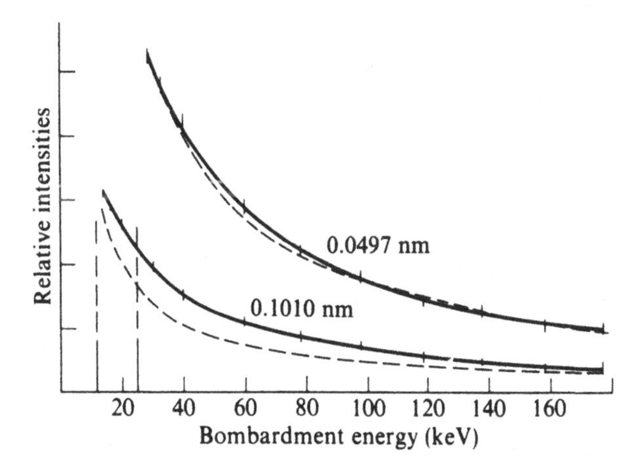

of voltage and wavelength. Accurate determinations of h/e require careful consideration of the exact relation between the applied voltage and the energy of the accelerated electrons at the moment of emitting a photon at the high energy limit. To do this it is necessary to consider the electron band structure of the cathode and anode materials. In the case when both cathode and anode are conductors, it is clear that the voltage indicated by an ordinary measuring device is in fact the potential difference between the Fermi surfaces of the metals (fig. 2.32). However, the electrons leaving the cathode start not from the Fermi surface but from states of higher energy, in fact as free electrons just outside the metal, having been lifted there by the process of thermionic emission. So, referring to fig. 2.32, we can write

$$eV_{\text{photon}} = eV_{\text{meter}} + e\phi_{\text{cathode}} \tag{2.52}$$

where ϕ_{cathode} is the work function of the cathode.

Sandström (1960), in the course of an investigation of the structure near the threshold of these isochromats (see below) found that the use of a cathode of tungsten oxide (as compared with metallic tungsten) required an additional potential difference of 2.8 volts across the X-ray tube to produce X-rays in a given waveband. This corresponds closely to the difference between the work functions of these two materials ($\phi_{\text{metal}} = 4.52$ volts, $\phi_{\text{oxide}} = 2.0$ volts approx.). Corrections for the work function of the cathode have been applied in most of the measurements from the time of DuMond and Bollman (1937) onwards.

Simple theory predicts a smooth rise of intensity as the applied voltage is raised above the threshold, but in practice a pronounced structure is

Fig. 2.32. See text.

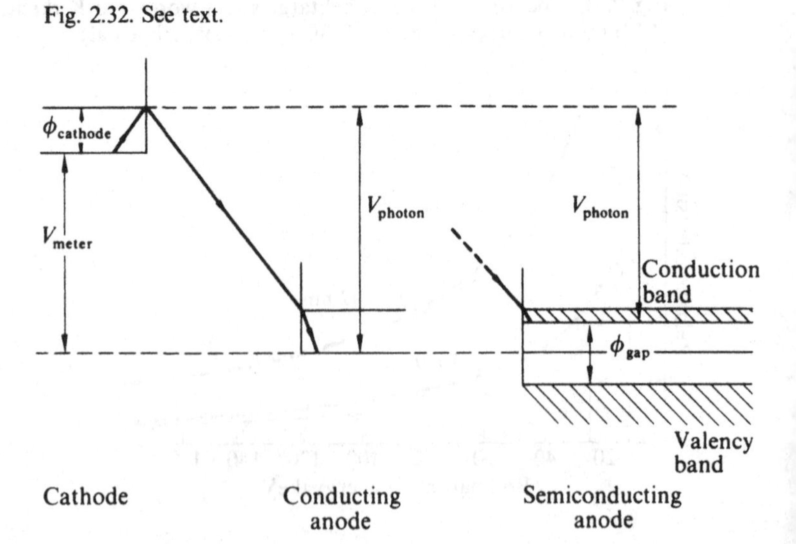

usually observed. This depends primarily on the distribution of final states available to the 'zero-energy' electron in the target after emitting the photon. In order to make an accurate determination of h/e, it is desirable to eliminate the isochromat structure altogether. This has been accomplished by Spijkerman and Bearden (1964) by the use of a target of mercury vapour. They measured the intensity as a function of voltage in a waveband which was sharply defined by means of a double crystal monochromator, and obtained an isochromat which was completely devoid of structural features. The shape of the profile is still, however, determined by the final states available to the electrons which in the case of a free electron gas is proportional to $(\Delta E)^{\frac{1}{2}}$, where ΔE is the amount of energy left over after emitting the photon.

It should be noted that whereas in a metal the electrons come to rest near the Fermi surface, the final states which are appropriate for a gaseous target are those for free electrons. This means that the work function of the anode now has to be subtracted from the right-hand side of (2.52). A further point arises from the fact that the mercury vapour approximates to a *thin* target, for which (2.50) has to be applied rather than (2.51). The curve does in fact fall off according to a $1/V$ law as predicted, once the threshold has been passed.

In this determination a value of 4.13561×10^{-15} Volt Hz^{-1} was obtained for h/e, with a probable error of 30 ppm (i.e., a standard deviation of approximately 45 ppm).

This experiment was originally based on a wavelength expressed in 'X-units', the result then being converted to absolute units using the best value for the conversion constant available at that time. The X-unit was originally introduced by M. Siegbahn, in order to relate X-ray wavelengths directly to the lattice constant of calcite, and is close to 10^{-11} cm. In fact, recent X-ray determination of h/e have sometimes been regarded as experiments to determine the ratio, Λ, of wavelength measured in Angstrom units (10^{-8} cm) to wavelength measured in 'kilo-X-units' (or kXu). For if h/e is calculated directly from values of h and e obtained independently, without recourse to X-ray methods, an X-ray experiment of the type carried out by Spijkerman and Bearden enables the wavelength of the X-ray isochromat to be *calculated*, in absolute measure (i.e. in metres or subdivisions of it) from the experimental observations, simply by the use of the equation $\lambda = hc/eV$. This value for the wavelength can then be compared with the wavelength in kXu. Spijkerman and Bearden obtained a new value for the ratio

$$\Lambda = 1.002024\ \text{Å kXu}^{-1},$$

with a probable error of 24 ppm.

A review of the history of the X-unit has been given by Thomsen and Burr (1968).

A new method of determining h/e directly, but by a non-X-ray method, has become available, and has been shown to be of much greater accuracy than any of the previous methods, including the X-ray method. This method makes use of the Josephson effect, in which the application of a voltage between two weakly coupled superconductors causes an alternating current to flow between them with a frequency proportional to voltage, the constant of proportionality being $2e/h$. For further details of this and other methods of determining fundamental constants, the reader is referred to the review by Taylor, Parker and Langenberg (1969). From the data presented in table 34 of their paper, it may be calculated that the best value of h/e available to them from studies of the Josephson effect was

$$4.1357040 \times 10^{-15}\ \text{Volt Hz}^{-1},$$

with a standard deviation of 3.4 ppm. This value and the X-ray value are thus consistent with each other, but the estimated error of the newer method is less by an order of magnitude.

In a conductor, an electron must reach its final state at the Fermi surface, as we have seen, but it was pointed out by Nijboer (1946), that a target of insulating material will require the electron to take up a final state in the conduction band. As this lies somewhat *above* the Fermi surface it follows that the threshold of the band isochromats will be shifted a little in the direction of higher applied voltages. Isochromats for tungsten, nickel, chromium, and chromium oxide were studied by Johansson (1960), using applied voltages in the region of 5.42 kV. The threshold for chromium oxide exceeded that for the metal by 1.8 volts, because of the energy gap between valence and conduction electrons in this semiconductor, yielding a gap width of 3.6 volts in Cr_2O_3.

When metallic targets are used, the spectrum is complicated by the characteristic energy losses which occur when electrons traverse the metal. This gives rise to peaks in the isochromats, which in Johansson's work occurred at 17 and 31 volts above threshold in tungsten, with similar effects in chromium and nickel. Below the first minimum, however, the shape of the curve gives the density of unfilled states immediately above the Fermi level. Sandstrom (1960) reports a variation in the relative heights of these peaks as the selected waveband is varied, and no explanation of this appears to be as yet forthcoming. Work on the isochromat structure of cadmium sulphide, cadmium telluride, germanium and silicon has been reported by Liden and Auleytner (1962), and by Bergwall and Elango (1967), who have studied a series of alkali halides. The first three peaks of their curves were attributed to the *s* and *d* states of the cations, and *d* state of the anions, and the *p* states of the cations respectively. Subsequent peaks were observed which were consistent with the known characteristic energy losses in the salts under investigation. We make further reference to isochromat structure in chapter 8.

The resolution obtainable in studies of this type depends upon instrumental factors such as the smoothness of the EHT supply, but there is an intrinsic effect due to the spread of thermal energies of the electrons leaving the cathode, amounting to 0.1 or 0.2 eV in practice.

At higher energies the nature of the spectrum near the high energy limit is of interest in connection with the interpretation of photo-nuclear reactions. A convenient way of studying the isochromats in this region is by the use of the gamma-ray resonance in ^{12}C, which occurs at 15.1 MeV (in the lab. system). Studies at this energy are reported by Hall, Hanson and Jamnik (1963).

2.11 Additional Bremsstrahlung processes

In principle, the electromagnetic interaction between any two particles can give rise to Bremsstrahlung, and its production by means other than electrons impinging on material targets is of interest. If we consider a beam of moderately energetic protons it is clear that for a given Coulomb force the deceleration of a proton is down by a factor of $\frac{m}{M_p}$: hence the electric field and the intensity of radiation are reduced by factors of $\frac{m}{M_p}$ and $\left(\frac{m}{M_p}\right)^2$ respectively, where m are M_p are the electron and proton masses respectively. We see immediately therefore that the efficiency of this process is several orders of magnitude less than for an electron of similar kinetic energy.

Although the proton and the nucleus have charges of similar sign, the proton–nucleus combination does nevertheless possess an electric dipole moment (along with higher moments) so the theory discussed above will still obtain in principle. In the case of $Z=1$, the dipole moment would however vanish, and higher approximations in the theory would be needed to achieve a non-vanishing cross-section for proton–proton Bremsstrahlung. Much more important than this primary process are the secondary processes by which the 'knocking-on' of electrons (during the slowing down of the protons) emits Bremsstrahlung by virtue of acceleration of these electrons during the collision process, and by which the slowing down of these 'knocked-on' electrons then makes a further contribution to the radiated energy.

Ogier *et al.* (1966) have discussed these processes in the following manner. If the radiation from the acceleration of the electron is viewed in centre-of-mass co-ordinates, the theoretical treatment is the same as for electron–nuclear Bremsstrahlung, with $Z=1$. Writing E_p (the proton energy) $=\frac{1}{2}M_p v_p^2$, we see that in C of M co-ordinates the electron velocity is $-v_p$ and hence radiates as if it were an electron of energy $\frac{1}{2}mv_p^2$. So the high energy limit of the spectrum is given by $E\left(\frac{m}{M_p}\right)$ in the case of a head-on collision, or $E_p/1837$ MeV, if E_p is in MeV.

In the laboratory system, the initially stationary electron will have been 'knocked-on', and its knock-on velocity may reach a value of $2v_p$, corresponding to a kinetic energy of $4(m/M_p)E_p$. The electron may then radiate photons up to a high-energy limit of this value, $4(m/M_p)E_p$, during its subsequent slowing-down.

For protons of 1 MeV, for example, we would therefore expect 2 additional contributions, both in the form of continuous spectra, with high energy limits of approximately 2 keV and 0.5 keV, as a result of these processes.

The low energy component ('acceleration radiation') would be difficult to observe, but the radiation due to the subsequent slowing down is now well-known, and is discussed in more detail in Chapter 6.

The *electron–electron* Bremsstrahlung has to be taken into consideration when calculating the X-ray yield under electron bombardment. In the dipole approximation the contribution from this source is zero, but it has been pointed out by Stabler (1965) that the quadrupole contribution is significant and indeed is comparable with the total cross-section expected from electron–proton Bremsstrahlung. This is of immediate interest in the field of plasma physics to which reference is made in chapter 8, but it will also increase the X-ray yield from ordinary target materials by a factor $\frac{Z+1}{Z}$ which is expected to be appreciable, at least in the case of the lighter elements.

A new method of investigating the continuous X-ray spectrum has been introduced by Nakel (1966), who has obtained angular distributions of the continuous spectrum in coincidence with the electrons emerging from a thin target at a defined angle. A magnetic spectrometer can be used to select electrons of a well-defined energy, thereby fixing the energy of the photons in coincidence with them. An example of this work is shown in fig. 2.33, in which the electron scattering angle is $+5°$.

The direction of the emitted photons may be understood by considering the manner in which the electron radiates during its hyperbolic path around the nucleus. Most of the radiation will be produced when the electron is closest to the nucleus (because its acceleration is at its maximum value there) and because of relativistic effects it will be emitted along a lobe substantially parallel to the direction of the deflected electron. The maximum in the X-ray intensity will therefore lie on the same side of the forward direction as the electron selected by the magnetic spectrometer. The transverse momentum necessary to balance this is taken up by the recoiling nucleus.

Fig. 2.33. Angular distribution of Bremsstrahlung in coincidence with electrons. Electron scattering angle, $+5°$; photon energy 130 keV; incident electron energy 300 keV (Nakel, 1967).

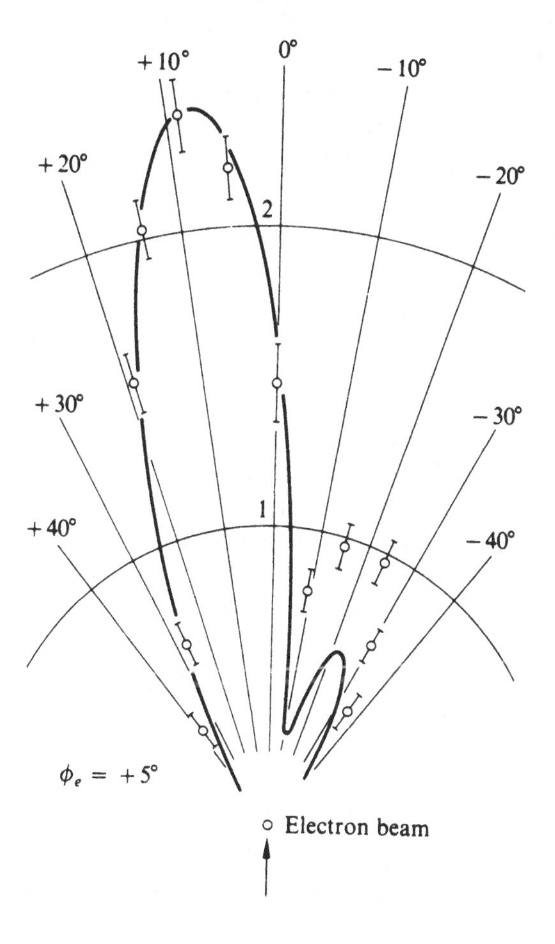

3

Characteristic X-rays

3.1 Energy levels and X-ray spectra from singly-ionized atoms

Soon after the discovery of X-rays, attempts were made to investigate the spectral characteristics of the radiation. Crystal diffraction had not yet been discovered, but by studying the absorption of the radiation by successive layers of material it was established that the radiation consisted of a continuous spectrum (discussed in the previous chapter), and also a monoenergetic component, or 'line' spectrum (soon to be found to consist of several lines) the penetrating power of which depended markedly on the target element. This 'characteristic' radiation has been the subject of much study, and can be understood in considerable detail from the Bohr theory of the atom as elaborated by Sommerfeld and others. The development of wave mechanics added to this understanding and enabled a detailed interpretation of the fine structure of spectra to be built up. The classification of the orbital electrons into shells and the designation of individual electrons by means of quantum numbers are well-known features of the Bohr theory. Fig. 3.1 shows how the orbital electrons may be grouped together into the K, L, M etc. shells of progressively decreasing binding energy, and it is seen that (with the exception of the K shell) each shell consists of several sub-shells of slightly different energy. Each shell is characterised by a Principal Quantum Number n, which has the values 1, 2, 3 etc. for the K, L, M etc. shells, and it is this number which features in the expression for the energy of electrons in the hydrogen-like atom (e.g. Bransden and Joachain, 1983)

$$E_n = -\frac{m_r}{2\hbar^2}\left[\frac{Z_{\text{eff}}e^2}{4\pi\varepsilon_0}\right]^2\frac{1}{n^2} \tag{3.1}$$

In this expression Z_{eff} represents the 'effective' atomic number, i.e. the atomic number (Z) modified by the screening effect of the orbital electrons (see below).

This expression is often written in the form

$$E_n = \frac{-RhcZ_{\mathrm{eff}}^2}{n^2}, \text{ where } R = \left(\frac{1}{4\pi\varepsilon_0}\right)^2 \frac{m_r e^4}{4\pi\hbar^3 c} \tag{3.2}$$

R is the Rydberg constant used in optical spectroscopy, m_r is the reduced mass of the orbital electron, and may be written as $m \cdot \dfrac{M}{M+m}$, where M is the atomic mass.

There are thus many distinct Rydberg constants, for nuclei of different mass, and these distinctions are observable in the optical spectra of light elements. However, the screening effect of orbital electrons imposes much greater departures from the simple formulae, so we shall take m_r in (3.1) to be simply the mass of the electron. The Rydberg constant would then be R_∞, appropriate for an infinitely heavy nucleus, and its value is $R_\infty = 1.09737 \times 10^7\ \mathrm{m}^{-1}$. With this approximation, E_n is often written as

$$-\tfrac{1}{2}mc^2 \frac{(Z_{\mathrm{eff}}\alpha)^2}{n^2} \tag{3.3}$$

where α is the fine structure constant, given by $\alpha = e^2/(4\pi\varepsilon_0)\hbar c$.

Next we consider the orbital angular momentum quantum number ℓ. To

Fig. 3.1. The Bohr atom–shells and sub-shells of krypton. In the designation $3d^6$ (for example), 3 refers to the principal quantum number n, d refers to the angular momentum quantum number ℓ ($\ell = 0, 1, 2, 3, \ldots$ is indicated by the letters $s, p, d, f, \ldots$) and the superscript indicates the number of the electrons in the sub-shell.

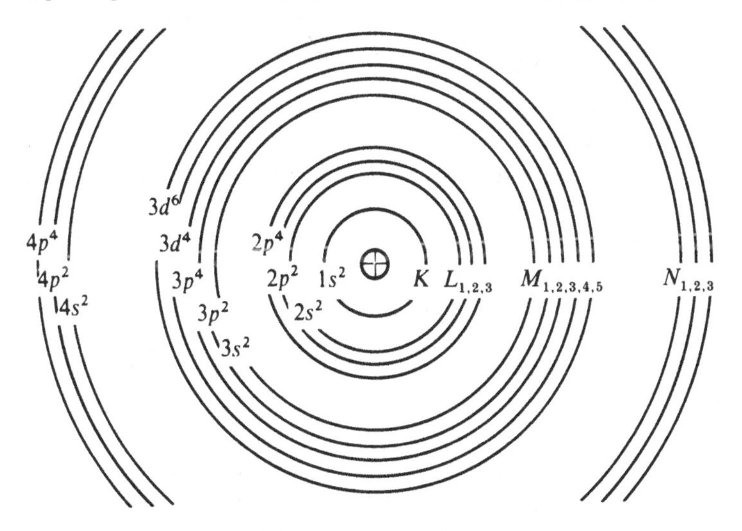

a first approximation this does not affect the energy level of the electron, because hydrogenic atoms are ℓ-degenerate, but in higher approximations its effect must be taken into consideration. In the Bohr–Sommerfeld theory it enters as a relativistic correction which is most important for the highly eccentric orbits, of low angular momentum, in which the electron spends appreciable time travelling very near to the nucleus at high velocity. This correction (neglecting screening for the moment) is given by

$$\Delta E_1 = -E_n \frac{(Z\alpha)^2}{n^2}\left[\frac{3}{4} - \frac{n}{\ell+\frac{1}{2}}\right] \tag{3.4}$$

Thirdly we consider the effect of electron spin–orbit coupling, which may be written as

$$\Delta E_2 = -E_n (Z\alpha)^2 \frac{\mathbf{L}\cdot\mathbf{S}}{n\ell(\ell+\frac{1}{2})(\ell+1)} \tag{3.5}$$

This term contains the vectors **L** and **S**; $\mathbf{L}\cdot\mathbf{S}$ may be written as

$$|\mathbf{L}|\,|\mathbf{S}|\cos(\mathbf{L},\mathbf{S}) \text{ or } \tfrac{1}{2}[j(j+1)-s(s+1)-\ell(\ell+1)]$$

j is given by $\ell \pm s$ where j is the *total angular momentum quantum number* and $s(=\frac{1}{2})$ is the spin quantum number for an individual electron. By making these substitutions, (3.5) may be reduced to the expression

$$\Delta E_2 = -E_n (Z\alpha)^2 \frac{1}{2n\ell(\ell+\frac{1}{2})(\ell+1)} \times \begin{cases} \ell, \text{ for } j=\ell+\frac{1}{2} \\ -(\ell+1), \text{ for } j=\ell-\frac{1}{2} \end{cases} \tag{3.6}$$

These terms may be combined to give

$$E = E_n + \Delta E_1 + \Delta E_2 = E_n \left\{1 - \frac{(Z\alpha)^2}{n^2}\left[\frac{3}{4} - \frac{n}{j+\frac{1}{2}}\right]\right\} \tag{3.7}$$

For the special case of $\ell=0$, the spin–orbit interaction vanishes and $\Delta E_2 = 0$. However, a third term ΔE_3 (the so-called Darwin term), given by

$$\Delta E_3 = -E_n \frac{(Z\alpha)^2}{n} \tag{3.8}$$

is operative when $\ell=0$. When this is included, (3.7) is once again obtained, and is therefore of general applicability. Equation (3.7) may be written out as

$$E_{n,\ell,j} = -Rhc\left[\frac{Z^2}{n^2} + \frac{\alpha^2 Z^4}{n^4}\left(\frac{n}{j+\frac{1}{2}} - \frac{3}{4}\right)\right] \tag{3.9}$$

Table 3.1 *Electronic orbits, with quantum numbers*

n	ℓ	j			
1	0	$\frac{1}{2}$	$1s$	K	$1s$
2	0	$\frac{1}{2}$	$2s\frac{1}{2}$	L_1	$2s$
2	1	$\frac{1}{2}$	$2p\frac{1}{2}$	L_2	$2p$
2	1	$\frac{3}{2}$	$2p\frac{3}{2}$	L_3	
3	0	$\frac{1}{2}$	$3s\frac{1}{2}$	M_1	$3s$
3	1	$\frac{1}{2}$	$3p\frac{1}{2}$	M_2	$3p$
3	1	$\frac{3}{2}$	$3p\frac{3}{2}$	M_3	
3	2	$\frac{3}{2}$	$3d\frac{3}{2}$	M_4	$3d$
3	2	$\frac{5}{2}$	$3d\frac{5}{2}$	M_5	
4	0	$\frac{1}{2}$	$4s\frac{1}{2}$	N_1	$4s$
4	1	$\frac{1}{2}, \frac{3}{2}$	$4p\frac{1}{2}, \frac{3}{2}$	$N_{2,3}$	$4p$
4	2	$\frac{3}{2}, \frac{5}{2}$	$4d\frac{3}{2}, \frac{5}{2}$	$N_{4,5}$	$4d$
4	3	$\frac{5}{2}, \frac{7}{2}$	$4f\frac{5}{2}, \frac{7}{2}$	$N_{6,7}$	$4f$

in order to compare the result with the older form given, e.g. by Compton and Allison (1935).*

In these earlier treatments, screening constants σ_s and σ_s' were introduced to allow for partial screening of the nucleus by the orbital electrons. Equation (3.9) therefore takes the form

$$E_{n,l,j} = -Rhc\left[\frac{(Z-\sigma_s)^2}{n^2} + \frac{\alpha^2(Z-\sigma_s')^4}{n^4}\left(\frac{n}{j+\frac{1}{2}} - \frac{3}{4}\right)\right] \quad (3.10)$$

The detailed significance of the screening constants will be discussed below.

We are now in a position to tabulate the sub-shells together with their quantum numbers n, ℓ, and j. We use the conventional X-ray notation L_1, L_2, L_3 and also the optical notation – s, p, d, etc. states – for comparison (table 3.1).

It should be added that each sub-shell can contain several electrons up to a maximum of $2j+1$. The so-called 'multiplicity' of the state arises from the fact that the total angular momentum vector (magnitude $[j(j+1)]^{\frac{1}{2}}$) can be

* The energy levels of (3.9) and (3.10) are often expressed in the form of 'term values' $\tilde{\nu}$, where $h\tilde{\nu}c = E$, the energy level. The *separation* between two states (e.g. the doublets tabulated in table 3.3) may be expressed as differences, $\Delta\tilde{\nu}$, between the term values of the 2 states in question. A *transition* between 2 states (e.g. the transitions indicated in figs 3.3 and 3.4) may also be expressed in this way, in which case $\Delta\tilde{\nu}$ is then the *wave-number* of the emitted radiation, and is equal to the reciprocal of its wavelength.

Fig. 3.2 Electronic levels in krypton.

	keV
$M_{4,5}$	0.091
M_3	0.210
M_2	0.215
M_1	0.289
L_3	1.67
L_2	1.73
L_1	1.92
K	14.32

Fig. 3.3. Energy levels for a singly-ionized atom. The diagrams contain all the lines for which specific symbols are listed by Bearden (1967), except a few weak quadrupole transitions. Many additional lines have been identified, especially in the heavier elements. (Dyson, 1975).

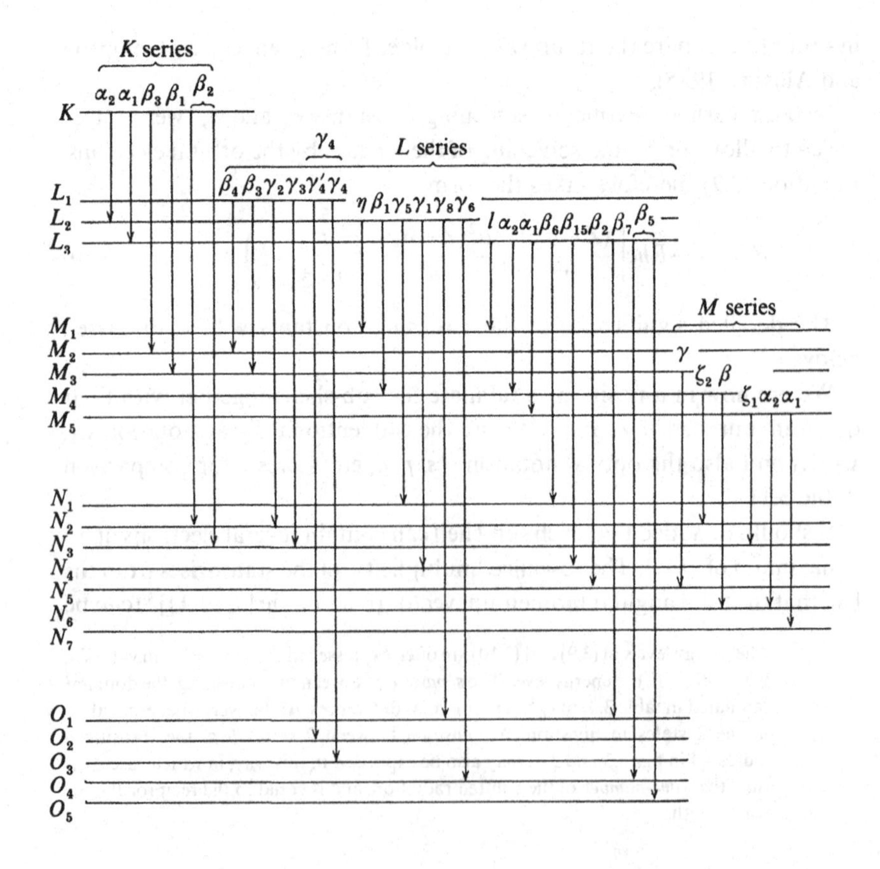

oriented in several directions according to the rules of spatial quantisation. An additional 'magnetic' quantum number m_j is necessary to characterise this feature, and a complete sub-shell will contain electrons with all possible values of $m_j(|m_j| \leqslant j)$. The application of a magnetic field would give rise, in principle, to a further splitting of levels, well-known as the Zeeman effect in optical spectroscopy. In the section dealing with line widths it will be seen that such additional splitting is not observable in X-ray spectra; but the multiplicity of levels is nevertheless important in that it affects the *intensity* of lines through its effect upon the *transition probabilities* between states.

It is a central feature of the interpretation of optical and X-ray spectra that observed lines correspond, in the main, to transitions between two states. We may draw an energy level diagram in which each combination of n, ℓ and j gives rise to a distinct level, and we are faced with two alternative ways of doing this. We may construct a diagram as in fig. 3.2 in which each group of levels represents the energy of a group of electrons, with the more strongly bound electrons near the bottom of the diagram. If an electron is removed, the atom becomes ionized, and the vacancy thus created may be filled by an electron from an outer shell, in accordance with certain *selection rules*. This movement of an electron from an outer shell to an inner shell gives rise, in principle at least, to the emission of a photon of characteristic X-radiation.

Alternatively we may depict the energy levels of the singly-ionized *atom* as in fig. 3.3. The removal of a strongly bound electron gives rise to a level at the top of the diagram in this case, and the emission of an X-ray photon corresponds once again to a transition from an upper level to a lower level. This mode of presentation is preferred by most authors, and has the advantage that the transitions between the excited (not necessarily ionized) states of optical spectroscopy can occupy positions at the foot of the diagram. States corresponding to the removal of an electron are known as 'hole states', and are denoted by terms of the form $1s^{-1}$, etc. A diagram giving the levels for cadmium (including optical levels), together with their wave-numbers, is shown in fig. 3.4.

The selection rules which govern 'allowed' transitions between states are

$$\Delta\ell = \pm 1; \; \Delta j = -1, 0, +1 \tag{3.11}$$

Δn may take any value.

$\Delta\ell = \pm 1$ corresponds to the emission of dipole radiation, and is the most likely class of transition. Transitions for which $\Delta\ell = \pm 2$ are occasionally observed, though with reduced intensity. These are known as electric quadrupole transitions.

Following the ionization, an electron has to 'fall in' from an outer shell, but it should be noted that this energy is not always emitted in the form of an X-ray photon. The alternative process, known as the Auger effect, is discussed in section 3.2. If a photon is emitted, there will then be a vacancy in the outer shell remaining to be filled. If an Auger electron has been emitted, there will be *two* vacancies which will be filled in turn. We see therefore that X-ray emission takes on the features of a cascade process of some complexity.

In our discussion of the detailed structure of X-ray spectra we shall refer to the higher state (i.e. ionization of an *inner* shell) as the *initial* state, and the *lower* energy state (when the vacancy has migrated to an outer shell) as the *final* state.

Fig. 3.4. Energy level diagram for cadmium. (White, 1934).

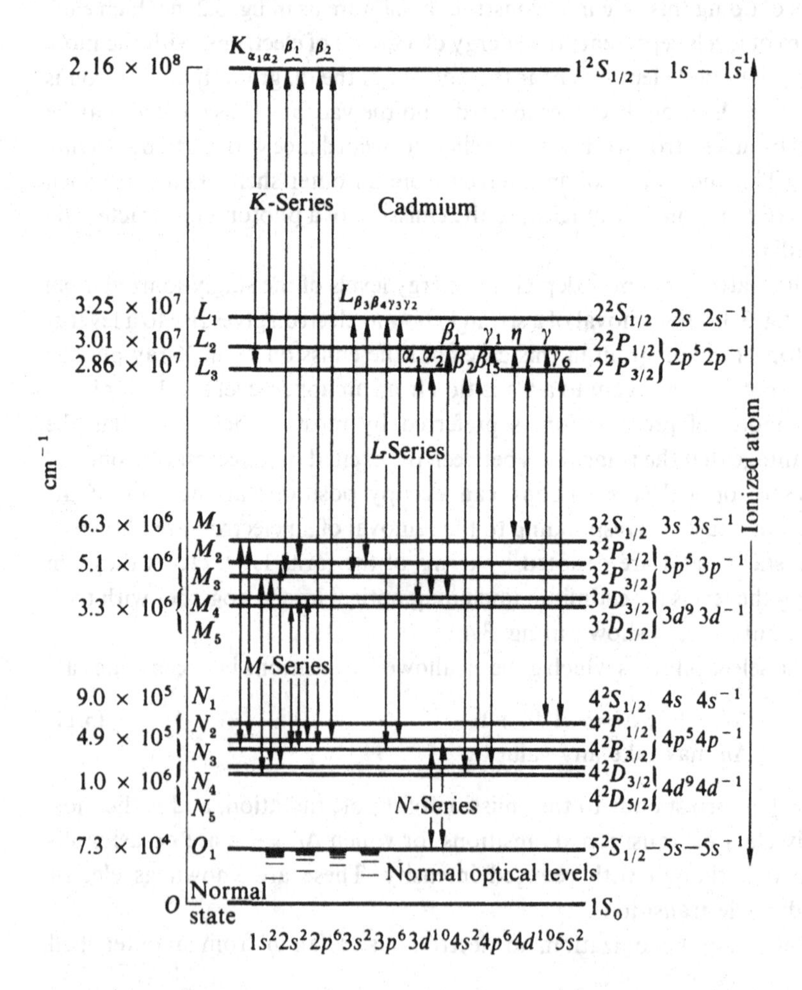

The main features of the K-series are the $K_{\alpha_1}K_{\alpha_2}$ doublet, and the K_{β_1} and K_{β_2} lines, also resolvable into doublet structures in the heavier elements. All these lines conform to the selection rules for dipole radiation, and their relative intensities (see below) can be understood in terms of the principles already discussed. The energies of the K_α doublet are given by the approximate formula

$$E = KRhc(Z - \sigma_m)^2, \text{ where } K = \tfrac{3}{4} \text{ and } \sigma_m \quad (3.12)$$

is approximately equal to 1. An expression of this type was first obtained experimentally by Moseley, and the approximate dependence upon Z^2 is known as Moseley's law. The constant $\frac{3}{4}$ can be written as $\left(\frac{1}{1^2} - \frac{1}{2^2}\right)$, by analogy with the Balmer series in the optical spectrum of hydrogen. A plot of the energies of the lines of the K series as a function of Z is shown in fig. 3.5. It should be remarked that Moseley's original work (1913, 1914) led for the first time to a systematic *physical* way of establishing the definitive

Fig. 3.5. K-series lines, illustrating Moseley's law. (White, 1934).

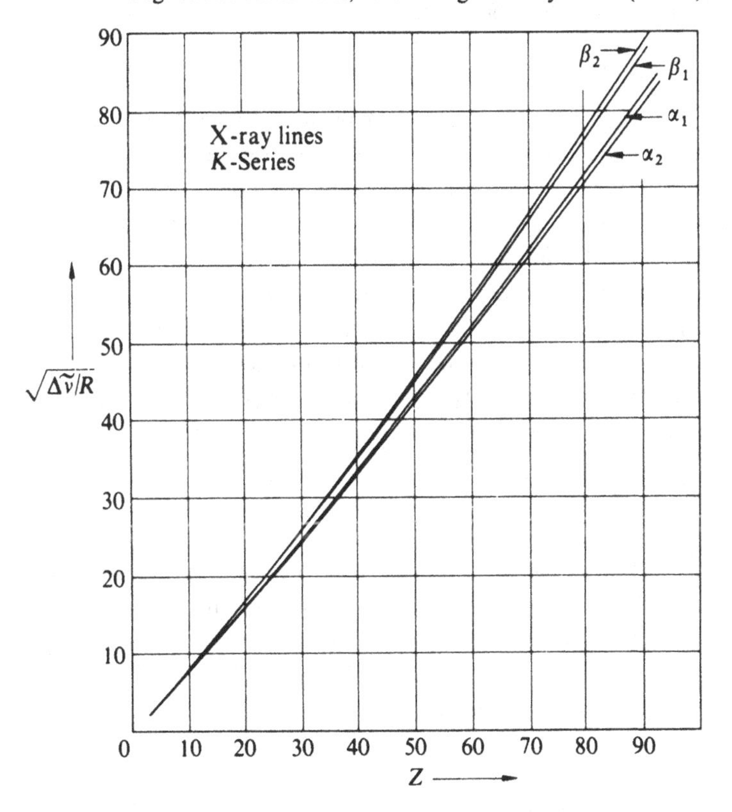

order of the chemical elements. Until this time, cobalt was placed *above* nickel (on account of the atomic weights of these elements being 58.94 and 58.69 respectively); the energies of their K_{α_1} lines are, however, 6.930 and 7.477 keV respectively, establishing clearly the accepted order cobalt ($Z=27$), nickel ($Z=28$). The K_β lines lie on the high energy side of the K_α lines by an amount which varies from 2.5% for sodium (the lightest element for which a K_β line is possible) to about 15% for uranium.

In the L series, which we shall now examine in detail, the strongest lines would be expected to be those which originate from the L_3 atomic energy level (since this contains four electrons, any one of which may be ejected by the initial ionization process) and which terminate in the M_4 and M_5 levels. These lines are designated L_{α_2} and L_{α_1} respectively. Their photon energies lie between one tenth and one sixth of the K spectra. These and other transitions are shown in figs. 3.3 and 3.4. Other strong lines are the L_{β_1} and L_{β_2} corresponding to atomic transitions $L_2 \rightarrow M_4$ and $L_3 \rightarrow N_5$ respectively. It should be noted that in cases where l and j both change they do so in the same sense, i.e. the electron spin remains unchanged in direction. This is a factor which is often associated with a high transition probability.

For the L_{β_1} line Moseley established a law similar to (3.12) with $\sigma_m = 7.4$ and K equal to $\frac{5}{36}$ $\left(\text{i.e. } \frac{1}{2^2} - \frac{1}{3^2}\right)$. It should be noted however that the screening constant can only be discussed realistically when *terms* rather than *transitions* are considered, so the values for σ_m quoted in Moseley's expressions should be regarded as empirical only. Detailed tabulations of accurate values of X-ray lines in the K, L and subsequent series are given by Sandström (1957) and by Bearden (1967).

For each value of l in any given group of levels there will always be two values of $j(l+\frac{1}{2}$ and $l-\frac{1}{2})$. Transitions to such a pair of levels will give rise to a doublet, and we can examine the expected doublet separation using (3.10). If we write this equation with two different values of j, differing by unity, and evaluate the difference in the energy levels, we obtain

$$\Delta E = Rhc\alpha^2 \frac{(Z-\sigma_s')^4}{n^3 l(l+1)} \tag{3.13}$$

A doublet of this kind is known as a *spin-relativity* doublet, and a table of spin-relativity doublets is given (table 3.2). Such doublets are characterized by a common initial *or* a common final level. The two levels comprising the doublet have the same l, and values of j which differ by unity.

We can write (3.13) in the form of $\Delta\lambda$, the wavelength difference:

$$\Delta\lambda = ch\Delta E/E^2.$$

Table 3.2 *Spin-relativity doublets*

K series			L series		
	Initial (atomic) level	Final (atomic) level		Initial level	Final level
$\alpha_1\alpha_2$	K	L_3L_2	$l\eta$	L_3L_2	M_1
$\beta_1\beta_3$	K	M_3M_2	$\alpha_2\beta_1$	L_3L_2	M_4
			$\beta_6\gamma_5$	L_3L_2	N_1
			$\beta_{15}\gamma_1$	L_3L_2	N_4
			$\gamma_8\beta_7$	L_2L_3	O_1
			$\beta_5\gamma_6$	L_3L_2	O_4
			$\beta_4\beta_3$	L_1	M_2M_3
			$\alpha_1\alpha_2$	L_3	M_5M_4
			$\gamma_2\gamma_3$	L_1	N_2N_3
			$\beta_{15}\beta_2$	L_3	N_4N_5

For the case of the $K \to L_2, L_3$ doublet we may substitute ΔE from (3.13) with $\ell = 1$ and $n = 2$. For E we may use the first term in (3.10) with $n = 1$. Hence

$$\Delta\lambda = \frac{\alpha^2}{16R} \frac{(Z-\sigma_s')^4}{(Z-\sigma_s)^4} \tag{3.14}$$

Neglecting, for the moment, the difference in behaviour between σ_s and σ_s', we might expect $\Delta\lambda$ to be approximately constant, for varying Z. This is approximately true for $Z > 50$, but the deviations at lower Z are sufficient to indicate that the two screening constants must be treated separately.

Detailed examination of the separation of the spin-relativity doublets has enabled the 'spin-relativity screening constant' σ_s' to be calculated from (3.7) for various doublets over a wide range of Z. Data for the doublet L_2L_3 are given in table 3.3. It is seen that σ_s' varies little with Z over this range. Sommerfeld examined about 40 elements in the range $Z = 41$ to $Z = 90$, and obtained an average value

$$\sigma_s' = 3.50 + 0.05$$

Wentzel gives data for the doublet M_2M_3 and finds σ_s' to be approximately independent of Z and equal to

$$8.5 \pm 0.4$$

Similar calculations for other doublets of this kind lead to the values of σ_s

Table 3.3 *Spin-relativity screening constants (L_2L_3 doublet) (after White)*

Element	Z	$\Delta\tilde{\nu}/R$	$Z-\sigma_s'$	σ_s'
Nb	41	6.90	37.53	3.47
Mo	42	7.70	38.50	3.50
Ru	44	9.49	40.54	3.46
Rh	45	10.48	41.53	3.47
Pd	46	11.57	42.55	3.45
Ag	47	12.69	43.52	3.48
Cd	48	13.96	44.52	3.48
In	49	15.29	45.50	3.50
Sn	50	16.72	46.51	3.49

Table 3.4 *Spin relativity screening constants σ_s (White, 1934)*

L_1	L_2L_3	M_1	M_2M_3	M_4M_5	N_1	N_2N_3
2.0	3.50	6.8	8.5	13	14	17

given in table 3.4. Values of σ_s' for s levels cannot be obtained by this method, but an indirect method of achieving this is described by White (1934) or Compton and Allison (1935) following Wentzel (1923). These additional values are given in the table 3.4.

An additional doublet structure arises from pairs of levels whose n and j values are the same, but whose l values are different. Equation (3.10) shows that an energy difference between such terms is only possible if the screening constants depend upon l. Such a doublet is called a screening doublet. The effect of a change in σ_s' will be small compared with that of a change in σ_s because of the presence of the fine structure constant in (3.10), so we may write, approximately,

$$\frac{\tilde{\nu}}{R}=\frac{(Z-\sigma_s)^2}{n^2} \tag{3.15}$$

and

$$\Delta\left(\frac{\tilde{\nu}}{R}\right)^{\frac{1}{2}}=\Delta\sigma_s/n \tag{3.16}$$

Screening doublets can arise as a result of transitions to or from the pairs of levels L_1L_2; M_1M_2; M_3M_4; N_1N_2, etc. It should be noted, however, that if j is the same for each member of a pair, and ℓ different, the selection rule $\Delta\ell = \pm 1$ would have to be violated for one of the two transitions forming the doublet. That is, one line of the pair would be 'forbidden' in the sense that it would be, e.g., a quadrupole, rather than a dipole, transition. By making absolute comparisons between observed term values and those calculated from the simple Balmer formula, values of the 'Balmer' screening constant σ_s can be calculated. A graph showing σ_s for each sub-shell as a function of Z is given in fig. 3.6. The curves are indeed parallel to each other within a given group, but σ_s is seen to depend markedly on Z and is also much larger than the spin-relativity screening constant.

Theoretical attempts to calculate the constants σ_s and σ_s' have been reported (e.g. Pauling, 1927). Calculations of σ_s are said to be in reasonable agreement with those obtained experimentally. Screening will depend

Fig. 3.6. Screening constant σ_s. (After A. Sömmerfeld *Atombau and Spekrallinien*, 1944, Braunschweig: Vieweg. For an English translation of an earlier edition of this text, see Sömmerfeld (1934)).

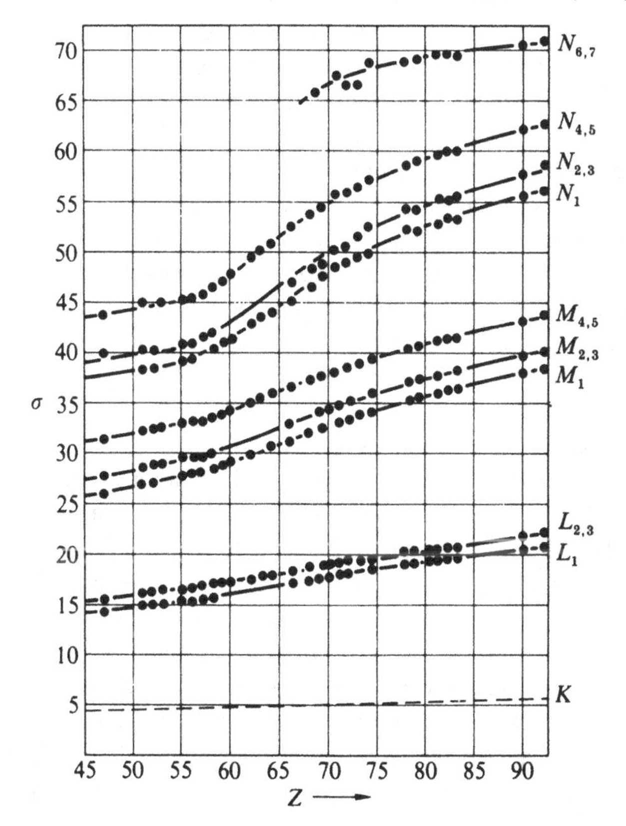

partly on the charge which is to be found at radii *less* than the shell being screened, and partly on the charge at *greater* radii. The former effect ('internal screening') will to a first approximation be dependent only on the total charge within the radius of the shell being screened and not on its distribution, whereas the latter effect, ('external screening'), will depend on the charge distribution – a shell of large radius will have less effect than a shell of radius little greater than the shell being screened.

If we write the potential at radius r_0 in the form

$$\frac{Ze}{r_0}-\frac{q}{r_0}-\int_{r_0}^{\infty}\rho(r)4\pi r^2\mathrm{d}r \tag{3.17}$$

where q is the 'internal' charge given by

$$\int_0^{r_0}\rho(r)4\pi r^2\mathrm{d}r,$$

we see that the internal and external screenings are given respectively by the 2nd and 3rd terms of this equation.

There still remains the problem of the difference between σ_s and σ_s'. Bohr and Coster (1923) have explained this by arguing that σ_s is given by the *total* screening, and σ_s' by *internal* screening only. Sommerfeld (1934) has given a discussion of this aspect of X-ray spectra. In more recent discussions, the distinction between internal and external screening is not made. Effective charges, Z^*, are derived from wave-functions obtained from the self-consistent field method. Alternatively, σ_s may be chosen so that a point nucleus of charge Z^* $(=Z-\sigma_s)$ binds a hydrogenic electron with the same energy as that actually observed for an electron in the real atom of nuclear charge Z. Further discussion (with references) of these more recent approaches is given by Bambynek *et al.* (1972).

3.2 The fluorescence yield and the Auger effect

We now consider the fluorescence yield of X-radiation associated with the return of the ionized atom to its ground state. When a vacancy in an inner shell is filled, the atom changes to a state of lower energy, and this energy may be released in two ways – an X-ray photon may be emitted, or, alternatively, a radiationless transition may take place in which the available energy is used to release an electron from an outer shell. This electron is known as an Auger electron, and the radiationless transition the Auger effect. It is readily seen that an atom which has emitted an Auger electron now has *two* vacancies, neither of which is the same as the vacancy created by the initial ionizing event. An Auger transition thus involves *two* orbital electrons and *three* electron orbits. If the K vacancy is filled by an electron from, say, the X_p sub-shell (where X represents the L, M etc. shell and p the sub-shell, following the notation of Bergström and Nordling (1965)) and an Auger electron is subsequently emitted from the Y_q sub-shell, we can describe the process by the symbol $K \rightarrow X_p Y_q$. The simplest situation is when X and Y both represent the L shell, when in principle *nine* transitions are possible. However the energy of the $K \rightarrow L_p L_q$ and $K \rightarrow L_q L_p$ electrons must be the same (and indeed there is no way of distinguishing between these two processes) so the number of discrete lines reduces to *six**. By similar arguments there are 15 lines in the $K \rightarrow L_p M_q$ series, 21 in the $K \rightarrow L_p N_q$, 15 in the $K \rightarrow M_p M_q$ series, and so on. Auger spectra may therefore be of considerable complexity.

The fluorescence yield is defined as $\omega_K = N_K/(N_K + A_K)$, for the K shell, where N_K is the number of emitted X-ray photons of the K series, and A_K is the number of K Auger electrons. Calculation of the K (or L etc.) fluorescence yield involves considering the relative probabilities of the two alternative processes. The radiationless transition probabilities vary relatively slowly with atomic number, and the probability of a radiative transition is approximately proportional to Z^4.

This latter result may be demonstrated by considering the radiating atom as a classical oscillator, losing energy by radiative damping. For a damped harmonic oscillator,

$$\begin{aligned} \text{relaxation time} &= \frac{\text{energy stored}}{\text{mean energy loss per unit time}} \\ &= \frac{W}{\langle \mathrm{d}W/\mathrm{d}t \rangle} \end{aligned} \tag{3.18}$$

where $W = \frac{1}{2} m v_0^2$ (v_0 = peak velocity).

In the case of an accelerated charge with a displacement f from the origin we may write (see for example section 2.7)

$$\langle \mathrm{d}W/\mathrm{d}t \rangle = \frac{1}{4\pi\varepsilon_0} \frac{2e^2 \langle \ddot{f}^2 \rangle}{3c^3} = \frac{1}{4\pi\varepsilon_0} \frac{2e^2}{3c^3} \frac{\ddot{f}_0^2}{2}$$

where $\ddot{f}_0$ is the peak acceleration.

If the damping is light, $\ddot{f}_0 = p v_0$, where p is the angular frequency $2\pi\nu$. So for $\langle \mathrm{d}W/\mathrm{d}t \rangle$ we have

$$\frac{1}{4\pi\varepsilon_0} \frac{2e^2}{3c^3} \frac{p^2 v_0^2}{2}$$

* It should be remembered that the selection rule which causes the $K \rightarrow L_1$ transition to be forbidden in X-ray emission, does not apply to radiationless transitions.

which is equal to

$$\frac{1}{4\pi\varepsilon_0}\frac{2e^2 4\pi^2 \nu^2}{3mc^3}W$$

Hence the relaxation time is given by

$$\frac{3mc^3\varepsilon_0}{2\pi e^2 \nu^2} \tag{3.19}$$

where ν is the frequency of the characteristic radiation. However, $\nu \propto Z^2$ approximately, from Moseley's law, so this leads to a relaxation time $\propto Z^{-4}$ and a transition probability $\propto Z^4$. So we may write, for the K fluorescence yield,

$$\omega_K = \frac{Z^4}{a_K + Z^4} \tag{3.20a}$$

This expression has been widely used with $a_K = 1.12 \times 10^6$ (Burhop, 1952).

A semi-empirical version

$$\left(\frac{\omega_K}{1-\omega_K}\right)^{\frac{1}{4}} = A + BZ + CZ^3 \tag{3.20b}$$

has been found useful which takes account of deviations at low and high Z.

A short account of the experimental methods of determining fluorescence yields is given in chapter 4, and values derived from the review by Bambynek *et al.* (1972) are to be found in appendix 5. Fig. 3.7 illustrates experimental values selected from this review together with a curve of best fit.

For high atomic numbers the fluorescence yield is high, this being essentially a consequence of the high transition energy. For light elements the fluorescence yield and the energy of the fluorescent X-ray photons are both low, giving rise to experimental difficulties. The measurement of the yield for the complementary process of Auger emission is no easier, in view of the very low energy of the Auger electrons in this region of Z.

When the fluorescence yield in the L shell is studied, it is found that the yields of the three sub-shells in general differ from each other. In circumstances where they are measured separately the results are independent of the method of initial ionization, but if the L fluorescence yield is measured as an average of the whole shell, the results depend markedly on the mode of ionization, because the ionization probabilities of the three shells are not the same. The fluorescence yield for the L_3 sub-shell follows

(3.20a), with a equal to 1.02×10^8. L_3 fluorescence yields are thus somewhat lower than yields in the K shell, and this is true also of L_2 and L_1 yields.

A detailed account of the Auger effect has been given by Burhop (1952), and a more recent review of the fluorescence yield and the Auger effect has been given by Fink *et al.* (1966). The article by Bergström and Nordling already referred to provides a useful introduction to the subject, and the review by Bambynek *et al.* (1972) gives a comprehensive discussion of the radiative and the Auger processes. The theory of the Auger effect has been treated in detail by Chattarji (1976), who also gives an account of the associated techniques of Auger electron spectroscopy.

Attempts have been made to describe the Auger effect as a 'two-stage' process, in which an X-ray photon is emitted and then subsequently reabsorbed before leaving the atom. However, this has been found to be

Fig. 3.7. Fluorescence yield (K-shell) as a function of atomic number (data from Bambynek *et al.*, 1972).

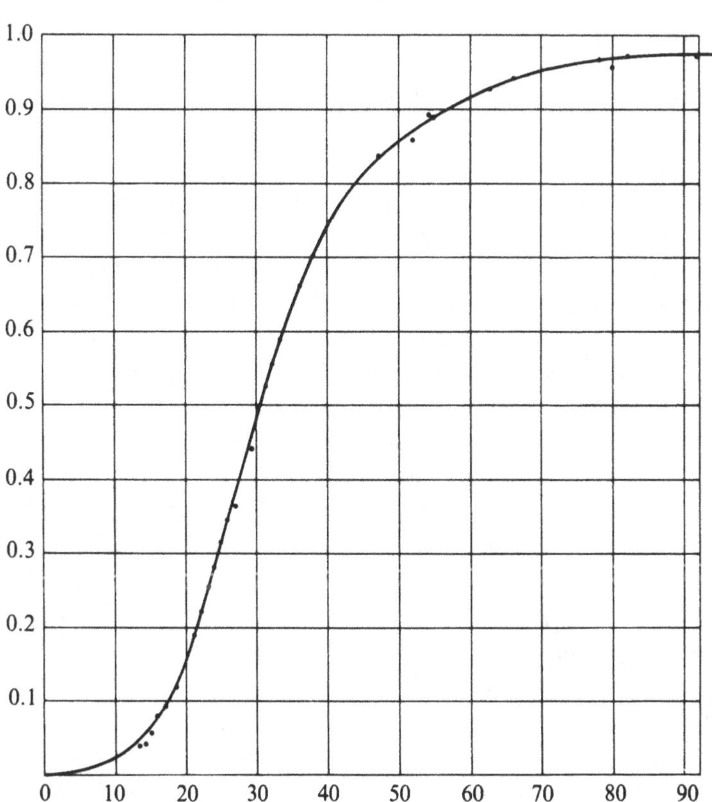

untenable, because of the presence in the Auger spectrum of quite strong lines (such as the $K \rightarrow L_1$ transition) which are forbidden in radiative transitions. Clearly in such cases an X-ray photon plays no part in the process, which must accordingly be a 'one-stage' process.

3.3 Relative intensities of characteristic X-ray lines

Having considered the probability with which a radiative transition will occur in a singly-ionized atom, we can now consider the relative probabilities of the various ways in which this radiative transition may occur. We distinguish between 'allowed' and 'forbidden' transitions. The selection rules for 'allowed' transitions are those for electric dipole transitions, for which

$$\Delta \ell = \pm 1; \ \Delta j = 0 \text{ or } \pm 1 \tag{3.21a}$$

Δn may take any value.
The most important of the forbidden transitions are the magnetic dipole transition for which

$$\Delta \ell = 0; \ \Delta j = 0 \text{ or } \pm 1 \tag{3.21b}$$

and the electric quadrupole transition for which

$$\Delta \ell = 0, \ \pm 2; \ \Delta j = 0, \ \pm 1, \text{ or } \pm 2. \tag{3.21c}$$

To calculate the relative intensities of lines emitted in allowed transitions we may make use of the 'sum rules' which state that the total intensity of all lines proceeding from a common initial level *or* to a common final level is proportional to the statistical weight $(2j+1)$ of that level. For a simple application of these rules we may consider the relative intensities of the K_{α_1} and K_{α_2} lines (atomic transitions $K \rightarrow L_3$ and $K \rightarrow L_2$ respectively). Referring to the energy level diagram in fig. 3.3 and remembering that an *electron* undergoing a transition proceeds from a *lower* to a *higher* level in this diagram, we see that the only electronic transitions proceeding from the L_2 and L_3 sub-shells are those giving rise to the lines in question, so the ratio of these line intensities is given simply by the statistical weights of the levels from which the electrons originate. The intensity ratio $K_{\alpha_2}:K_{\alpha_1}$ is thus expected to be simply 1:2. Experimental determinations of this quantity are illustrated in fig. 3.8(a). We see that the intensity of the K_{α_2} line increases progressively with an increase of Z, but that at low Z the agreement with the simple theory is good.

Experimental data for the ratio of $K_\beta(K_{\beta_1}+K_{\beta_3})$ to $K_{\alpha_1}+K_{\alpha_2}$ intensities is illustrated in fig. 3.8(b). Turning to weaker lines, experimental data for $K_{\beta_2}:K_{\alpha_1}$ ratios is given in fig. 3.9. The gradual fall of this ratio to

Fig. 3.8*a*. $K_{\alpha_2}:K_{\alpha_1}$ intensities (expressed as a ratio of numbers of photons) as a function of Z (after Rao *et al.*, 1972).

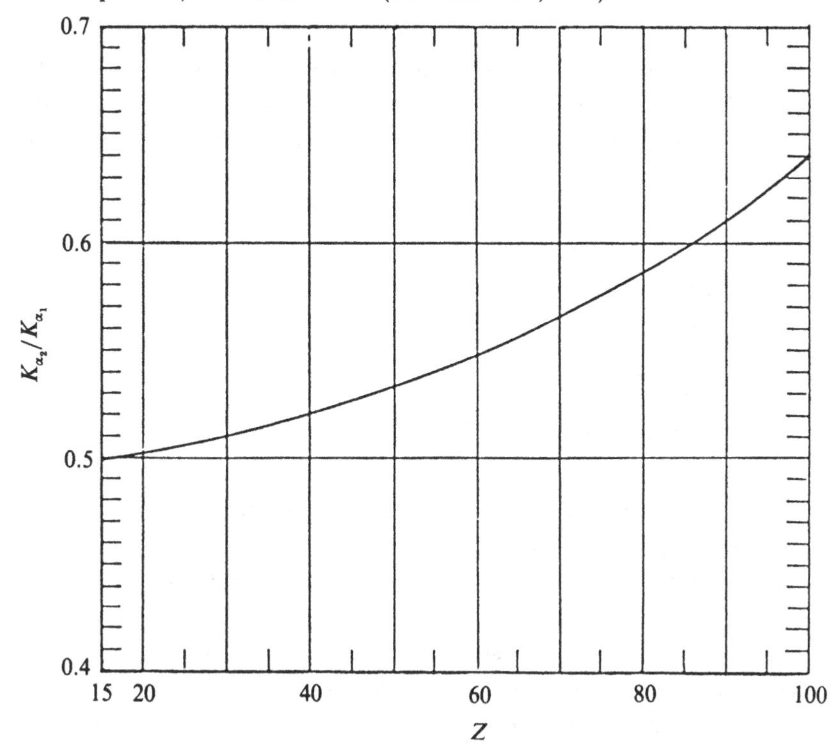

Fig. 3.8*b*. K_β: $K_{\alpha_{1+2}}$ intensities (expressed as a ratio of numbers of photons) as a function of Z. ($K_\beta = K_{\beta_1} + K_{\beta_3}$) (after Rao *et al.*, 1972).

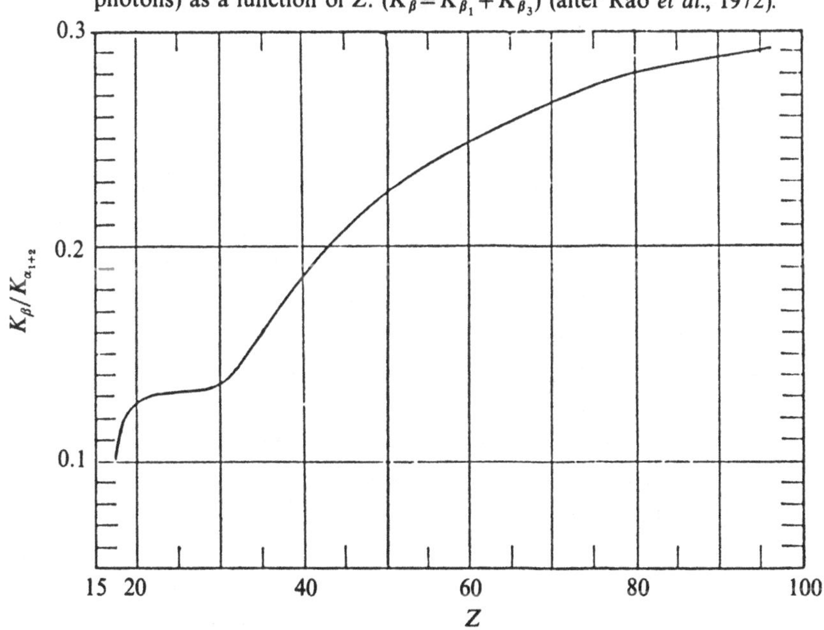

near zero as Z falls towards 30 is associated with the gradual emptying of the N_2 and N_3 sub-shells.

The intensity of forbidden lines is much less than for the lines already considered. The $K \rightarrow L_1$ magnetic dipole transition has been observed by Beckman (1955), and fig. 3.10 illustrates his observation on gold. Data for the electric quadrupole lines K_{β_5} and K_{β_4} are given in fig. 3.11(*a*) and 3.11(*b*).

We may apply the sum rules to the strongest lines of the L series. Referring to fig. 3.3 we may write

$$\frac{L_{\alpha_1}+L_{\alpha_2}}{L_{\beta_1}}=2 \qquad \frac{L_{\alpha_1}}{L_{\beta_1}+L_{\alpha_2}}=\frac{3}{2}$$

from which we find

$$L_{\alpha_1}:L_{\beta_1}:L_{\alpha_2}=9:5:1.$$

Before the predictions can be compared with experimental determinations, further considerations are necessary. In the case of the lines which are produced by transitions from the same *initial* state of ionization, we would expect agreement with the statistical arguments given above; but when two lines stem from *different* states of ionization it is necessary to take into account the other processes of de-ionization which are in competition with the radiative process. The Auger effect may be different for the two states; and furthermore there is an important type of radiationless

Fig. 3.9. Relative intensities of $K_{\beta_2}:K_{\alpha_1}$ lines as a function of z. X and * are theoretical points. The remainder are experimental determinations by Beckman and others (Beckman, 1955).

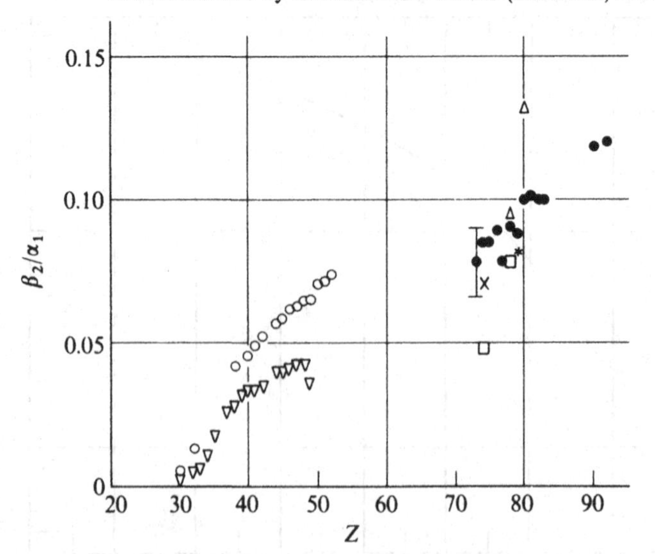

transition which may take place (often with high probability) between sub-shells. These transitions are known as Coster–Kronig transitions and may cause a radical redistribution of electrons in, for example, the L shell, prior to radiation. In the L shell, vacancies may undergo radiationless transitions from L_1 to L_2, from L_2 to L_3, and from L_1 to L_3. Energy becomes available as a result of this, and will normally cause the ejection of an electron from any outer shell whose binding energy is less than that made available by the Coster–Kronig redistribution.

For example, an $L_1 \rightarrow L_3$ radiationless transition will eject an electron from the M_3 sub-shell, in atoms for which Z is less than 37. This would be represented as an atomic transition $L_1 \rightarrow L_3 M_3$. This has three consequences – first, lines of the L_3 spectrum will be enhanced at the expense of those of the L_1 spectrum. Secondly, these transitions will be taking place in an atom which is already ionized in the M_3 sub-shell, and will be slightly displaced in energy, therefore appearing as satellites of the main lines. Thirdly, the extra vacancy in the M shell will enhance the lines of the M spectrum, but these transitions will be modified because the vacancy in the L shell reduces the screening effect by approximately 1 unit of charge, causing the new M lines to correspond to element $Z+1$ instead of Z. Victor (1961) has discussed these effects, and established that the ratio of total intensities of lines in the L_3 spectrum to those in the L_2 spectrum varies with Z. This is interpreted as due primarily to variation of the relative probability of Coster–Kronig transitions.

A further factor affecting the relative intensities of lines associated with different sub-shells is the energy of the bombarding electrons. This effect

Fig. 3.10. The $K \rightarrow L_1$ magnetic dipole transition in gold (Beckman, 1955).

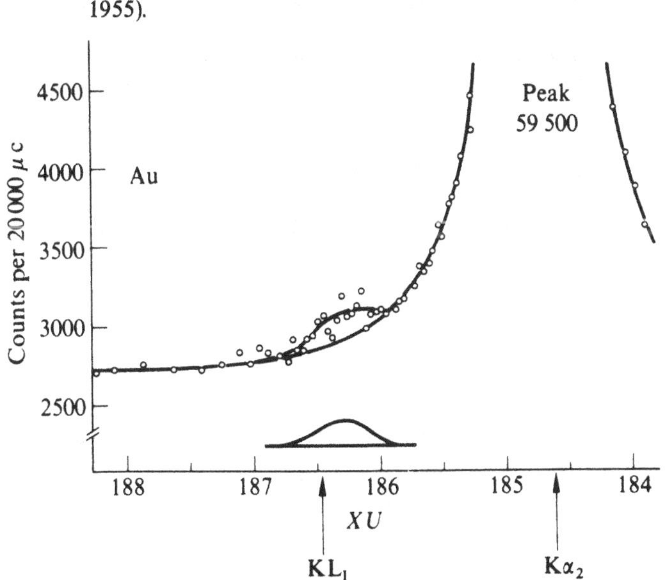

will be most marked near the excitation thresholds (when indeed the L_2 and L_1 spectra will not appear at all unless the appropriate excitation energy is exceeded). For example, Hirst and Alexander (1935) have reported that the $L_{\alpha_2}:L_{\alpha_1}$ and $L_{\beta_2}:L_{\alpha_1}$ intensity ratios are independent of bombarding energy, as would be expected from their common association with the L_3 level, but that the $L_{\beta_1}:L_{\alpha_1}$ ratio rises with bombarding energy, showing that the $L_2:L_3$ ionization probability rises similarly. This ratio is found to approach a constant value, for bombarding energies of about 4 times the excitation energy, and it is under these conditions that the detailed measurements of Victor (1961) just referred to were obtained. Further data is given by Goldberg (1961), reproduced here in table 3.5. It may be seen

Fig. 3.11(a). The intensity of the $K_{\beta_5}(K\rightarrow M_{4,5})$ quadrupole line compared with $K_{\beta_1}(K\rightarrow M_3)$ (Beckman, 1955).

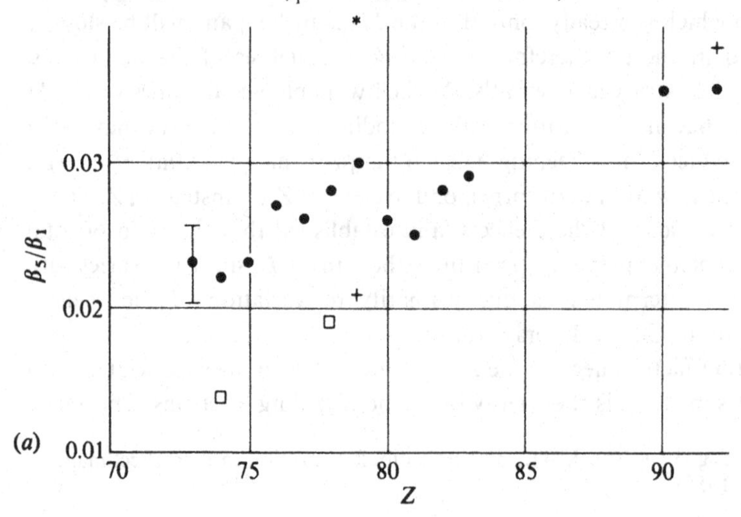

Fig. 3.11(b). The intensity of the K_{β_4} $(K\rightarrow N_{4,5})$ quadrupole line compared with K_{β_2} $(K\rightarrow N_{2,3})$ (Beckman, 1955).

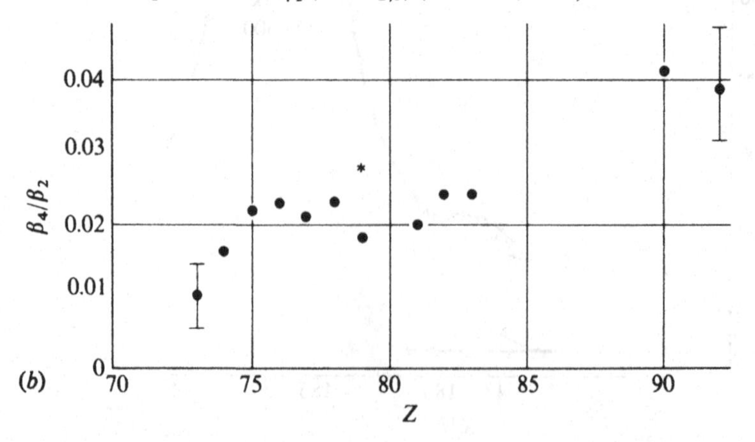

Table 3.5 *Relative intensities of lines in the L spectra of heavy elements (Goldberg, 1961)*

Initial level		Final level	73 Ta	74 W	75 Re	78 Pt	79 Au	80 Hg	81 Ti	82 Pb	83 Bi	90 Th	92 U
L_3	α_1	M_5	100	100	100	100	100	100	100	100	100	100	100
L_3	α_2	M_4	12	11.7	11.7	11.8	11.5	11	10.5	11.9	11.3	10.7	10.6
L_2	η	M_1	1.1	1.1	0.8	0.9	0.8	1.1	0.9	1.1			0.9
L_2	β_1	M_4	45.5	46.3	44.1	45.5	42.4	45.3	46.4	41.6	40.5	39.9	38.6
L_3	$\beta_{15}+\beta_2$	$N_{4,5}$	20.2	22.2	21.4	23.2	25.9	22.9	26.8	22.4	24	28.4	26.8
L_1	β_3	M_3	8.6	6.7	6.1	4.7	4.7	5.7	4.5	4.8	4.3	3.4	3.7
L_1	β_4	M_2	6.5	5.3	4.4	3.7	3.6	3.8	4.3	4.3	4.1	**3.3	3.6
L_3	β_5	$O_{4,5}$	0.28	0.25	0.53	1.7	2.1	2.3	3	3.2	4	**5.7	5.7
L_3	β_6	N_1	1.2	1.2	1.8	1.5	1.3	1.3	1.7	1.7	1.8	1.9	1.7
L_3	β_7	O_1	0.33	0.21		0.32	0.36	0.35	0.5	*0.9	*1.1		0.42
L_3	β'_7	$N_{6,7}$		0.07			0.16						0.22
L_1	β_9	M_5	0.48	0.34	0.5	0.32	0.4		0.5				0.4
L_1	β_{10}	M_4	0.26	0.3		0.27	0.3		0.3				0.25
L_2	γ_1	N_4	9.6	12	12.6	14.7	13.9	13.3	13.7	12.8	11.6	12.2	10.8
L_1	γ_2	N_2	1.9	1.9	1.3	1.6	1.8	0.95	1.6	1.3	1.3	1.4	1.05
L_1	γ_3	N_3	3	2.8	2	2.2	2.3	1.5	1.8	1.6	1.7	1.75	1.15
L_1	$\gamma+\gamma_4$	O_2O_3	0.78	0.83	0.7	1	1	1	0.9	0.7	0.6		0.77
L_2	γ_5	N_1	0.32	0.39	0.5	0.45	0.65	0.5	0.65	0.6	0.45		0.34
L_2	γ_6	O_4	0.33	0.27	0.58	1	1.6	0.8	1.8	1.9	1.3	2.2	2.2
L_2	γ_8	O_1	0.16	0.06					0.28				

*$\beta_7+\beta'_7$
**Interpolated

that the ratio $L_{\alpha_2}:L_{\alpha_1}$ agrees closely with the predicted value of 1:9, but that the ratio $L_{\beta_1}:L_{\alpha_1}$ is appreciably less than the predicted value 5:9 due to the Coster–Kronig transition $L_2 \rightarrow L_3$.

The ionization probabilities of the L sub-shells, and hence the relative intensities of associated lines, depend critically on the mode of excitation. This has been discussed by Burhop (1952) and Victor (1961). For excitation by electron bombardment at energies well above the excitation thresholds, the ratio of cross-sections is approximately $\sigma_{L_1}:\sigma_{L_2}:\sigma_{L_3}=1:1:2$, but for excitation by fluorescence the ratio is more nearly 1:2:3. For ionization caused by internal conversion of a γ-ray emitted in radioactive decay the ionization probabilities depend on the character (electric or magnetic) and the multipolarity of the radiation. This is referred to in chapter 7.

3.4 Characteristic X-ray emission from thin targets

(a) *Cross-sections for inner-shell ionization*

The starting point for most treatments of the ionization cross-section is the expression derived by Bethe (1930) and given in a convenient form* by Mott and Massey (1949).

* Some understanding of the form of this expression may be obtained by relating it to the radius, r_n, of the nth shell. This is given by

$$r_n=\frac{a_{\text{H}}n^2}{Z_{\text{eff}}}$$

where a_{H} is the Bohr hydrogen radius

$$\left(4\pi\varepsilon_0\frac{\hbar^2}{m^2e}\right).$$

E_n (3.1) may be written as

$$E_n=\frac{-1}{4\pi\varepsilon_0}\frac{1}{a_{\text{H}}{}^2}\frac{e^2}{2}\frac{Z_{\text{eff}}^2}{n^2}=\frac{-1}{4\pi\varepsilon_0}\frac{e^2}{2}\frac{Z_{\text{eff}}}{r_n}.$$

For values of $\frac{1}{2}mv^2\approx|E_{nl}|$,

$$\sigma_{nl}\approx\frac{e^4}{(4\pi\varepsilon_0)^2}\frac{1}{E_{nl}{}^2}$$

which is

$$\approx\frac{4r_n{}^2}{Z_{\text{eff}}^2}$$

i.e. the cross-section is of the order of the mean square shell radius divided by Z_{eff}^2. Viewed classically, the incident electron needs to approach the orbital electron sufficiently closely for the interaction energy to be dominant in comparison with the interaction energy between the orbital electron and the nucleus. The reduction factor of Z_{eff} in the impact parameter, and Z_{eff}^2 in the cross-section, expresses this requirement.

$$\sigma_{nl} = \frac{1}{(4\pi\varepsilon_0)^2} \frac{2\pi e^4 Z_{nl}}{m_0 v^2 |E_{nl}|} b_{nl} \ln \frac{2m_0 v^2}{B_{nl}} \tag{3.22}$$

Here, σ_{nl} is the cross-section for ionization in the (n,l)th shell, Z_{nl} is the number of electrons in that shell, and e, m_0, and v are respectively the charge, rest-mass, and velocity of the incident electrons. $|E_{nl}|$ is the shell binding energy, b_{nl} is a numerical constant, and B_{nl} is a function of the electronic binding energy. Values of 0.35 for b_K, for the K shell and 1.65 E_{nl} for B_{nl}, are given by Mott and Massey.

For an approximate formulation at relatively low electron energies, we may use the non-relativistic electron energy $E = \frac{1}{2}m_0 v^2$. If, in addition, we introduce the 'excitation ratio' $U_{nl} = E/E_{nl}$, and the additional quantity*

$$c_{nl} = 4E_{nl}/B_{nl}$$

(3.22) may be written

$$\sigma_{nl} E_{nl}^2 = \frac{1}{(4\pi\varepsilon_0)^2} \frac{\pi e^4 Z_{nl}}{U_{nl}} b_{nl} \ln c_{nl} U_{nl}$$

Rearranging terms,

$$(4\pi\varepsilon_0)^2 \frac{\sigma_{nl} E_{nl}^2 U_{nl}}{\pi e^4 Z_{nl}} = b_{nl} \ln U_{nl} + A_{nl} \tag{3.23}$$

where

$$A_{nl} = b_{nl} \ln c_{nl}$$

The left-hand side is thus a linear function of $\ln U_{nl}$. Experimental data may be plotted in this form, which is known as a Fano plot, in order to obtain experimental values of the parameters b_{nl} and A_{nl} by a least squares fit. This formulation has been elaborated by Powell (1976).

Characteristic X-rays are produced as a result of an inner-shell electron being ejected by a swiftly-moving incident electron. This process is not possible unless the incident electron energy exceeds the shell binding-energy of the relevant bound electrons. The cross-section for inner-shell ionization must therefore be zero for $E < E_{nl}$, and will rise initially as E is increased beyond this value. Equation (3.22) can therefore be satisfactory only if the quantity B_{nl} is allowed to approach $4E_{nl}$ for excitation energies near the excitation limit.

Worthington and Tomlin (1956) have devised an empirical formula which approaches the original expression (3.22), with $b_K = 0.35$, and B_K

* We shall assume positive values for E_{nl}, E_K, etc., from now on.

approaching 1.65 E_K in the logarithmic term for large excitation energies, but which allows B_K to approach $4E_K$ for excitation energies just exceeding the excitation limit. Their expression for B_K is

$$B_K = [1.65 + 2.35 \exp(1 - U_K)]E_K \quad (3.24)$$

where U_K is the excitation ratio E/E_K. This expression for B_K has to be inserted into (3.22). Confining ourselves to the K shell ($Z_{nl}=2$), (3.22) becomes

$$\sigma_K = \frac{1}{(4\pi\varepsilon_0)^2} \frac{2\pi e^4}{EE_K} b_K \ln 4E/B_K \quad (3.25a)$$

or

$$\sigma_K E_K{}^2 = \frac{1}{(4\pi\varepsilon_0)^2} \frac{2\pi e^4}{U_K} b_K \ln 4E/B_K \quad (3.25b)$$

A simpler approach may be adopted, with $B_K = 4E_K$ for all values of U, in which case (3.22) may be written as

$$\sigma_K = \frac{1}{(4\pi\varepsilon_0)^2} \frac{2\pi e^4}{EE_K} b_K \ln U_K \quad (3.26a)$$

or

$$\sigma_K E_K{}^2 = \frac{1}{(4\pi\varepsilon_0)^2} \frac{2\pi e^4}{U_K} b_K \ln U_K \quad (3.26b)$$

We see that $\sigma_K E_K{}^2$ is the same function of U_K for all elements, and that this is true whether we take $B_K = \text{const.} \times E_K$, or B_K as given by (3.24).

Green and Cosslett (1961), following Worthington and Tomlin, have developed this simpler approach, choosing $b_K = 0.35 \times 1.73$ to fit certain experimental data (referred to below) arbitrarily at $U=3$. We should note that to obtain a better approximation at higher energies, a somewhat higher value of b_K than this is required. A value of 0.35×2 is adopted for the comparison illustrated by the dotted line in fig. 3.13. Burhop (1940) has made more detailed calculations, and fig. 3.12 (full line) illustrates his calculations for a thin silver target.

(b) *Experimental data obtained using thin targets*

Up to the time of the calculations of Worthington and Tomlin not a great deal of experimental data was available for comparison, but information for nickel and silver had been reported, consisting in each case of a series of relative measurements standardised by a single absolute observation*.

In fig. 3.12 the calculations of Burhop for silver are compared with experimental data taken from the paper of Clark (1935), from which it is clear that the experimental data fall off less rapidly than expected when the bombarding energy is increased. The experimental methods used for determinations of this type are described in chapter 4, but it should be mentioned here that Clark claimed only 33% accuracy for his absolute measurement.

It is interesting to attempt to approximate to these data by means of a simple expression of the form of (3.26a). By putting $b=0.7$ to obtain the correct scale we get the curve shown by the dotted line in fig. 3.12. ($Z_{n,l}$ in (3.22) has been put equal to 2, the number of electrons in the K shell.) A reasonable fit is obtained.

Fig. 3.13 illustrates the experimental data on nickel and silver to which reference has already been made, with a theoretical curve derived by Worthington and Tomlin from (3.24) and (3.25b) (with $b_K=0.35$) for comparison. In this diagram the units of E_K are electron volts. We show also the theoretical prediction using (3.26b), with $b_K=0.7$.

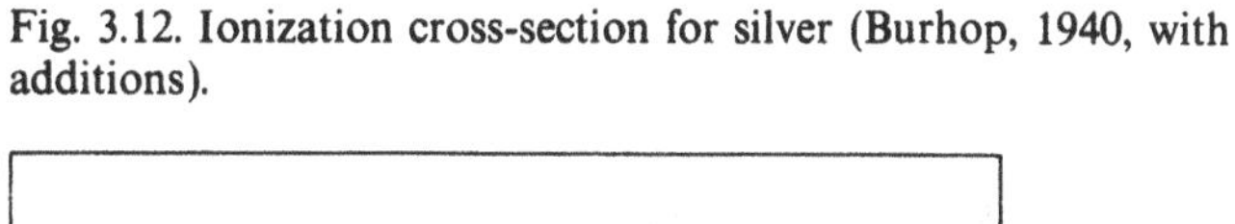

* The relative measurements were by Pockman *et al.* (1947) for nickel, and Webster *et al.* (1933a) for silver. The absolute measurements were by Smick and Kirkpatrick (1945) and Clark (1935) respectively.

Fig. 3.12. Ionization cross-section for silver (Burhop, 1940, with additions).

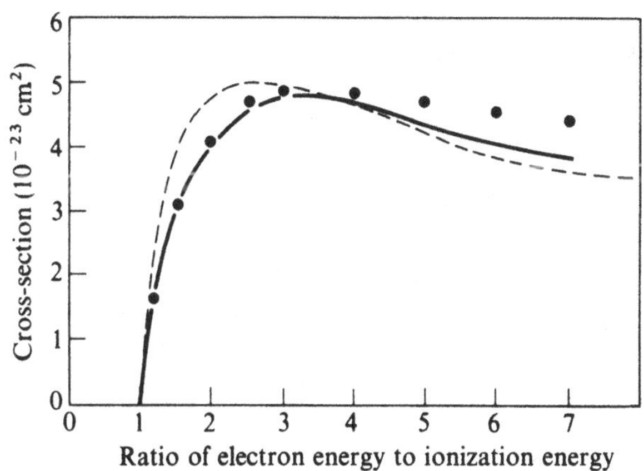

If E_K and σ_K are expressed in electron volts and cm² respectively, and with $b_K=0.7$, (3.26b) becomes

$$\sigma_K E_K{}^2 = 9.12 \times 10^{-14} \frac{\ln U_K}{U_K} \text{cm}^2\,\text{eV}^2 \tag{3.27}$$

Using the value of 0.35×1.73 for b_K, Green and Cosslett (1961) obtained the equation

$$\sigma_K E_K{}^2 = 7.92 \times 10^{-14} \frac{\ln U_K}{U_K} \text{cm}^2\,\text{eV}^2, \tag{3.28}$$

which has been widely used.

Turning to more recent data, cross-sections for aluminium for $1 < U < 20$ have been published by Hink and Ziegler (1969), and a series of measurements on carbon has been reported by Hink and Paschke (1971a, 1971b). Some data for light elements has been reviewed by Powell (1976) and is reproduced here in fig. 3.14.

When presented in the form of a Fano plot, these data appear as in fig. 3.15, and the Bethe parameters b_K and c_K deduced from these data are shown in table 3.6.

A certain amount of experimental data are available for ionization in the L shell. Data for the combined L_2 and L_3 shells are summarised in fig. 3.16. The curves follow the same general pattern as those for the K shell illustrated in fig. 3.14, but lie somewhat higher in absolute value. Bearing in

Fig. 3.13. Ionization cross-section for nickel and silver (the dotted line is from (3.26b) with $b=0.7$). The full line is by Worthington and Tomlin, from (3.25b) with B given by (3.24) and $b=0.35$. σ_K is in cm², and E_K in electron volts. (After Worthington and Tomlin, 1956).

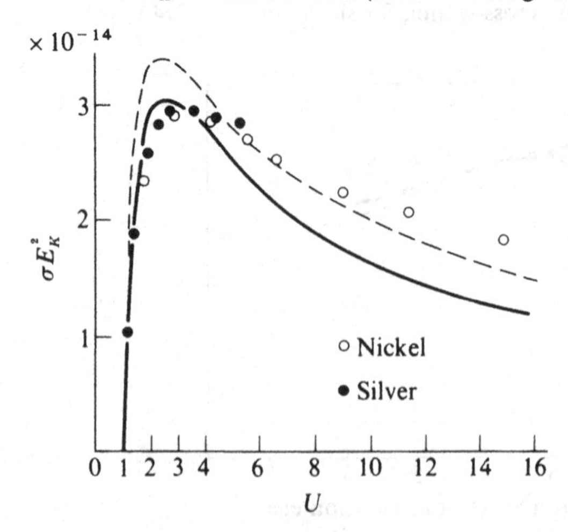

Table 3.6 *Values of Bethe parameters* b_K *and* c_K *and their probable errors obtained from the linear least-squares fits to measured* K-*shell ionization cross-sections for the range of* U_K *indicated.* (*Powell, 1976.*)

Author	Element	Range of U_K	b_K	c_K
Pockman *et al.* (1947)	Ni	5.5–22	1.05 ±0.03	0.51±0.05
Hink and Ziegler (1969)	Al	4.4–19.2	0.90 ±0.01	0.79±0.02
Glupe and Melhorn (1967, 1971) and Glupe (1972)	C	4.2–16.4	0.887±0.004	0.62±0.01
	N	4.4–25.5	0.970±0.004	0.63±0.01
	O	4.3–23.9	0.908±0.003	0.63±0.01
	Ne	4–12	0.932±0.003	0.67±0.01

Fig. 3.14. Experimental values of $\sigma_K E_K{}^2$ as a function of U_K. Data of Glupe and Melhorn (1967, 1971), Glupe (1972) and Bekk (1974) for carbon (triangles), neon (crosses), nitrogen (squares) and oxygen (circles). (Powell, 1976).

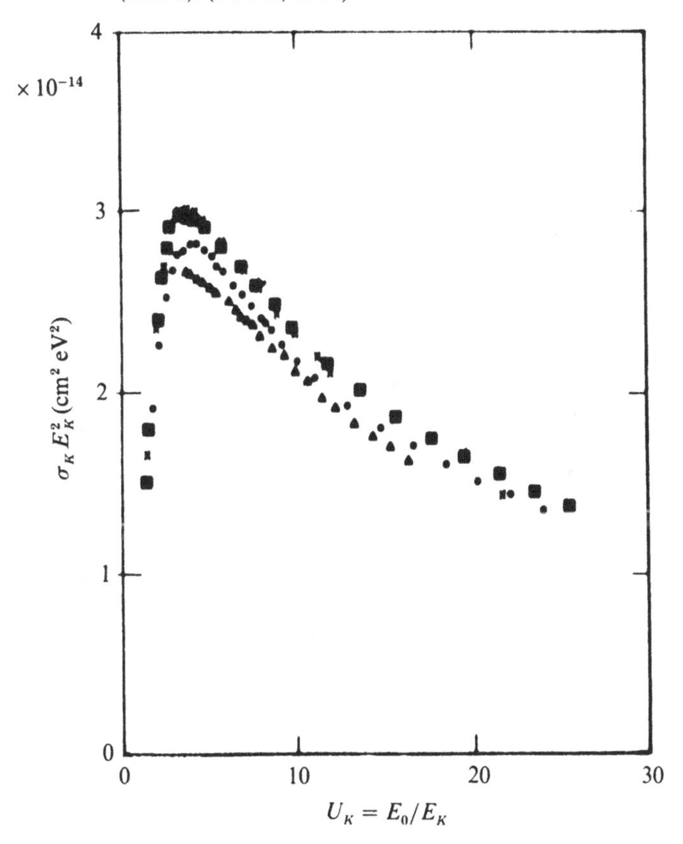

mind that $Z_{nl}=6$ for the L_2 and L_3 sub-shells combined ($n=2$, $\ell=1$), as compared with 2 for the K shell, the data lead to values of b_{nl} which are rather *less* than those for the K shell. $b_{L_{23}}(=b_{21}$ in the '$b_{nl'}$ notation) lies generally in the region of 0.5, as compared with ≈ 1 (table 3.6) for the K shell. These values are both appreciably higher than the values of 0.25 (b_L) and 0.35 (b_K) which were adopted by Mott and Massey.

Much interest is attached to cross-section measurements extending to

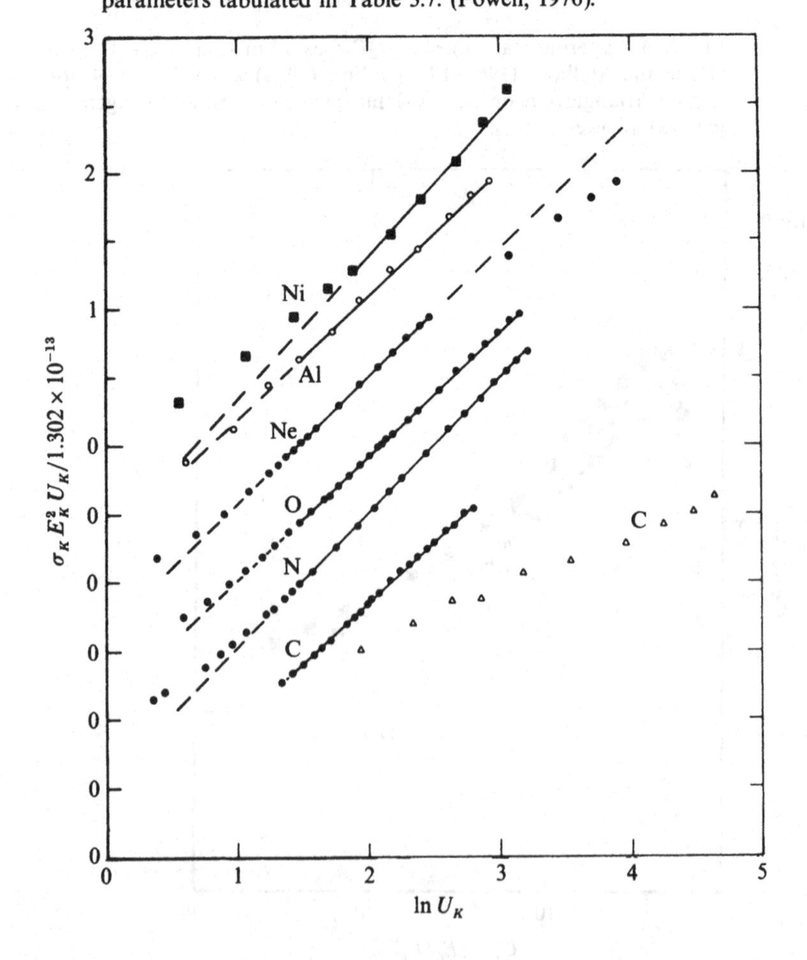

Fig. 3.15. Plot of experimental values of $\sigma_K E_K{}^2 U_K/1.302\times10^{-13}$ (LHS of (3.23) with $Z_{nl}=2$) v. ln U_K. Data as follows:
solid squares: Ni (Pockman *et al.*, 1947)
open circles: Al (Hink and Ziegler, 1969)
solid circles: date of fig. 3.14.
open triangles: C (Hink and Paschke, 1971a,b)
Successive plots have been displaced vertically for clarity. Solid lines (with dashed extrapolation) represent least-squares fit, with the parameters tabulated in Table 3.7. (Powell, 1976).

higher electron bombarding energies and heavier elements. Measurements have been made by Motz and Placious (1964), Hansen *et al.* (1964), Hansen and Flammersfeld (1966), and Dangerfield and Spicer (1975). The parameter $\sigma_K E_K^2$ is no longer the same function for all elements and the plots of cross-section v. electron energy do not always have the same general shape as those for lighter elements. The $m_0 v^2$ in the denominator of (3.22) approaches constancy as $v \rightarrow c$ at relativistic energies; hence the cross-section, instead of reaching a maximum and then declining, (as exhibited by

Fig. 3.16. Experimental values of $\sigma_{L_{23}} E^2{}_{L_{23}}$ as a function of $U_{L_{23}}$. Data of Vrakking and Meyer (1974) for P(squares), S(circles) and Cl (triangles); data of Orgurtsov (1973) for Ar (crosses); data of Christofzik (1970) for Ar (open circles). (Powell, 1976).

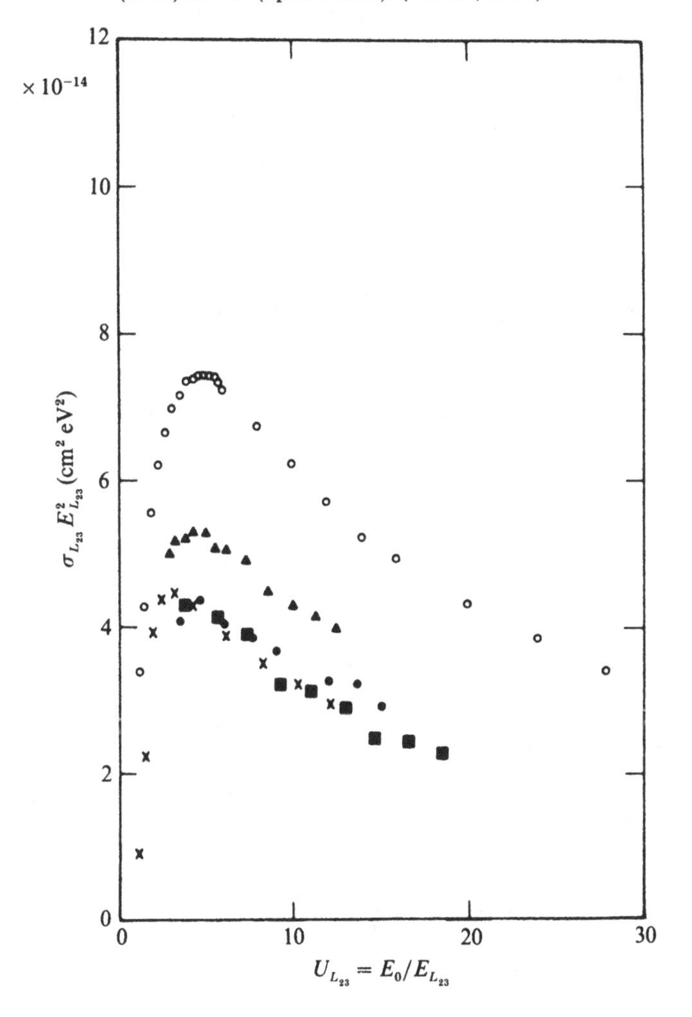

the data of fig. 3.14) reaches a steady value, and remains approximately constant – or exhibits a continued slow rise – as the electron energy is further increased (fig. 3.17). The cross-sections for the heavier elements approach values which are considerably greater than those expected from (3.25a) and (3.25b). Motz and Placious show that their data are in good agreement with relativistic calculations of Arthurs and Moiseiwitsch (1958), and a discussion of the relativistic cross-section for *K*-ionization, including calculations for mercury and nickel, is given by Perlman (1960).

Dangerfield and Spicer (1975) have summarised the behaviour of the σ and σE_K^2 curves by pointing out that, in general, the cross-section rises to a maximum at about four times the shell ionization energy, and then

Fig. 3.17. *K*-ionization cross-section for (*a*) Ag, (*b*) Sn, and (*c*) Au as a function of incident electron kinetic energy. (Rester and Dance, 1966).

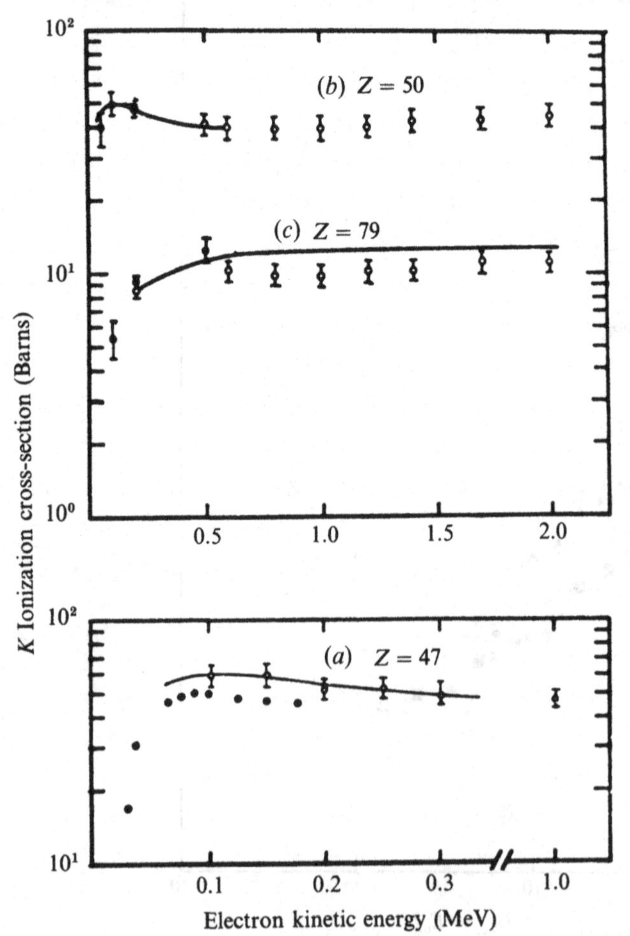

decreases steadily to a minimum when the bombardment energy reaches approximately 1 MeV. The cross-section then *rises* logarithmically (i.e. $\sigma \propto \ln E$) due to contributions from distant collisions. For the higher atomic numbers, the minimum is not observed, due to the preponderance of the steady logarithmic rise. This behaviour is seen in the data reproduced in fig. 3.18.

Fig. 3.18. *K*-ionization cross-sections for (*a*) nickel, (*b*) silver and (*c*) gold, at higher electron energies (Dangerfield and Spicer, 1975).

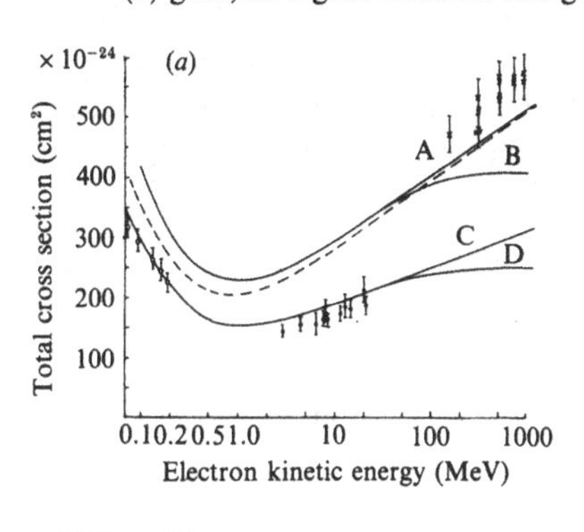

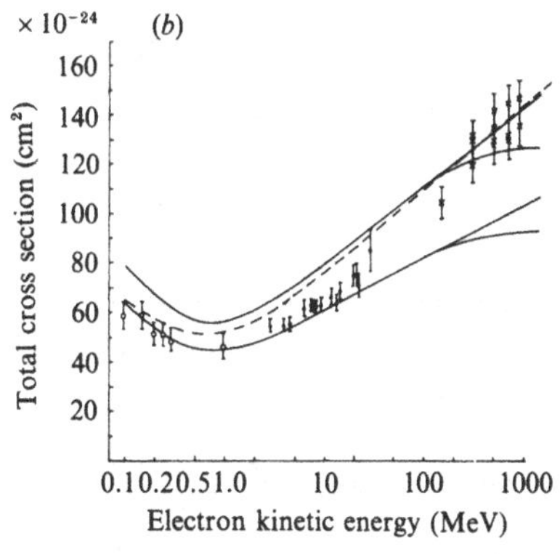

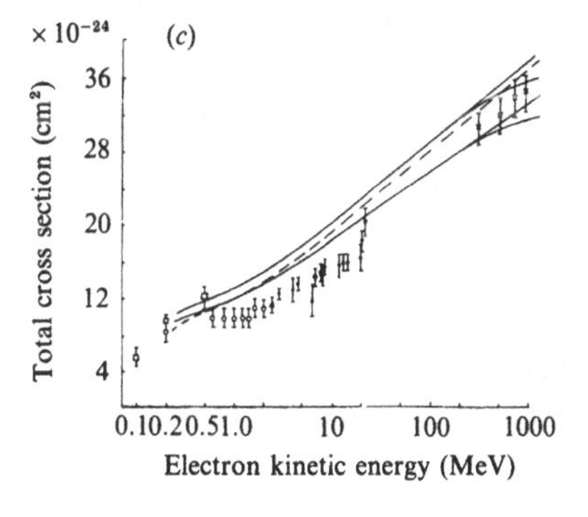

The data of Hansen *et al.* (1964) include measurements obtained with beams of positrons, obtained from radioactive sources. The cross-sections for inner-shell ionization do not differ, within the experimental error, from those obtained by electron bombardment.

3.5 The production of characteristic radiation from thick targets

(a) *Integration of the cross-section*

When electrons enter the target they gradually lose energy, and, to obtain the yield of characteristic X-rays, the cross-section requires an integration over the distance of penetration, using a suitable energy-loss expression, between the incident electron energy E_0, and the shell excitation energy E_{nl}. Confining ourselves to the K-shell, we may write, for the number of K X-rays emitted per incident electron

$$N_K=\omega_K\int_{E_0}^{E_K}\sigma_K n\left(\frac{dE}{dx}\right)^{-1}dE \tag{3.29}$$

where ω_K is the fluorescence yield, σ_K the ionization cross-section, n the number of atoms per unit volume, and $\frac{dE}{dx}$ the *linear stopping power* of the material for electrons. This quantity is given by the Bethe–Bloch formula, which for the present purposes may be written (e.g. Segrè, 1977) as

$$\frac{dE}{dx}=-\frac{1}{(4\pi\varepsilon_0)^2}\frac{4\pi e^4 nZ}{m_0 v^2}\ln\left(\frac{e}{2}\right)^{\frac{1}{2}}\frac{m_0 v^2}{2\bar{J}} \tag{3.30}$$

A discussion of the extensive literature relating to this expression would be out of place here, but it should be noted that this expression is non-relativistic, and that there is some uncertainty in the logarithmic part of this expression; Evans (1955), for example, has $2^{\frac{1}{2}}$ instead of $\left(\frac{e}{2}\right)^{\frac{1}{2}}$, and Mott and Massey (1949) the factor 2. In (3.30), m_0 is the electronic rest-mass, v the velocity, and $\bar{J}$ is a geometric mean of the excitation and ionization potentials of the atom, averaged with due regard to their weighting. Some information regarding the range, energy-loss and scattering of electrons is to be found in appendix 1. Equation (3.30) may be written as

$$\frac{dE}{dx}=\frac{-1}{(4\pi\varepsilon_0)^2}\frac{2\pi\varepsilon^4 nZ}{E}\ln\left(\frac{e}{2}\right)^{\frac{1}{2}}\frac{E}{\bar{J}} \tag{3.31}$$

An approximate calculation may be usefully carried out by noting that the logarithmic term will vary only slowly with E and Z. If this variation is neglected, the stopping power may be written as

$$\frac{\mathrm{d}E}{\mathrm{d}x}=\frac{-1}{(4\pi\varepsilon_0)^2}\frac{knZ}{2E} \tag{3.32}$$

where k is constant.

This equation is a useful approximation for electron energies less than 40 keV, and as such is applicable to the situation obtaining in, e.g., the electron microprobe, and in X-ray tubes for crystallography, or diagnostic radiology.

Reverting to (3.29) and using (3.26a) for σ_K, and (3.32) for $\frac{\mathrm{d}E}{\mathrm{d}x}$, we obtain

$$N_K=\frac{\omega_K 4\pi e^4 b_K}{kZE_K}\int_{E_K}^{E_0}\ln\frac{E}{E_K}\,\mathrm{d}E \tag{3.33}$$

which becomes

$$N_K=\frac{\omega_K 4\pi e^4 b_K}{kZ}(U_0\ln U_0-(U_0-1)) \tag{3.34}$$

for the thick target yield of characteristic K X-rays. This expression has been derived explicitly because several features of importance may be deduced from it. First, it may be seen that the variation of yield with energy is the same for all elements. Secondly, we may deduce the more complicated type of variation with Z – for light elements, the variation is dominated by the Z^4 dependence of ω_K, leading to a Z^3 dependence. As Z approaches middle range, ω_K rises less rapidly and its effect on N_K is almost offset by the Z term in the denominator. At still higher Z, the yield begins to rise again because of the increasing indirect contribution due to fluorescence induced by the continuum. This is not included in (3.34) but is discussed in detail in the following section.

(b) *The indirect production in thick targets*

When a thick target is bombarded by electrons, some of the radiation of the continuous spectrum is absorbed in the target, and if the electrons are incident at angles approaching normal incidence, much of the radiation emitted in the forward hemisphere will be absorbed in this manner. Because the radiation is produced slightly below the surface, there will be some degree of absorption of the backward radiation. Approximately half the energy of the continuous spectrum will therefore be absorbed, and much of this will ultimately cause fluorescence of the target material, and a corresponding enhancement of the emitted characteristic radiation, to an extent which will depend on the number of photons in the continuous spectrum with energies greater than the appropriate excitation energy and the fluorescence yield of the target material.

The contribution due to the indirect radiation has been studied by several workers using X-ray tubes in which the target is covered by a thin layer of material with a lower atomic number, thick enough to stop all the incident electrons, but thin enough to allow some of the characteristic radiation from the underlying target to emerge. Using a target of this type, the radiation from the underlying target is wholly indirect, (i.e. none is produced by electron bombardment) and by comparison with a massive target of ordinary construction the direct and indirect effects can be completely separated. The method has been used by several workers for the study of K-shell X-radiation from silver, palladium, copper and gold.* The interesting result was obtained, in the case of the first three elements named, that the ratio

$$P = \frac{\text{number of } K\text{-ionizations produced by electron impact, } Y_K}{\text{number of } K\text{-ionizations produced by fluorescence, } F_K} \tag{3.35}$$

was very nearly constant over a wide range of accelerating voltages. The ratio varied strongly with Z, the indirect contribution being more important for the heavier elements. In the case of gold, the ratio varied appreciably with voltage, but approached a constant value as the accelerating voltage was increased. A further quantity (which we shall denote by S) was introduced and was defined as

$$S = \frac{\text{number of } K\text{-ionizations produced by electron impact, } Y_K}{\text{number of photons of the continuous spectrum with } h\nu > E_K,\ N_c} \tag{3.36}$$

It was evaluated by dividing the directly produced K ionizations by the number of photons in the continuous spectrum with energies above E_K obtained from (2.2). This quantity also was found to be constant. The data were presented as a function of E_0 or of U_0, and are reproduced in table 3.7.

We can study the theoretical behaviour of S by means of the following argument.

First we obtain Y_K, the number of K ionizations per electron, by writing $Y_K = n\int_0^{x_K} \sigma_K \, dx$, where n is the number of atoms per unit volume of target material, σ_K is the ionization cross-section discussed previously, and x_K is the distance for slowing down to the K excitation limit.

* Respectively by Webster (1928a, b), Hansen and Stoddard (1933), Stoddard (1934) and Stoddard (1935).

Table 3.7 *Direct and indirect production of characteristic K radiation in thick targets*

			(Silver $E_K = 25.535$ keV)					
U_0	1.37	1.96	2.54	3.13				
P	1.83	1.89	1.96	1.96				
S	0.915	0.92	0.925	0.90				
			Palladium ($E_K = 24.356$ keV)					
U_0	1.65	2.46	3.3	4.1	4.9	5.75	6.6	7.4
P	1.95	1.98	2.05	2.03	2.12	2.09	2.06	2.13
S	0.86	0.83	0.85	0.84	0.88	0.87	0.85	0.89
			Copper ($E_K = 8.982$ keV)					
U_0	2	4	6	8	12	14.7	17.4	
P	6.63	6.71	6.87	6.83	7.0	7.1	7.2	
S	3.67	3.58	3.50	3.50	3.5	3.6	3.9	
			Gold ($E_K = 80.713$ keV)					
U_0	1.3	1.6	1.9	2.2				
P	0.195	0.225	0.255	0.27				
S	0.07	0.088	0.103	0.11				

We have

$$Y_K = \frac{1}{(4\pi\varepsilon_0)^2} \frac{2\pi e^4 b_K n}{E_K} \int_{x=0}^{x=x_K} \frac{1}{E} \ln \frac{4E}{B_K} \mathrm{d}x$$
$$= \frac{1}{(4\pi\varepsilon_0)^2} \frac{2\pi e^4 b_K n}{E_K} \int_{E_0}^{E_x} \ln \frac{4E}{B_K} \frac{1}{E} \frac{\mathrm{d}x}{\mathrm{d}E} \mathrm{d}E \tag{3.37}$$

If we use the Thomson–Whiddington Law in the form of (3.32), we must substitute

$$\frac{\mathrm{d}x}{\mathrm{d}E} = -(4\pi\varepsilon_0)^2 \frac{2E}{knZ}$$

and thus obtain

$$Y_k = \frac{4\pi e^4 b_K}{Zk} \left(U_0 \ln \frac{4U_0}{a} - (U_0 - 1) - \ln \frac{4}{a} \right), \tag{3.38}$$

having put $B_K = aE_K$ and $E_0/E_K = U_0$.

In section 2.8 we have evaluated the continuous spectrum in terms of the power radiated per unit current. This would of course be equal to the energy radiated per unit charge. It is, however, sometimes more convenient to work in terms of the energy radiated per incident electron, and we adopt this arrangement here.

From (2.2), omitting the small term in Z^2, we obtain, for the energy E_ν radiated per unit frequency interval per incident electron

$$E_\nu = A_c e Z(\nu_0 - \nu), \tag{3.39a}$$

and we now write $A_c e$ as C*. So, from equation (3.39a) we get

$$n_\nu = \frac{CZ(\nu_0 - \nu)}{h\nu} \tag{3.39b}$$

$$\therefore N_c = \int_{\nu_K}^{\nu_0} n_\nu \, d\nu = \frac{CZ}{h}\left[\nu_0 \ln \frac{\nu_0}{\nu_K} - (\nu_0 - \nu_K)\right]$$

$$= \frac{CZE_K}{h^2}[U_0 \ln U_0 - (U_0 - 1)] \tag{3.40}$$

Dividing (3.38) by (3.40), we obtain

$$S = \frac{4\pi h^2 e^4 b_K}{CZ^2 E_K k} \frac{(U_0 \ln \frac{4}{a} U_0 - (U_0 - 1) - \ln \frac{4}{a})}{U_0 \ln U_0 - (U_0 - 1)} \tag{3.41}$$

This expression becomes independent of U_0 if a is put equal to 4. *The experimentally observed constancy of S as U_0 is varied has therefore provided us with useful information about the constant B in the Mott–Massey expression for ionization cross-section.*

With a equal to 4, S then becomes

$$\frac{4\pi h^2 e^4 b_K}{CZ^2 E_K k}. \tag{3.42}$$

If we substitute $E_K \propto (Z-1)^2$, the result is obtained that S is approximately proportional to Z^{-4}. In practice, the experimental data suggest a Z^{-3} dependence, but as information on only a few elements is available, there is naturally some uncertainty in this.

We plot the available experimental data for P, S, and $(P+1)/P$, as a function of Z in figs. 3.19(a)–(c). $(P+1)/P$ is the factor by which the directly

* This achieves uniformity of notation with the first edition of this book. We may further note that

$$\frac{C}{h^2}\left(= \frac{A_c e}{h^2}\right)$$

is *twice* the continuous efficiency constant denoted as K^1 in section 2.8, expressed in reciprocal energy units.

produced radiation is enhanced due to indirect production. We note the rapidly increasing rôle of indirect production as Z is increased. We shall have occasion to refer again to fig. 3.19 when discussing the ratio of characteristic to continuous radiation in section 3.5.

Little information is available for P and S in the case of L radiation, but Burbank (1944) obtained data for L_3 in thorium. P was found to be 3.1 at 40 kV and 1.98 at 100 kV.

It is useful to summarize at this stage the several factors which have to be included in a complete expression for the number of photons in the K series emitted per electron incident on the target:

$$N_K = \omega_K n \left[\int_{E_K}^{E_0} \sigma_x \frac{dx}{dE} dE \right] \frac{P+1}{P} R f(\chi) \tag{3.43}$$

In this expression, ω_K is the fluorescence yield, n the number of atoms per unit volume, and the integral represents integration of the ionization cross-section along the electron path. The factor following this term represents the correction necessitated by indirect production. This is followed by two further factors, representing respectively the reduction in output due to backscattering of the electrons and to self-absorption of the emergent X-rays. These are discussed below.

The theoretical behaviour of the number of ionizations produced indirectly, F_K, is examined in the context of practical X-ray production in the following section.

(c) *Studies of the intensity of characteristic radiation in practical situations*

Amongst the earliest discoveries in the study of X-ray emission was that the characteristic radiation did not appear until a certain threshold voltage was reached, which varied from element to element, and that the intensity became greater with further increase of the accelerating voltage. Many observers have attempted to fit the rise of intensity with voltage to a simple power law, and the measurements of Webster *et al.* (1933b) are important amongst the earlier studies. In this work it was concluded that the intensity from a thick target of silver, when corrected for absorption in the target, could be expressed quite accurately in the form

$$I \propto (V - V_K)^{1.65} \tag{3.44}$$

and that if a correction for electron rediffusion (backscattering) was incorporated the relation had to be modified slightly to

$$I \propto (V - V_K)^{1.71} \tag{3.45}$$

Fig. 3.19 (*a*),(*b*),(*c*) P,S, and $(P+1)/P$, as a function of Z. (see text).

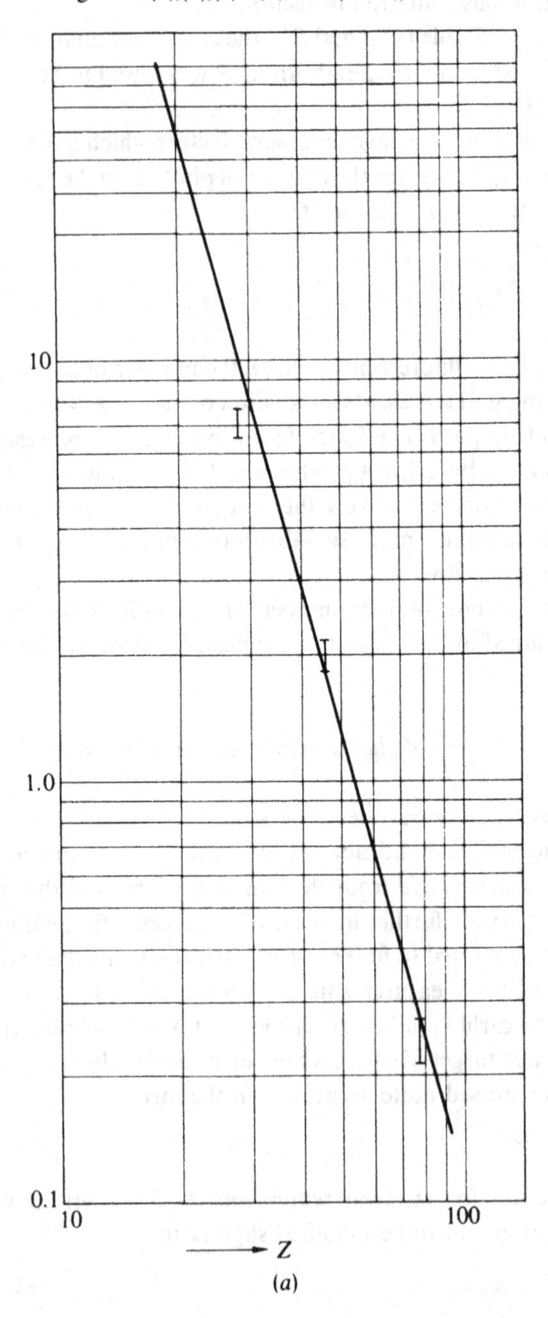

More recent studies have been directed towards three main ends – to establish theoretical grounds for an expression of this form, to study the intensity on an absolute basis as a function of energy and atomic number, and to study the angular distribution of the radiation with particular reference to the intensity at small take-off angles in the region of 2–5 degrees, such as are commonly used in tubes for X-ray diffraction, etc.

Worthington and Tomlin (1956) have used the ionization cross-sections for thin targets of nickel and silver referred to in section 3.3 and have integrated these data using the Bethe–Bloch energy-loss equation (3.30) in the form given by Mott and Massey (1949). The number of K-ionizations per electron in a thick target may be written, as before, as

$$Y_K = n\int_0^{x_K} \sigma_K \, dx \text{ or } n\int_{E_0}^{E_K} \sigma_K \frac{dx}{dE} \, dE \qquad (3.46)$$

It was found that values of Y_K obtained in this way for silver are in good agreement with the absolute measurements of Kirkpatrick and Baez (1947).

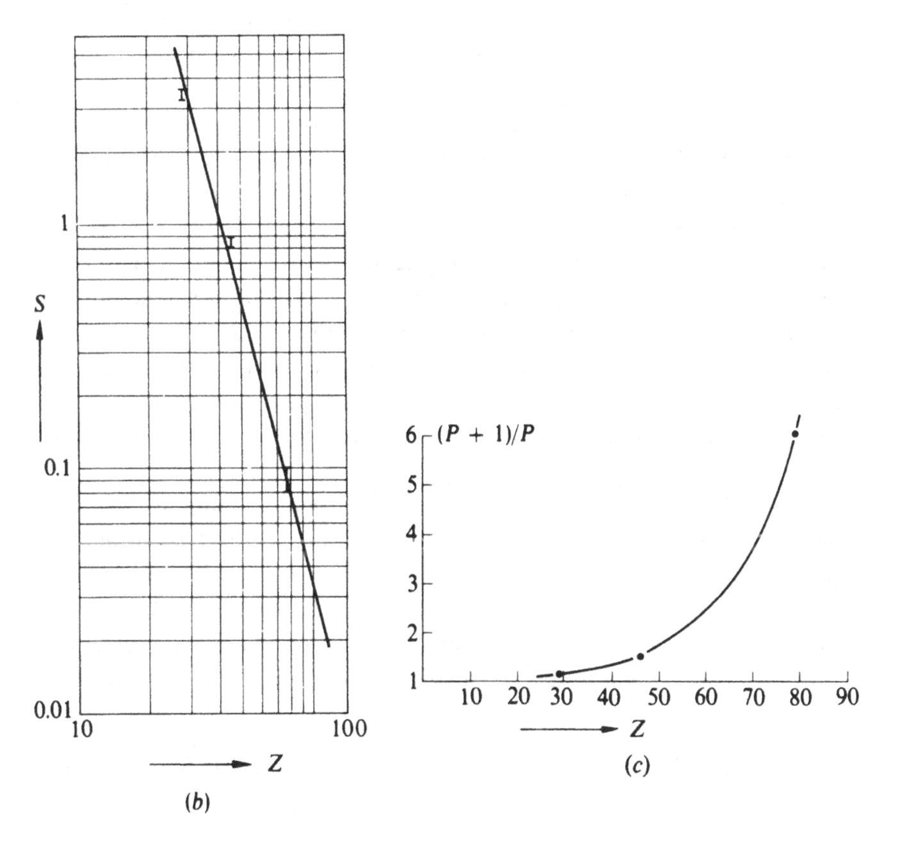

Worthington and Tomlin then evaluated a quantity N_ϕ, such that $\frac{N\phi}{4\pi}\mathrm{d}\omega$ is the number of K_α quanta emitted into a solid angle $\mathrm{d}\omega$ at an angle ϕ to the surface of the target, where

$$N_\phi = K_\omega \int_0^{x_K} \exp(-\mu x \operatorname{cosec}\phi) n\sigma_K \mathrm{d}x \tag{3.47}$$

In this expression the factor K_ω allows for the fluorescence yield, the proportion of $K_\alpha(=K_{\alpha_1}+K_{\alpha_2})$ photons in the K spectrum, and the effect of back-diffusion of electrons. N_ϕ was calculated for silver, copper, and chromium using experimental values for σ_K*, and a calculation was also performed for aluminium using expressions (3.24) and (3.25a) for B_K and σ_K, with $b_K=0.35$.

These calculations have been subjected to extensive experimental verification by Metchnik and Tomlin (1963), using these four elements. The theory of Worthington and Tomlin was modified by taking into account the scattering of the electrons during the slowing-down process. This causes much of the characteristic radiation to be produced considerably nearer the surface than would be the case if the electrons travelled along a rectilinear path. Adoption of this latter assumption by Worthington and Tomlin suggests that their original predictions might be somewhat low, especially at low take-off angles, when the absorption of the outgoing radiation is considerable. The mean depth of production is in fact found to be only about one-sixth of the electron range measured along the track.

Metchnik and Tomlin's experimental data are shown in figs. 3.20(*a*)–(*c*), and a comparison with theory in figs. 3.21(*a*)–(*c*). At moderate angles the agreement between theory and experiment is good, but at smaller angles there are divergences due to the reason already discussed. It seems that the treatment of Metchnik and Tomlin does slightly over-allow for the effect of electron scattering, yielding predictions which are high, and that the experimental data lie between the extremes represented by the two theories.

At small take-off angles the increasing effect of absorption as the mean depth of production rises causes the observed intensity to pass through a maximum at a voltage which depends upon take-off angle. This maximum occurs at lower voltages the smaller the take-off angle but the available absolute intensity is of course greater at larger take-off angles. In tubes for practical use a compromise is struck between the need for a greatly

* In the absence of thin-target data for copper and chromium, the experimental data for nickel were used.

Fig. 3.20,(*a*)-(*c*). Production of K_α radiation in (*a*) silver, (*b*) copper and (*c*) chromium ($N\phi/4\pi$ = number of quanta emitted per unit solid angle in the direction ϕ per incident electron). (Metchnik and Tomlin, 1963).

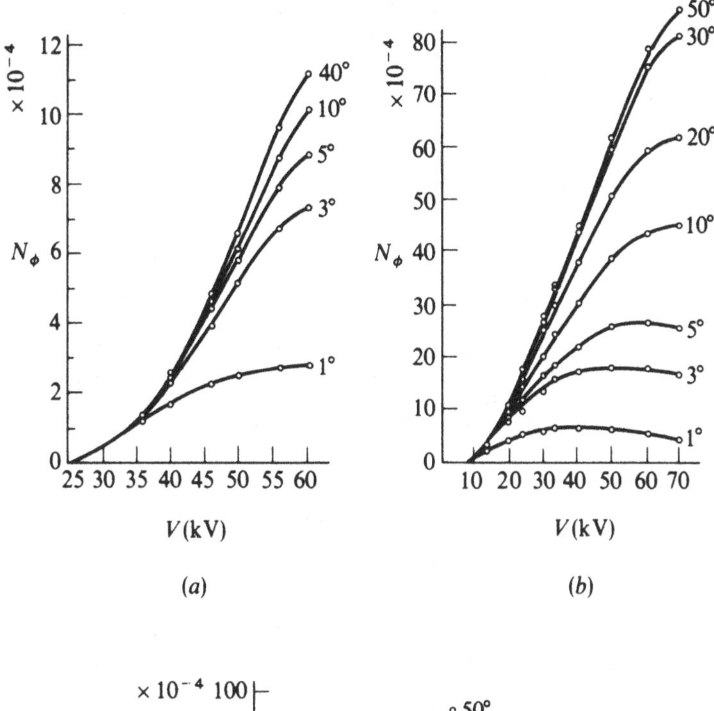

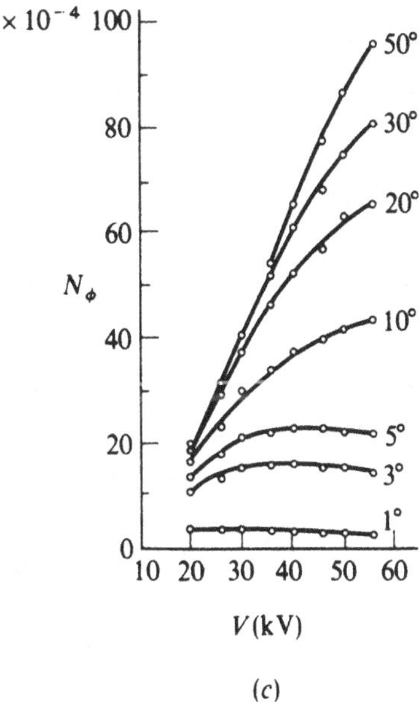

foreshortened apparent source size and the need for a high emitted intensity.

A different approach is taken by Green and Cosslett (1961), for purposes of numerical calculation. They prefer to use a simpler energy-loss expression in the form of (3.32), and, for the ionization cross-section (3.25a) is used with $B=4E_K$ and $b=0.35\times 1.73$. They introduce the electron backscattering factor R, but omit the X-ray self-absorption factor $f(\chi)$.

For the directly-produced component,

$$Y_K=Rn\int_{E_0}^{E_K}\sigma_K\frac{\mathrm{d}x}{\mathrm{d}E}\mathrm{d}E. \tag{3.48}$$

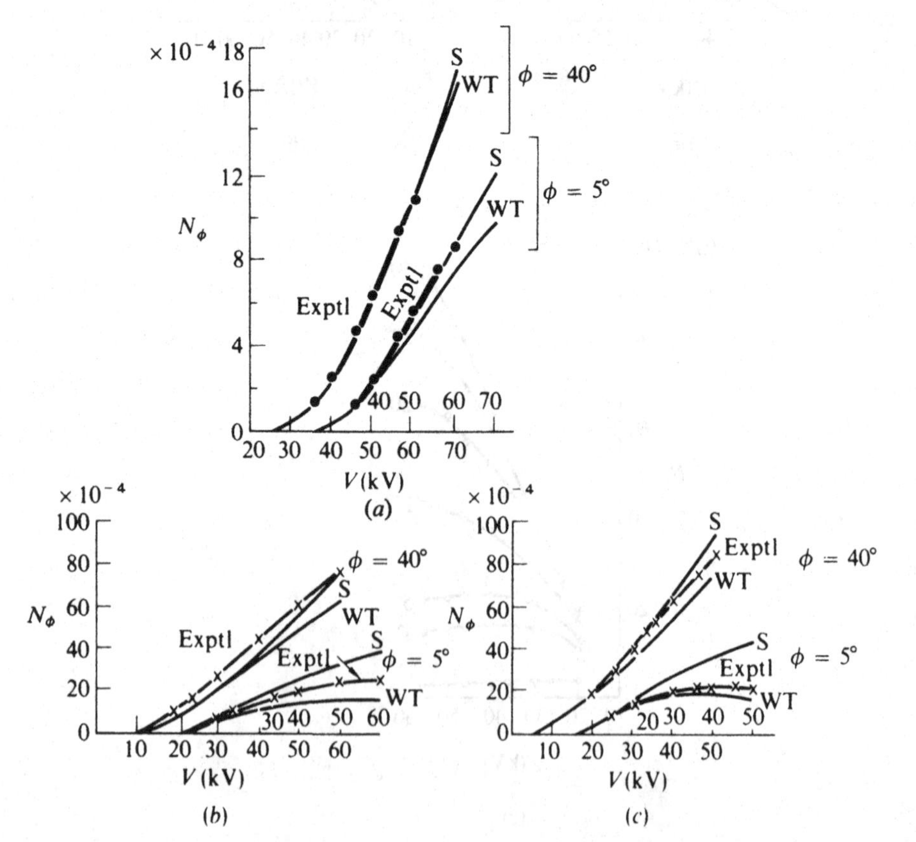

Fig. 3.21, (*a*)-(*c*). Production of K_α radiation in (*a*) silver, (*b*) copper and (*c*) chromium. (Metchnik and Tomlin, 1963). (WT = theoretical calculation of Worthington and Tomlin, S = theoretical calculation of Metchnik and Tomlin, allowing for electron scattering.)

Table 3.8 *The parameter c in the Thomson–Whiddington Energy-loss Equation* (Green and Cosslett, 1961)

E (keV)	1	2	5	10	20	50
c(keV^2 cm^2 g^{-1})	1.0×10^5	1.3×10^5	1.8×10^5	2.3×10^5	2.9×10^5	4.4×10^5

We may substitute as follows:

$$n=N_A\rho/A$$

$$\sigma_K=\frac{1}{E_K{}^2}7.92\times10^{-20}\frac{1}{U}\ln U$$

(from (3.28), but with E_K now in keV; σ_K is in cm^2, as before) and

$$\frac{\mathrm{d}x}{\mathrm{d}E}=-\frac{2E}{c\rho},$$

in which the constant c may be related to our (3.32) by putting

$$c=\frac{1}{(4\pi\varepsilon_0)^2}N_Ak\frac{Z}{A}.$$

Hence,

$$Y_K=\frac{2RN_A}{cA}\int_{E_K}^{E_0}7.92\times10^{-20}\frac{1}{UE_K{}^2}\ln U E\,\mathrm{d}E \tag{3.49a}$$

Inserting the numerical value of 6.02×10^{23} for N_A, we obtain

$$Y_K=9.54\times10^4\frac{R}{cA}\int_1^{U_0}\ln U\,\mathrm{d}U$$

$$=9.54\times10^4\frac{R}{cA}[U_0\ln U_0-(U_0-1)] \tag{3.49b}$$

Y_K being the total number of directly-produced K-shell ionizations in a thick target, per incident electron.

R is shown graphically in fig. 3.22. c is not in fact constant – because we have seen that (3.33) is only approximate – but it varies only relatively slowly with electron energy (table 3.8). Equation (3.49b) is thus appropriate for approximate numerical evaluation of emitted X-ray intensities, if the quantity Y_K is multiplied by the fluorescence yield for the K shell, ω_K.

Allowance must now be made for the indirectly produced ionization F_K. This is excited by those photons of the continuous spectrum which have photon energies in excess of E_K. Of these photons, approximately one half will be absorbed in an X-ray target of conventional shape, and a proportion $f_K=(r_K-1)/r_K$, (where r_K is the K absorption discontinuity ratio – see section 5.2) of these absorptions occurs in the K shell.

So we may write

$$F_K=\tfrac{1}{2}N_{(E>E_K)}f_K \tag{3.50}$$

For the continuous spectrum, from (2.2), we may write

$$n_\nu=\frac{CZ(\nu_0-\nu)}{h\nu},$$

or

$$n_E=\frac{CZ(E_0-E)}{h^2E} \tag{3.51}$$

Fig. 3.22. Fraction of ionization remaining in target after backscattering, as a function of Z and U. (Duncumb and Reed, 1968).

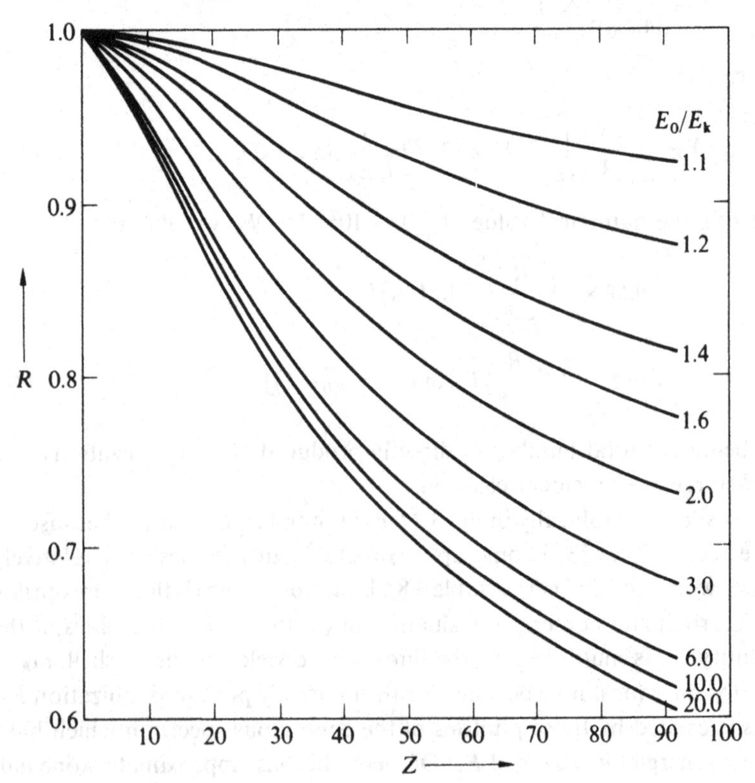

where $\frac{C}{h^2}$ (or $A_c e/h^2$) is *twice* the continuous efficiency constant K' (section 2.8). For $2K'$, Green and Cosslett adopt a value of $2.76 \times 10^{-6}\,\text{keV}^{-1}$, i.e.

$$n_E = 2.76 \times 10^{-6} Z(E_0 - E)/E \tag{3.52}$$

photons electron $^{-1}$ keV^{-1} energy interval. Hence,

$$N_{(E>E_K)} = 2.76 \times 10^{-6} Z \int_{E_K}^{E_0} (E_0 - E)/E \, \mathrm{d}E \tag{3.53a}$$

$$= 2.76 \times 10^{-6} Z E_K (U_0 \ln U_0 - (U_0 - 1)) \tag{3.53b}$$

For E_K, the empirical formula

$$E_K = 1.263 \times 10^{-2} (Z-2)^2 \text{ keV} \tag{3.54}$$

is used, and

$$\frac{r_K - 1}{r_K}$$

is taken as approximately 0.85 for $30 < Z < 80$. Substituting this value and the result of (3.53b) and (3.54) into (3.50), they obtain

$$F_K = 1.46 \times 10^{-8} (Z-2)^2 Z (U_0 \ln U_0 - (U_0 - 1)). \tag{3.55}$$

Adding (3.49b) and (3.55), and multiplying by the fluorescence yield ω_K, they then obtain, for the number of K photons produced per electron incident upon the target,

$$N_K = \omega_K \left(9.54 \times 10^4 \frac{R}{cA} + 1.46 \times 10^{-8} (Z-2)^2 Z \right) (U_0 \ln U_0 - (U_0 - 1)). \tag{3.56}$$

The quantity $U_0 \ln U_0 - (U_0 - 1)$ may be related to empirical expressions of the form of (3.44) by graphical plotting; from this it has been found that $U_0 \ln U_0 - (U_0 - 1)$ may be taken as $0.365(U_0 - 1)^{1.67}$ for $1.5 < U < 16$ with an error of less than 10%. The theoretical calculations therefore link up with the early experimental observations. For the final result for $N_K/4\pi$, the number of photons per steradian per electron, they write

$$\frac{N_K}{4\pi} = \omega_K \left(2.80 \times 10^3 \frac{R}{cA} + 4.27 \times 10^{-10} (Z-2)^2 Z \right) (U_0 - 1)^{1.67} \tag{3.57}$$

Green (1963b) has observed the absolute intensity of generation of K radiation from targets of several elements and finds that (3.57) is a useful fit over the whole range of parameters investigated. To calculate the emission at a given angle, (3.57) has to be multiplied by $f(\chi)$ and those data can then be compared with, e.g., the experimental determinations of Metchnik and Tomlin. Good agreement is found at moderately large take-off angles; at small take-off angles the agreement is less good, though still satisfactory. Green and Cosslett (1968) continue this treatment on the basis of (3.57) (with values of the numerical coefficients modified very slightly), and extend it to the L shell yield.

We may conclude that for elements in the range $20<Z<50$ the absolute intensity is satisfactorily accounted for by the theoretical and experimental data so far reviewed.

In the case of the lighter elements ($Z<20$) the situation is less good. Green finds that there is good agreement between his theory and his measurements, but the theoretical calculations of Worthington and Tomlin for aluminium lead to predictions which exceed Green's observations by a factor of between 3 and 5 in the range $2<U<8$, and by an even larger factor at lower voltages. The measurements of Dolby (1960) on the total K radiation ($K_\alpha+K_\beta$) from aluminium are in good agreement with those of Green. Further, an alternative theoretical treatment by Archard (1960) which gives reasonable agreement with the work already cited in the range $20<Z<50$ gives for aluminium a prediction which is lower than the observations of Green and of Dolby, by a factor of 2.

For still lighter elements there is a comprehensive series of measurements by Campbell (1963) on targets containing beryllium, boron, carbon, oxygen, fluorine, and also aluminium. For aluminium however, the intensities are only about one half of those obtained by Green and by Dolby, being thus quite near to the theoretical predictions of Archard. Data for carbon are shown in fig. 3.23. The experimental differences between the data are even greater. Archard's theoretical prediction is again low.

One reason for the discrepancies between different theoretical treatments is probably the uncertainty in the fluorescence yields for elements of Z less than about 15, and until more experimental data on this is available these differences will remain unresolved. A further difficulty is the uncertain nature of the energy-loss expressions for electron energies below about 5 keV. This affects the integration of (3.48), etc., and also introduces uncertainty into estimates of the mean depth of production of the radiation and therefore of the absorption correction. Campbell considers that the difference between the two sets of experimental data for carbon may be due to uncertainties in the large absorption correction for the window of the counter in Dolby's measurements.

Very little data is available on the production of characteristic radiation in thick targets for elements of Z greater than 50. In the case of L or M radiation, information is sparse over the whole range of elements. Green has obtained experimental data and theoretical expressions relating to L_3 ionization and the intensity of the L_{α_1} line. The ionization probability by the direct process is comparable with the K shell, for similar values of the excitation ratio; but indirect ionization appears to be of much less importance; and the lower fluorescence yields lead to observed quantum efficiencies which are considerably lower than those observed for K radiation. For further details the reader is referred to the original literature already cited.

Finally we show (figs. 3.24(a), (b)) the experimental and theoretical efficiencies for K_α production for which the generated intensity may be obtained at any electron energy and for any elements ($Z < 50$) by application of the empirical power-law in (3.57).

Fig. 3.23. Production of K radiation in carbon. × Campbell (1963); ○ Dolby (1960).

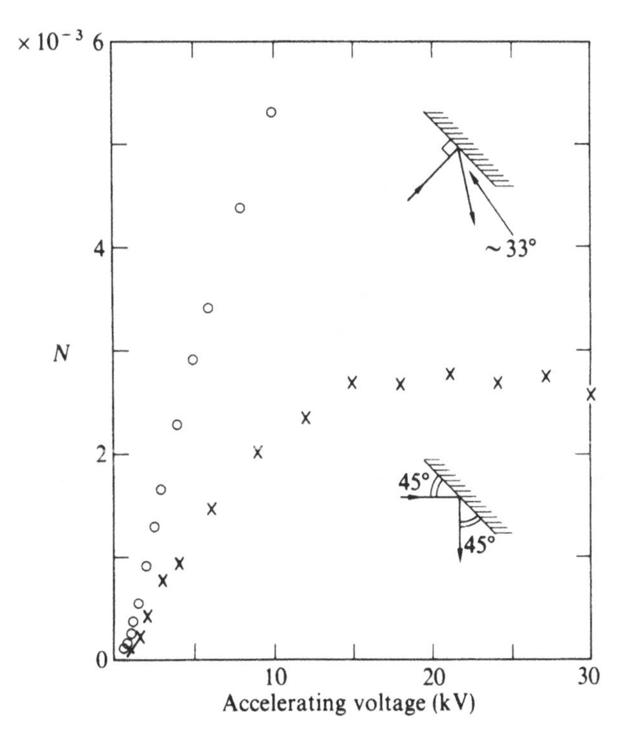

(d) *The variation of characteristic X-ray production with depth, and the self-absorption effect*

If the energy of electrons (of initial energy E_0) has a value of E after travelling a distance x within the target, we may write, for the number of ionizations,

$$Y_K \mathrm{d}x = \sigma_K(E) n \,\mathrm{d}x$$

By integration of (3.32) we may readily obtain

$$E = E_0(1 - x/x_0)^{\frac{1}{2}}$$

Fig. 3.24, (*a*). Efficiency of characteristic K X-ray production for $U_o = 2$. (Green, 1963.) (*b*) Efficiency of characteristic K X-ray production for $E_0 = 2.5$, 10 and 40 KeV. (Green, 1963b).

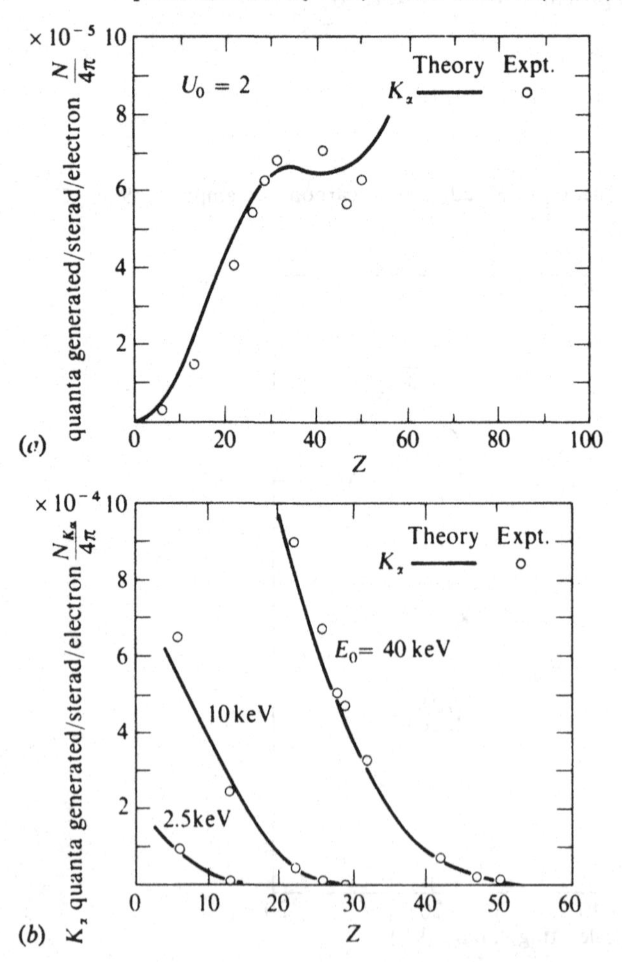

Using (3.26a) for $\sigma_K(E)$, we obtain

$$Y_K \, \mathrm{d}x = \frac{1}{(4\pi\varepsilon_0)^2} \frac{2\pi e^4 n b_K}{U_0 E_K{}^2 (1 - x/x_0)^{\frac{1}{2}}} \ln U_0 (1 - x/x_0)^{\frac{1}{2}} \, \mathrm{d}x \tag{3.58}$$

where $U_0 = E_0/E_K$. Y_K is plotted for $U_0 = 4$ in fig. 3.25. It is seen that the number of ionizations per electron per unit path length rises at first with increasing penetration, and then falls, but remains appreciable until the electron energy falls below the critical excitation energy; this point may be very near to the full range x_0 of the electrons, for values of U_0 commonly used in practical situations, e.g. in the electron microprobe (see chapter 8).

A major factor affecting the depth distribution of inner-shell ionization is the backscattering of electrons within the target. This will reduce the production at greater depths, and will increase the production at and near the point of entry. The overall effect of backscattering, however, will be to reduce the total production because of loss of electrons from the surface of the target, by the factor which we have denoted by R (see fig. 3.22).

To examine the production in more detail, we must distinguish between the increments of path length dx, and increments of *depth*, dz, within the target. As the depth within the target increases, the path lengths in a given increment of z will become progressively greater, because of increasing obliquity of travel. This also will increase the production at moderate depths over and above the rise deduced from the simple treatment

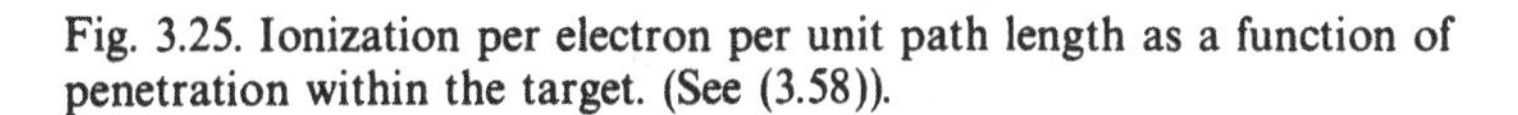

Fig. 3.25. Ionization per electron per unit path length as a function of penetration within the target. (See (3.58)).

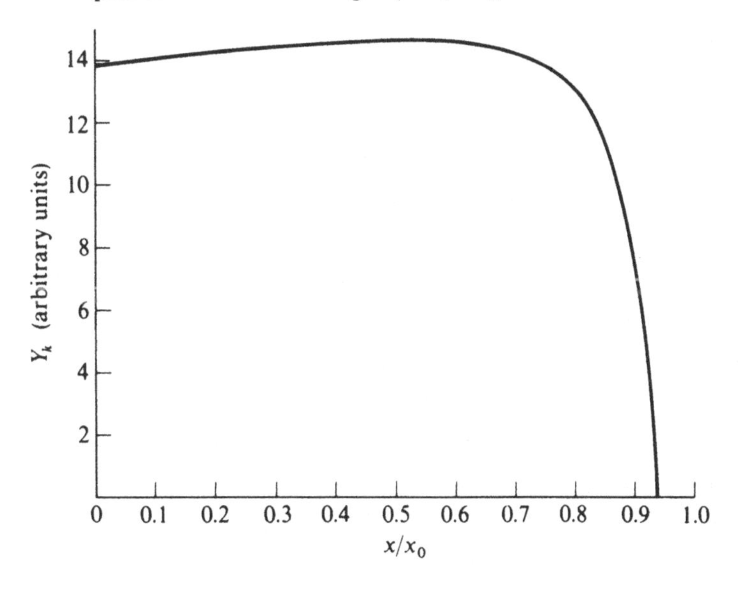

illustrated in fig. 3.25. An experimental approach to this problem has been undertaken by Vignes and Dez (1968), in connection with the evaluation of self-absorption corrections required in quantitative analysis using the electron microprobe. They used the 'tracer-layer' technique of Castaing and Descamps (1955). This consists of taking a block of the element under investigation (element A, atomic number Z) and placing in front of it a thin layer of the tracer element (element B, usually with atomic number $Z+1$). In front of the tracer layer varying thicknesses of element A are deposited. The radiation from the thin layer represents the intensity which would be produced at that depth in a solid target of element A, as modified by the effect of backscattering, obliquity, and energy-loss. In order to obtain the efficiency of production as a function of depth, a correction is applied to the emitted intensities to allow for attenuation in the known thickness of

Fig. 3.26. Distribution in depth of K_α production in titanium at 29 keV. (Vignes and Dez, 1968).

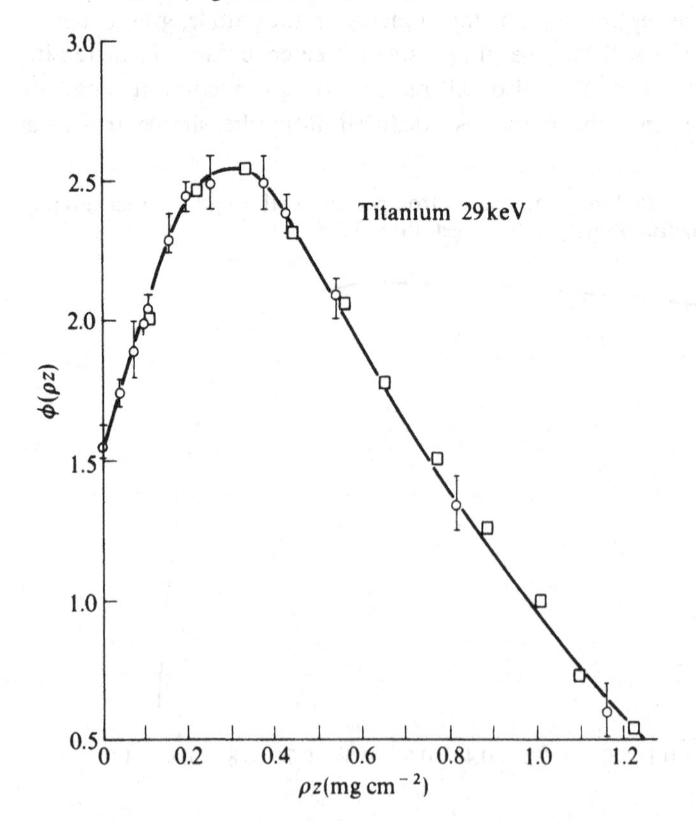

Table 3.9 *Surface ionization* ϕ_0

Elements	Accelerating voltages (kV)											
	10	13.4	15	17	18.2	20	23.1	24	27.6	29	33	35
Aluminium	1.26		1.38			1.45		1.51		1.52		
Titanium				1.42		1.45		1.52		1.55		1.57
Nickel				1.31		1.39		1.46		1.57		1.58
Copper		1.273			1.428		1.537		1.571	1.50		
Gold										1.714		
Lead										1.70	1.80	

overlying material. Data obtained in this way for titanium (K-radiation) at 29 keV are shown in fig. 3.26. These data include X-rays produced indirectly by fluorescence caused by the continuous radiation which is produced at the same time; this explains why some characteristic radiation is generated at depths greater than the electron range. This effect may be considerable in the heavier elements. The enhancement of X-ray production in the surface layer caused by electron backscattering within the target has already been noted. This also was investigated by Vignes and Dez, who compared the yield in the surface-layer with the yield obtained from an *isolated* layer of the same thickness. The 'surface ionization' for several elements and electron energies is shown in table 3.9, taking as unity the intensity emitted by the foils isolated in space. This table also includes data by Castaing and Descamps (1955), on whose method the later work is based.

If $I(\rho z)\,\mathrm{d}(\rho z)$ represents the intensity (per electron, or per unit of current) generated in a layer of thickness $\mathrm{d}(\rho z)$, as shown in fig. 3.26, the data can be used to calculate the total absorption correction which would have to be applied to the radiation *emitted* from the surface of a solid target of element A, in order to calculate the *generated* radiation, which would be required, for example, in the processing of data obtained in electron microprobe analysis. The ratio of *emitted* to *generated* radiation is given by

$$f(\chi)=\frac{\displaystyle\int_0^\infty I(\rho z)\exp(-(\mu/\rho)(\rho z\operatorname{cosec}\theta)\,\mathrm{d}(\rho z)}{\displaystyle\int_0^\infty I(\rho z)\,\mathrm{d}(\rho z)}, \tag{3.59}$$

where θ is the 'take-off' angle of the emerging X-rays. The quantity

Table 3.10 *The absorption correction factor $f(\chi)$*, given by (3.59), obtained experimentally by Vignes and Dez (1968)*

	Absorption correction values $f(\chi)$				
	Titanium			Lead	
	Accelerating voltages (kv)				
χ	20	25	29	29	33
500	0.839	0.809	0.777	0.784	0.758
1000	0.714	0.669	0.621	0.635	0.597
1500	0.615	0.562	0.507	0.528	0.485
2000	0.535	0.480	0.423	0.448	0.405
2500	0.470	0.416	0.359	0.387	0.345
3000	0.416	0.364	0.310	0.338	0.299
3500	0.372	0.322	0.270	0.300	0.263
4000	0.334	0.288	0.239	0.268	0.233
4500	0.302	0.259	0.213	0.241	0.209
5000	0.275	0.235	0.191	0.219	0.189
5500	0.252	0.214	0.173	0.200	0.172
6000	0.231	0.196	0.158	0.184	0.158
6500	0.214	0.181	0.145	0.170	0.145
7000	0.198	0.167	0.133	0.158	0.134
7500	0.184	0.155	0.123	0.147	0.125
8000	0.172	0.145	0.115	0.137	0.116

*$f(\chi)$ is the factor by which characteristic radiation is attenuated on emerging from the target. The point of origin of the radiation is distributed in depth according to the experimental data of fig. 3.26.

$\chi = \left(\frac{\mu}{\rho}\right)$cosec θ, where θ is the take-off angle.

ρz cosec θ is the path, in the target, of the X-rays generated at a depth z. The quantity (μ/ρ) cosec θ is denoted by χ. The factor $f(\chi)$ is tabulated in table 3.10.

The study of the depth distribution of characteristic X-ray production has been refined by the setting up of suitable theoretical models of electron scattering and diffusion within the target. These are reviewed by Green (1963a), who also developed a Monte Carlo method of treating the problem. A detailed application of this method, for the case of 29 keV electrons incident on a copper target, has been given by Bishop (1965).

The function $f(\chi)$ may be studied by a different method, by observing the angular distribution of characteristic radiation, and noting that departures from isotropy are due entirely to absorption effects of the type already discussed. This approach has been adopted by Green (1964), who presents graphs of $f(\chi)$ as a function of atomic number and take-off angle. He shows that $f(\chi)$ is nearly independent of Z for a given electron energy. This implies that, although the range measured along the electron path increases somewhat with Z, it is offset by the greater amount of scattering as Z is increased, causing the actual penetration to be almost independent of Z. Because of this it is possible to construct generalised $f(\chi)$ graphs from which the behaviour over a wide range of Z ($5 < Z < 30$) can be predicted by linear

Fig. 3.27. $f(\chi)$ (Green, 1964) (*a*) $Z = 10$, (*b*) $Z = 30$.

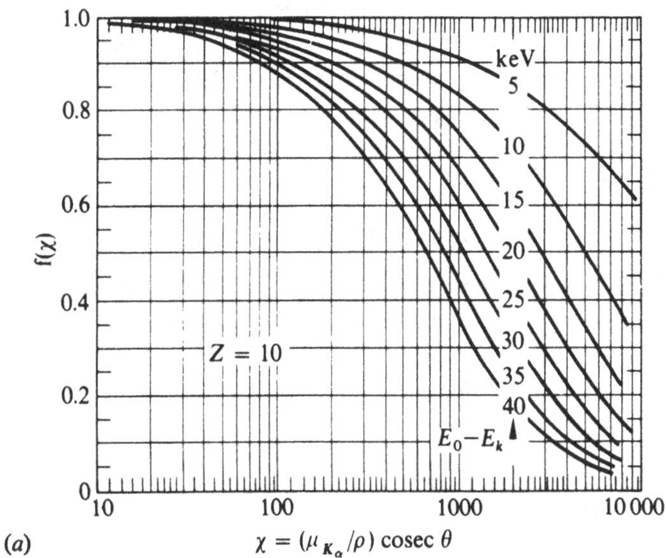

(*a*)

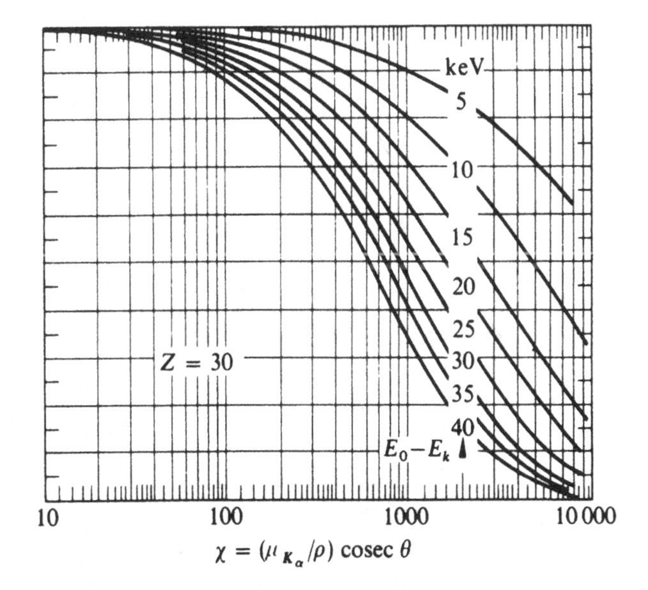

(*b*)

interpolation from just two sets of data for $Z = 10$ and $Z = 30$ (fig. 3.27). It should be noted that the values of Vignez and Dez for $f(\chi)$ in table 3.10 are appreciably higher than those of Green in fig. 3.27.

The problem of the depth of origin of characteristic X-rays and the influence of absorption upon the intensity of the emergent beam has assumed considerable practical importance in the field of X-ray emission microanalysis. A detailed knowledge of the self-absorption effect is required, in pure standards and also in the specimens undergoing analysis. This is in order to convert the observed intensities into generated intensity (by dividing by $f(\chi)$) to enable a quantitative analysis to be carried out. Those procedures are extensively discussed by Reed (1975).

The mean depth $\bar{z}$ at which characteristic radiation is produced is found to vary approximately linearly with excess $E - E_K$. Kirkpatrick and Hare (1934) quote the relation for silver

$$\bar{z} = 0.8(U - 1), \text{ in micrometres of silver} \tag{3.60}$$

Green obtained data for copper from his angular distribution measurements, and his values for $\bar{z}$ in this element are shown in fig. 3.28. A simple way of understanding this is to suppose that after reaching the depth at which 'full diffusion' takes place, the electron then moves in a 'random walk'

Fig. 3.28. Electron range and mean depth of K_α production in copper. (Green, 1964).

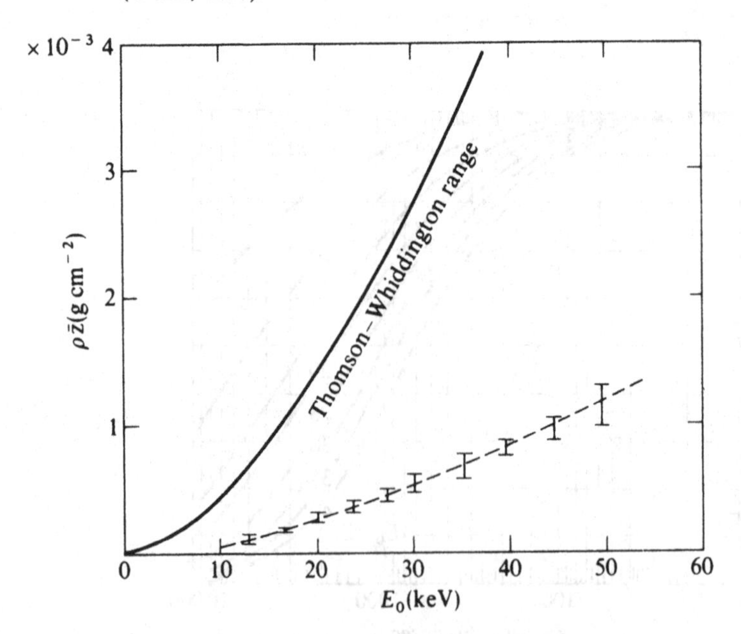

manner until reaching thermal velocities. If there are M scattering events, associated with a mean free path λ_e, we may write

$$x = M\lambda_e = kE^n,$$

where x is the range *measured along the track*, and

$$z = \lambda_e \sqrt{M} \text{ or } (\lambda_e kE^n)^{\frac{1}{2}} \tag{3.61}$$

where z is the penetration. If n is taken to be 2, we see that any further penetration into the target after full diffusion is reached could follow an approximately linear law of variation with energy. The random-walk approach is likely to become progressively better at higher values of Z, because of the increasing rôle of nuclear scattering (i.e. through large angles) in the heavier elements.

A quantity of interest is the mass-absorption coefficient of the target for its own characteristic radiation. This is illustrated in appendix 2 (fig. A2.1). In this region of energy the absorption is primarily photoelectric, so we may write, for the photoelectric absorption coefficient

$$(\mu/\rho) = \text{const.}\, Z^{9/2} \frac{N_A}{A} \frac{1}{E^3} \tag{3.62}$$

The energy of the K_α characteristic radiation is given by Moseley's law (3.12)

$$E = \text{const.}\, (Z-1)^2$$

Hence the absorption coefficient for K_α radiation is given by

$$\left(\frac{\mu}{\rho}\right)_{K_\alpha} \propto Z^{\frac{9}{2}} \frac{N_A}{A} \frac{1}{(Z-1)^6}$$

or, approximately, by putting $A \propto Z$ and $Z - 1 \approx Z$,

$$\left(\frac{\mu}{\rho}\right)_{K_\alpha} \propto Z^{-\frac{5}{2}} \tag{3.63}$$

The data in appendix 2 follow this law of variation quite closely.

A useful parameter is the quantity $(\mu/\rho \,.\, \rho z_0)$ where z_0 is the (full) depth of penetration of electrons in the target material, and μ/ρ is the absorption coefficient for K_α radiation, for it gives an index of the amount of absorption to be expected for characteristic radiation emerging perpendicularly from the target. We may consider two cases – first, for a *constant bombarding energy*, and secondly, for a constant value of E_0/E_K. Case 1 is illustrated in fig. 3.29(*a*) for constant bombarding energies of 40 and 100 keV. The dotted line follows the relation

$$\left(\frac{\mu}{\rho}\right)_{K_\alpha} \propto Z^{-\frac{5}{2}}$$

and the full line is derived from the values for range tabulated by Nelms (see appendix 1), and from the mass absorption coefficients tabulated in

Fig. 3.29(*a*). Graph of $(\mu/\rho)\rho z$ for K_α radiation (z = electron penetration depth) as a function of Z for electron bombarding energies of (*a*) 40 keV, (*b*) 100 keV.

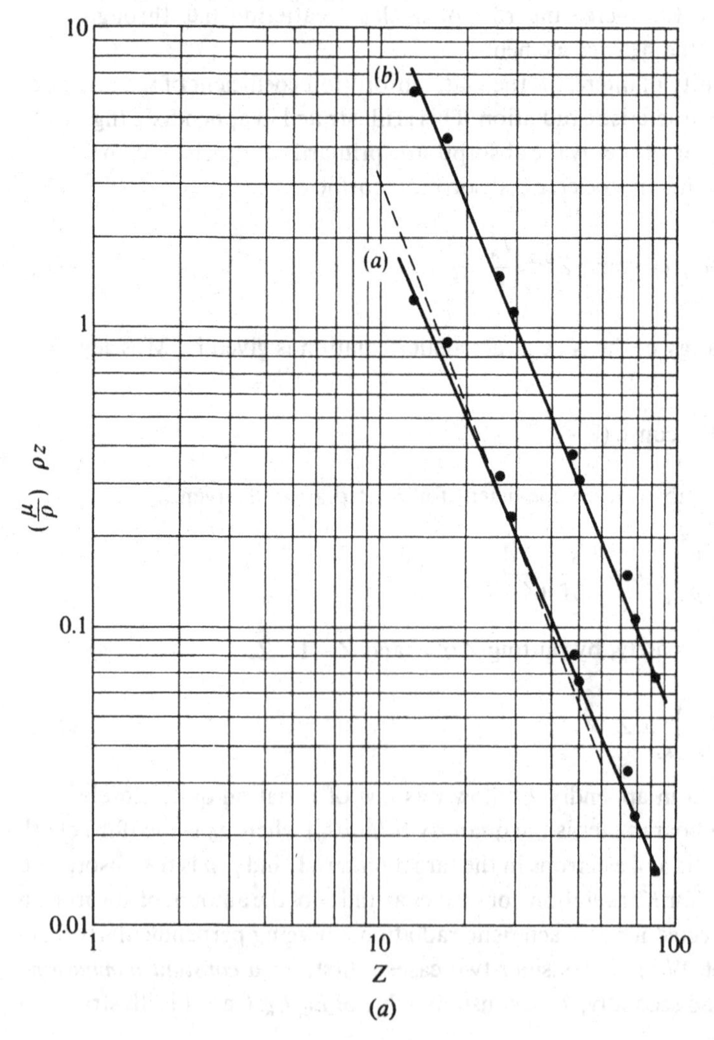

appendix 2. Case 2 is illustrated in fig. 3.29(*b*), for the case where $U_0 = 4$, using data derived from the same sources.

A simple power-law type of calculation is not possible for case 2, because of the more complex variation of range with electron energy over the wide span of the latter variable necessitated by the condition that E/E_K is constant at a value of 4; but the trends are clear – for a constant accelerating voltage the effects of self-absorption are most marked for light elements, decreasing strongly with increasing Z. For an accelerating voltage which

Fig. 3.29(*b*). Graph of $(\mu/\rho)\rho z$ for K_α radiation (z = electron penetration depth) as a function of Z for $E/E_K = 4$.

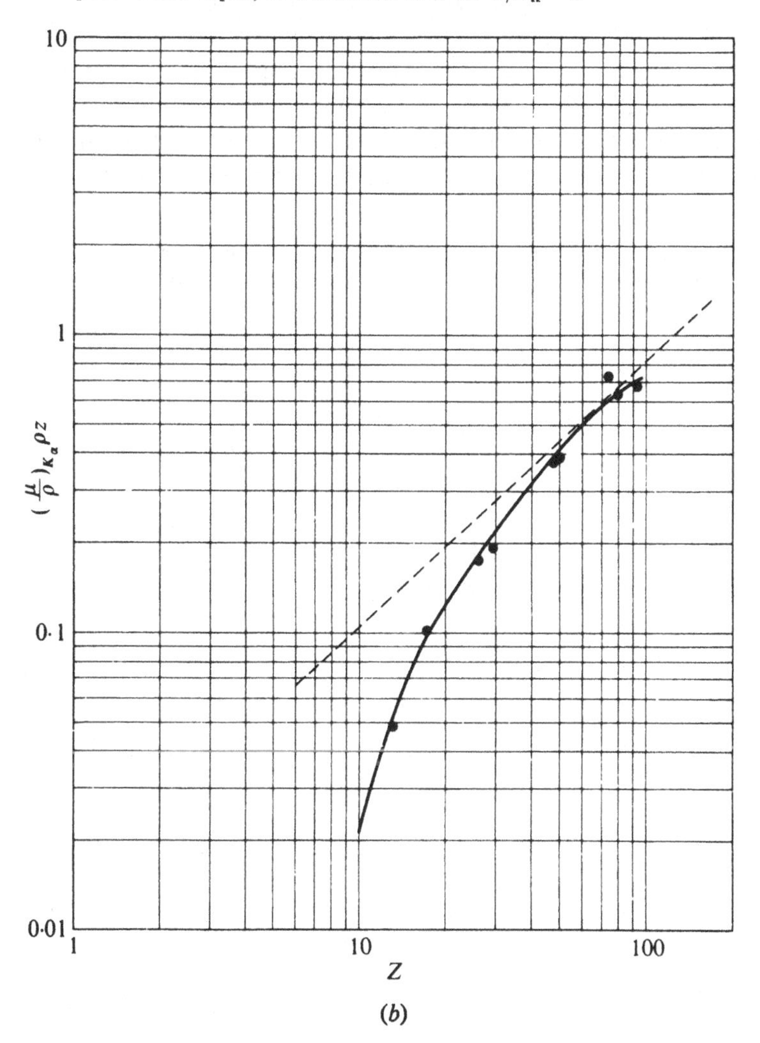

increases in step with E_K so as always to be near the optimum for the target in question, the self-absorption is most important for the heavier elements, although the variation with Z is less than for case 1.

3.6 The ratio of characteristic to continuous radiation from thick targets

This quantity can be readily estimated from the data already discussed, but in view of its importance in connection with, for example, the ratio of peak to background intensities in X-ray spectra and X-ray diffraction measurements it is as well to consider the ratio in more detail.

To calculate the ratio of characteristic $K_\alpha(=K_{\alpha_1}+K_{\alpha_2})$ to continuous radiation we may take the experimental data of Green (fig. 3.24) and combine it with calculations of the continuous spectrum intensity derived from (2.2). This has been carried out for copper and is shown in fig. 3.30(a). It is seen that the ratio is low for accelerating voltages not greatly in excess of the critical excitation potential, and that it rises to a broad maximum at which the intensities of the two spectra are approximately equal. The

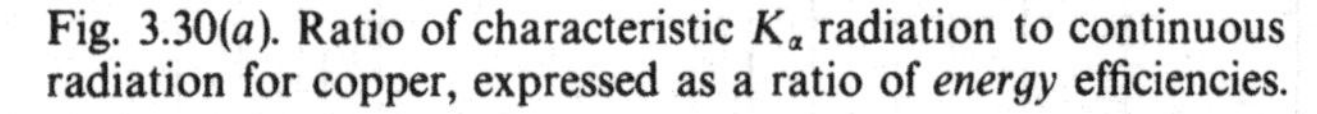
Fig. 3.30(a). Ratio of characteristic K_α radiation to continuous radiation for copper, expressed as a ratio of *energy* efficiencies.

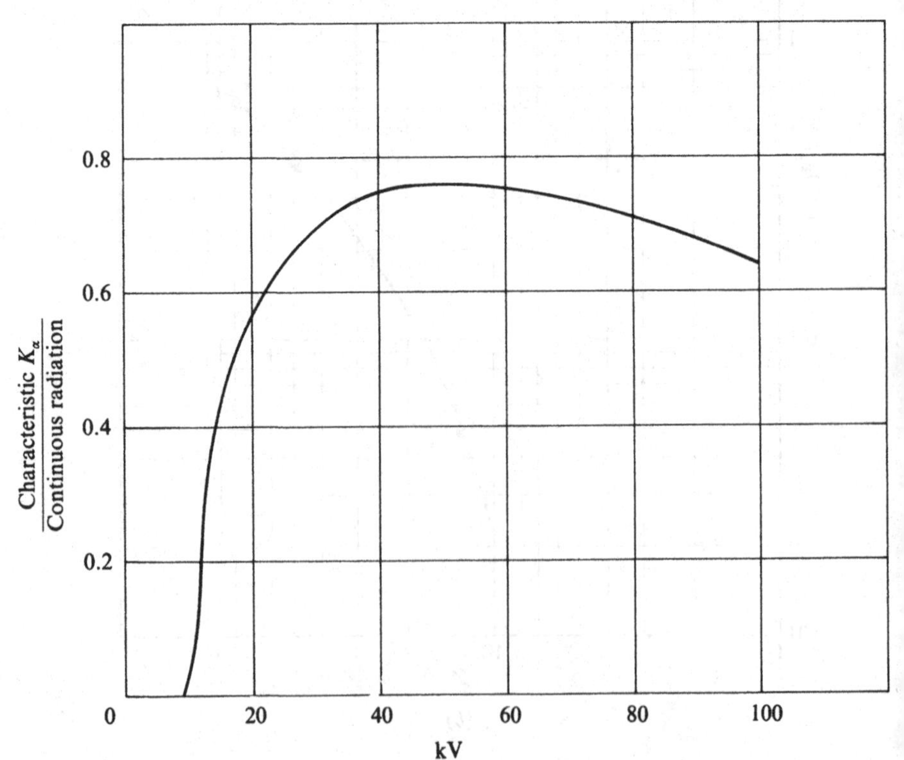

Fig. 3.30(*b*). Ratio of characteristic K_α to continuous radiation for tungsten. (After Tothill, 1968).

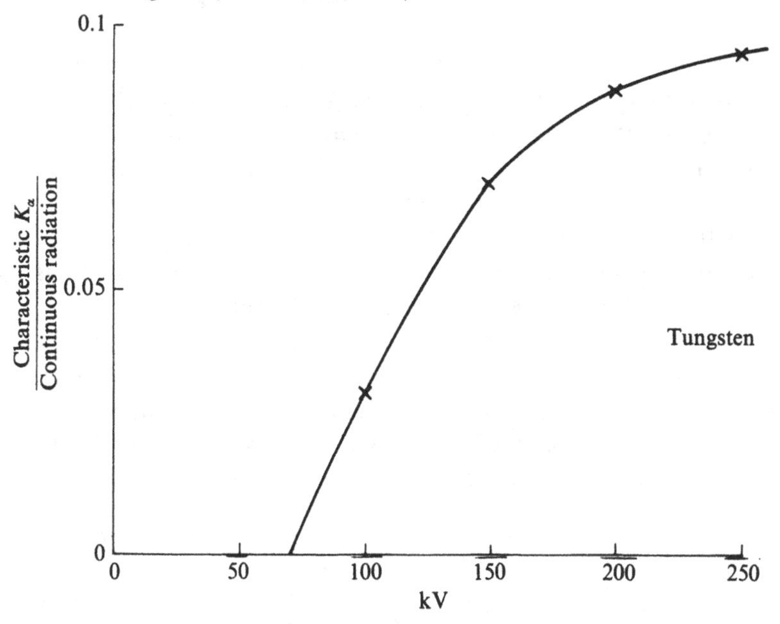

Fig. 3.30(*c*). Ratio of characteristic L to continuous radiation for tungsten. (Data of Unsworth and Greening, 1970a,b).

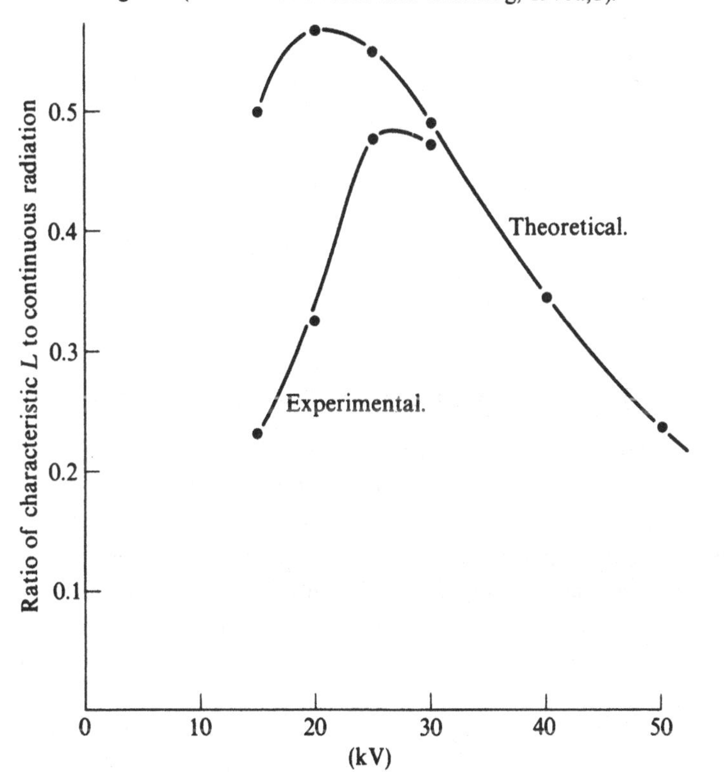

position of the maximum will be sensitive to the detailed geometry of the situation, but the graph will provide us with a rough guide as to the magnitude of the ratio.

To determine this ratio for heavier elements is less easy on account of the lack of absolute data for the characteristic 'thick target' intensity for Z greater than 50. But for tungsten ($Z=74$) we may quote the data of Tothill (1968) which we reproduce (in a form comparable with fig. 3.30(*a*) in fig. 3.30(*b*). The ratio for the same values of the excitation ratio U is seen to be much less than for copper.

Some data for the ratio of characteristic L radiation to continuous radiation for tungsten has become available and is shown in fig. 3.30(*c*).

The quantity which we defined as S in section 3.3 can readily be converted to the ratio J_d of the intensities of *directly* produced characteristic K_α radiation to total continuous radiation by multiplying by the fluorescence yield ω_K, the photon energy $E_{K\alpha}$ of the K_α radiation, the fraction f_α of K_α photons in the K spectrum, and by the factor

$$\frac{\text{probability of producing a photon in the continuous spectrum with } h\nu > E_K}{\text{total intensity in the continuous spectrum}}$$

This last factor is given, from (3.40) and integration of $h\nu n_\nu$ in (3.39b), by

$$\frac{E_K(U_o \ln U_o - (U_o - 1))}{h^2 \int_0^{\nu_0} (\nu_o - \nu)\mathrm{d}\nu}$$

Hence

$$J_d = 2Sf_\alpha \omega_K E_{K\alpha} E_K \{U_0 \ln U_0 - (U_0 - 1)\} E_0{}^2$$

or $\qquad$ (3.64)

$$2Sf_\alpha \omega_K \frac{E_{K\alpha}}{E_K} (U_0 \ln U_0 - (U_0 - 1)) U_0{}^2$$

This needs to be increased by a factor $\dfrac{P+1}{P}$, from fig. 3.19(*c*), to allow for indirectly produced radiation. This correction is large for heavy elements.

Equation (3.64) is plotted (using the previously cited experimental values for S) in fig. 3.31, with and without the correction for indirectly produced radiation, and is compared with values calculated directly from fig. 3.24 and (2.2), for a fixed value of U_0 equal to 2. We also include values for copper and tungsten from fig. 3.30. A strong inverse dependence on Z is noted, becoming less strong for the lighter elements, on account of the rapidly

Fig. 3.31. Ratio of characteristic to continuous radiation as a function of Z, for $U=2$.
* From equation (3.64) (i.e. J_d, the ratio of directly produced characteristic to continuous radiation, based on experimental values of S).
(1)———Straight line fit to points *.
(2)· · · ·$J_d \times (P+1)/P$(i.e. includes indirectly produced characteristic radiation).
(3)———from fig. 3.24(a) and (2.2) (i.e. based on experimental data which includes indirectly produced radiation).
●from figs. 3.30.
---extrapolation of (3) to pass through Tothill's experimental point (which includes indirectly produced radiation) for tungsten.

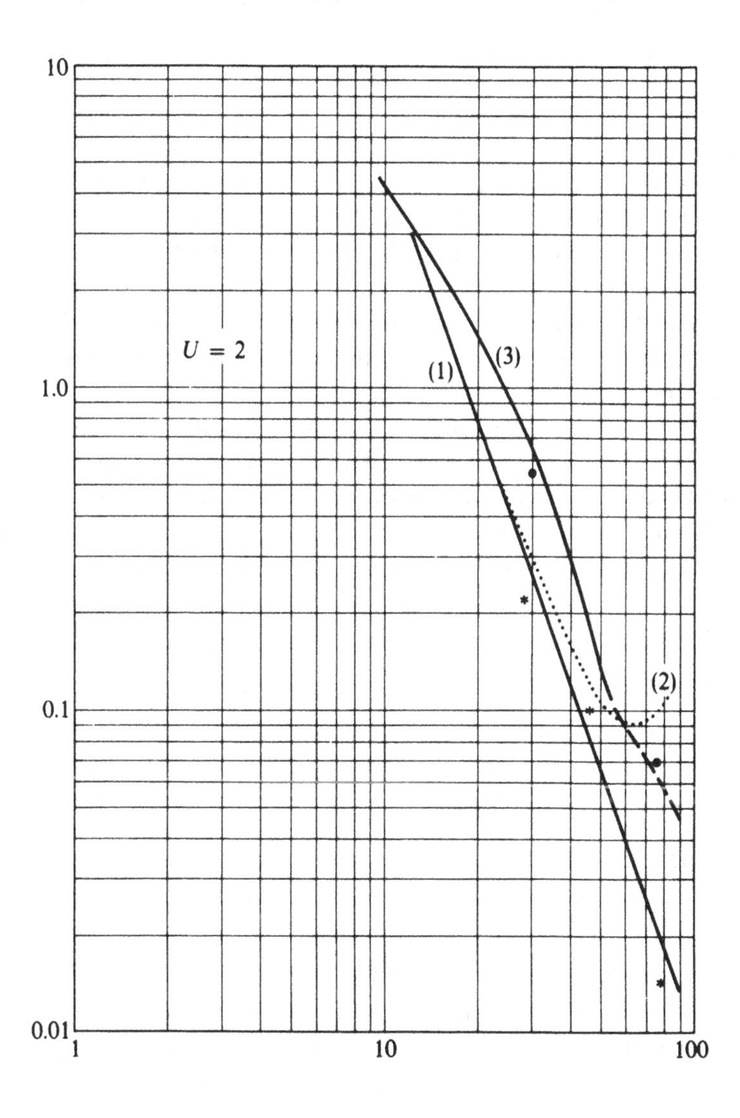

varying fluorescence yield for small Z. The ratio varies approximately as Z^{-3}, but this is modified by the effect of the fluorescence yield (particularly at medium and low Z), and by the effect of indirect production (the latter becoming important at high Z).

3.7 Satellites

In addition to the characteristic spectrum described in section 3.1 a number of additional lines appear close to, and usually on the high energy side of, the principal lines. Those lines cannot be explained in terms of energy level diagrams of the type illustrated in figs. 3.2 and 3.3, and are known as 'satellites' or 'non-diagram' lines. Satellites of the K series are considerably weaker than the lines they accompany, but in the L series the satellites may have an intensity which is comparable with that of their parent lines. The origin of satellites is explained by the occurrence of double and multiple ionization of the emitting atoms, and this can occur in several distinct ways: double ionization by a single incident electron, ejecting one K and one L electron, is feasible and is the mechanism by which satellites of the K series are produced. This mechanism is available for the production of L satellites also, but in this series considerable enhancement in satellite intensity is observed as a result of the additional multiple ionizations produced by the Auger effect and by radiationless transitions of the Coster–Kronig type.

Consider an atom ionized in the K shell, the energy of which may be written as K_Z. The removal of a second electron, e.g. from one of the L sub-shells, would require an energy somewhat greater than that required for L ionization in an otherwise unionized atom, because of the reduced screening of the nucleus in the K ionized atom. This reduction in screening would amount to the order of one electronic charge in this example. So we could write

$$(KL)_Z = K_Z + L_{Z+1}$$

Suppose now that the K vacancy is filled by an electronic transition from the M shell. This would result in the atom being doubly ionized in the L, M shells, for which we may write

$$(LM)_Z = L_Z + M_{Z+1}$$

Hence

$$(KL)_Z - (ML)_Z = (K_Z - M_Z) + (L_{Z+1} - L_Z) - (M_{Z+1} - M_Z) \quad (3.65)$$

The terms on the right-hand side of (3.65) are all positive, showing immediately that the satellite lies on the high energy side of the parent, with

Table 3.11 *The $K_{\beta'''}$ satellite ($KL \rightarrow ML$). (Druyvesteyn, 1927)*

Element	Z	Observed $\Delta\tilde{v}/R$			Calculated	
					$L_{2,3}$	L_1
Al	13	1.79		2.23	1.82	2.34
Si	14	2.07		2.74	2.11	2.63
P	15	2.38		2.97	2.40	2.92
S	16		2.97		2.70	3.22
Cl	17		3.27		2.97	3.49
K	19		3.59		3.55	4.07
Ca	20		3.83		3.84	4.36
Sc	21		3.77		3.53	4.05
Ti	22		3.72		3.73	4.25
V	23		3.92		3.95	4.47
Cr	24		4.00		4.16	4.68
Mn	25		4.34		4.38	4.90
Fe	26		4.69		4.58	5.10

a separation which increases with increasing Z. The $K_{\beta'''}$ satellite (parent K_{β_1}) has been examined by Druyvesteyn (1927) for the elements Al to Fe and the results are shown in table 3.11. In order to give rise to a satellite of the K_{β_1} line ($K \rightarrow M_2, M_3$) the L shell may be ionized in any of its three sub-shells, so a triplet structure might be expected. However, the L_2L_3 interval is small compared with L_1L_2; moreover the M_2 and M_3 ionizations do not differ in energy sufficiently to resolve the K_{β_1} parent into its expected fine structure. So in these measurements for $K_{\beta'''}$ satellite structure reduces to a doublet (for Al, Si, P) or to a single line (S–Fe). The origin of this satellite is shown diagrammatically in fig. 3.32, and it should be noted that we have not needed to invoke the concept of a *double transition* (i.e. two simultaneous transitions between states) in our explanation.

The satellites of the K_α lines have been studied in some detail. Their origins are explained satisfactorily by the theory of double ionization, but in order to explain the details of the fine structure the splitting of the KL_1, KL_2, and KL_3 levels has to be considered.

The allowed transitions of the type $K \rightarrow L_3$ (i.e. associated with the K_{α_1} line) are tabulated, along with the accepted symbols for the satellites, as follows. The j values for each level are given in parentheses.

$KL_1 \rightarrow L_3L_1$

α' $\quad 1s_{1/2}2s_{1/2}(1) \rightarrow 2p_{3/2}2s_{1/2}(2)$

$KL_2 \to L_3L_2$

α_0'' $1s_{1/2}2p_{1/2}(1) \to 2p_{3/2}2p_{1/2}(1)$

α_4 $\begin{cases} 1s_{1/2}2p_{1/2}(1) \to 2p_{3/2}2p_{1/2}(2) \\ 1s_{1/2}2p_{1/2}(0) \to 2p_{3/2}2p_{1/2}(1) \end{cases}$

$KL_3 \to L_3L_3$

α_3' $1s_{1/2}2p_{3/2}(1) \to 2p_{3/2}^2(2)$

α_3 $1s_{1/2}2p_{1/2}(2) \to 2p_{3/2}^2(2)$

α_3'' $1s_{1/2}2p_{3/2}(1) \to 2p_{3/2}^2(0)$

If the possibility of double transitions is admitted, further transitions become possible, e.g.,

$KL_1 \to L_3L_2$ or L_2L_3

α_0' $1s_{1/2}2s_{1/2}(0) \to 2p_{3/2}2p_{1/2}(1)$

These last two transitions are equivalent, in that the initial and final states are the same in each case. They involve transitions of the Coster–Kronig type.

Many other transitions are possible in principle, including several involving $K \to L_2$. These would give satellites of the K_{α_2} group, which would be of even lower intensity than the foregoing satellites of the K_{α_1} group and have not, as yet, been observed.

As examples of the K satellite group, data for copper and iron are given in table 3.12 (after Sandström). Detailed studies of K satellites in the region $Z = 25$–30 have been reported by Edamoto (1950) and discussed by

Fig. 3.32. Origin of K_{β}''' satellites.

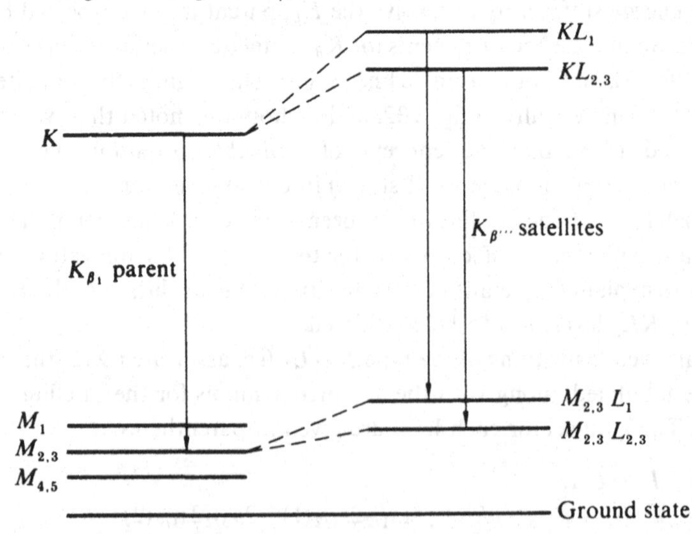

Table 3.12 *K satellites in* Cu *and* Fe (*after Sandström, 1957*)

	Copper		Iron	
	XU	$\Delta\tilde{v}/R$	XU	$\Delta\tilde{v}/R$
α'	1533.01	594.431	1926.05	473.128
α_4	1530.70	595.328	1922.76	473.937
α_3	1531.57	594.989	1923.94	473.647
$\alpha_3{}'$	1529.87	595.651	1922.16	474.085
$\alpha_3{}''$			1924.92	473.406
(α_1 parent	1537.396	592.734	1932.06	471.656)

Candlin (1955). The structure of the K_α satellites in calcium is illustrated in fig. 3.33 (Parratt, 1936a).

It should be noted that the double ionization is produced by a single process. The possibility that the ionization takes place in two successive stages can normally be discounted because of the shortness of the lifetime of a singly ionized atom in relation to the availability of electrons in the incident beam; an exception to this situation is discussed in section 3.8.

The excitation energies for satellites in the K series have been investigated, and have been found to be in general somewhat greater than for the parent line, because of the need for a double ionization. Calculation gives general agreement with experiment, and, in particular, the possibility that the production of these satellites involved the ejection of two K electrons, as propounded in some earlier theories, has been discounted.

Satellites of the L series follow a more complicated pattern. Double initial ionization of the type LL, LM, followed by a single transition are possible in a manner analogous to those which give rise to the satellites of the K series. However, when the K shell is not involved, double ionization can arise from a *singly*-ionized state followed by a Coster–Kronig process, as discussed in section 3.3. For example, an atom ionized initially in the L_1 shell may undergo an $L_1 \rightarrow L_2$ atomic transition associated with the emission of a second electron from an outer shell, normally the N shell in this case. We now have a doubly ionized atom of the L_2N type.

Remembering that the N ionization energy of such an ion would be that appropriate to an atom of atomic number higher by one, we may compare the $L_1 \rightarrow L_2$ transition energy with the N ionization energy of the various N subshells to determine exactly which double ionizations are energetically possible.

We may summarize the situation as follows:

(1) The immediate effect of the Coster–Kronig transition $L_1 \rightarrow L_2$ is to

enhance those diagram lines of the L spectrum which result from *electronic* transitions to the L_2 sub-shell, at the expense of those of the L_1 series.

(2) The effect of the associated emission of an electron from the N shell is to *displace* these diagram lines and give rise to satellites. As the probability of this is high, following the internal reorganization of that L shell, we see that satellites in the L series may have an intensity comparable with that of the parent line, a situation which would not occur for satellites of the K series.

If the energy of the incident electrons exceeds the K excitation potential, there exists the third possibility that the atom may be ionized in the K shell, and that the K vacancy may be filled by a $K \rightarrow L_2$ atomic transition which, if radiationless, would emit an Auger electron from, say, the N shell, giving an L_2N doubly ionized atom. An increase in the number of L satellites is in fact normally observed when the K excitation energy is exceeded.

We thus recognise three distinct ways in which the (L_2N) doubly ionized state can arise as a result of electron bombardment – (a) direct double ionization, (b) single L_1 ionization followed by a Coster–Kronig process $L_1 \rightarrow L_2$ together with emission of an N electron, and (c) K ionization followed by a $K \rightarrow L_2$ transition together with Auger emission from the N shell.

The atomic transition $L_1 \rightarrow L_3$ may release sufficient energy to cause ionization in an M sub-shell in the region of Z below 50 and greater than 73 (M_5), 77 (M_4) or 90 (M_3) but not in the region between these limits (fig. 3.34). The L satellites are in fact observed to be weak or non-existent in the

Fig. 3.33. Calcium K_α lines and satellites (Parratt, 1936a).

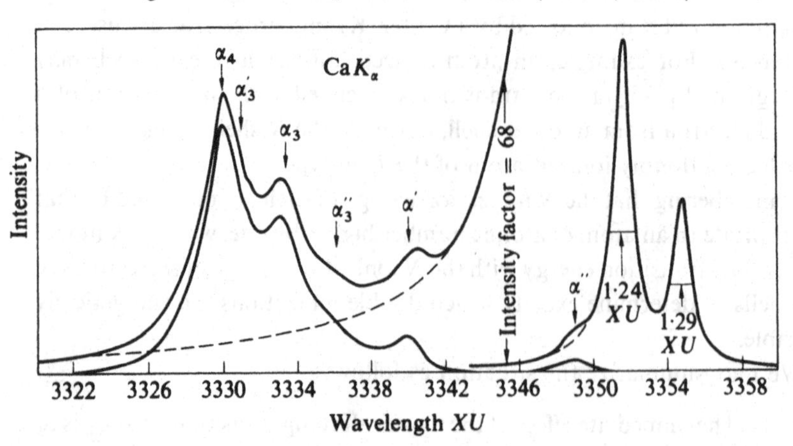

'forbidden' region of Z, and this lends strong support to the foregoing theory of their origin.

The possibilities of emitting L satellites are thus very rich. A complete list of Auger and Coster–Kronig transitions relevant to the production of L satellites is given by Burhop (1952).

Fig. 3.34. Energy of $L_1 \rightarrow L_3$ transitions in relation to M_3, M_4, M_5 binding energies in the L ionized atom.

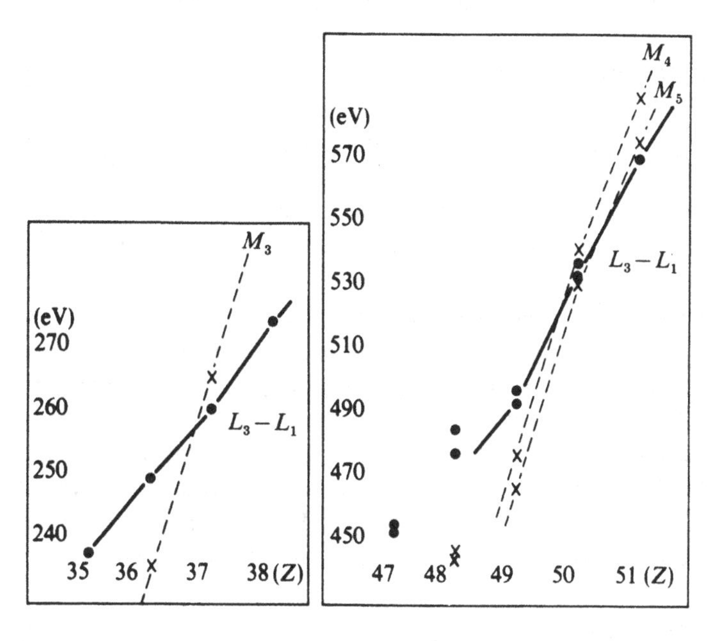

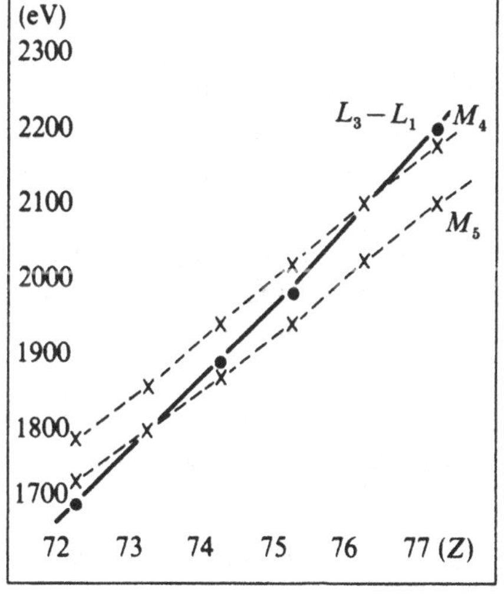

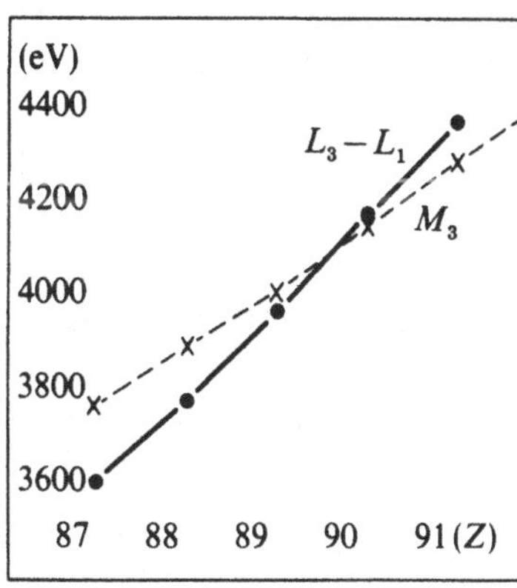

Additional processes which could in principle give rise to satellite lines include simultaneous inner-shell ionization and an optical excitation, followed by a double transition; further, in any lattice structure a vacancy in the inner shell of one atom can in principle be filled by an electron from an outer shell of an atom of different kind (e.g. $L_{Na^+} \rightarrow K_{Cl^-}$). Both these processes will give rise to lines different in energy from any so far considered. In the former case the transition energies could presumably be closely correlated with the energies of optical transitions.*

3.8 Line width

The width of the lines of the characteristic X-ray spectrum has been examined as a function of atomic number and is found to exhibit several interesting features. The width of a state and its lifetime are related by the Heisenberg uncertainty principle $\Gamma\tau = \hbar$, where Γ is the width in energy units and τ the mean life of the state. If p is the probability per unit time of the transition we can write $\tau = p^{-1}$ and $\Gamma = \hbar p$. In the case of a transition from a state of inner-shell ionization, the probability per unit time of radiative and non-radiative (Auger) transitions occurring may be written as p_r and p_n, and the mean life will be given by $(p_r + p_n)^{-1}$. The presence of competing processes thus reduces the mean life of the state and must therefore, through the application of the uncertainty principle, increase the width of the state. We may define the total width Γ_t as the sum of two partial widths Γ_r and Γ_n, where $\Gamma_r = \hbar p_r$ and $\Gamma_n = \hbar p_n$. The distribution of probability density p in the state is given by the expression

$$p\mathrm{d}E = \frac{\Gamma_t \mathrm{d}E}{2\pi\left[(E - E_0)^2 + \left(\frac{\Gamma_t}{2}\right)^2\right]} \tag{3.66}$$

where E_0 is the mean energy of the state. This is the so-called Lorenzian distribution and represents a resonance process similar to that encountered in the absorption of slow neutrons by atomic nuclei, when the expression is known as the Breit–Wigner formula.

Moreover, it has been shown that the width of a line emitted during a transition between two states is given by the sum of the widths of the two states. So, in principle, the widths of states may be obtained directly from experimental measurements of line profiles.

An alternative method of obtaining level widths would be to examine the shape of the critical absorption edges. This might be thought to have the

* For additional reading on X-ray satellites a further paper by Parratt (1936b) may be consulted, also Blokhin (1957).

advantage that only one inner-shell level is involved. In fact, however, absorption profiles are considerably affected by the state of chemical combination of the absorbing atom (see section 8.2), so are not much used in the study of level widths.

The transition probability for a state of inner-shell ionization is given by $p_t = A + BZ^4$, where the terms relate to the non-radiative and the radiative parts of the process, respectively. Hence we may write

$$\tau = p_t^{-1} = (A + BZ^4)^{-1} \tag{3.67}$$

K level widths have been studied as a function of Z and this general trend is confirmed (fig. 3.35(*a*), (*b*)).

For L states the situation is rendered more complex by the possibility of radiationless transitions of the Coster–Kronig type, which cause migration of L_1 vacancies into the L_2 and L_3 shells, whenever this is energetically possible.

The widths of the L_1 lines would be expected to increase in those regions of Z where the Coster–Kronig transitions are most probable, and this would be superimposed on the 'normal' variation given by (3.67). We have already noted the appearance of L satellites when Z is less than 50 or greater than 73 and it would also be expected that the width of the L_1 lines would increase in these circumstances.

The widths of the L_1 lines do in fact rise relatively to those of the L_2 lines, for Z greater than about 75. Below $Z = 50$, however, the expected effect is

Fig. 3.35(*a*). Width of K_α lines as a function of Z (Brogren, 1963). (● = K_{α_2} lines; ○ = K_{α_1} lines).

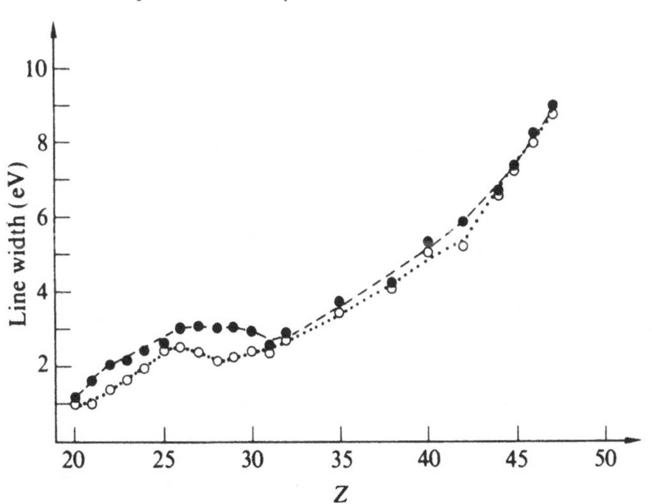

Table 3.13 *Level widths in gold (Ramberg and Richtmeyer, 1937)*

Level	K	L_1	L_2	L_3	M_1	M_2	M_3	N_1
Width (eV)	54	8.7	3.7	4.4	15.5	10.7	12.1	11.7
Lifetime (units of 10^{-18}s)	12.2	80	178	150	42.6	61.7	54.5	56.4

Fig. 3.35(*b*). *K*-level width as a function of atomic number (Bambynek *et al.*, 1972). (A full list of references is given in this review.)

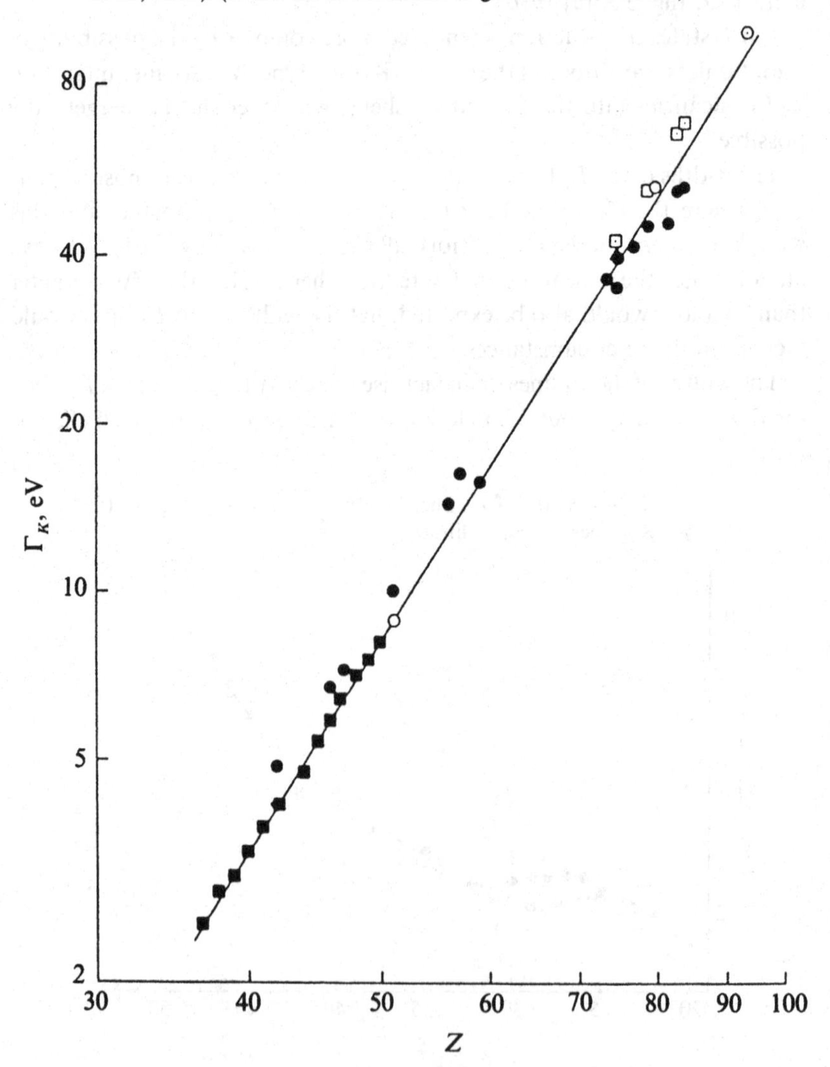

obscured, and this is discussed by Burhop (1952). A contributory factor appears to be irregular variations in the width of the *final* states due to radiationless transitions between sub-levels in the M shell.

Data for level widths of gold are given in table 3.13. The excess width of the L_1 level due to radiationless transitions is clearly seen. Lifetimes have been calculated from the experimental values using the relation $\tau = \frac{\hbar}{\Gamma}$.

Some very interesting effects have been observed in the so-called 'flash spectra' obtained from X-ray tubes in which the current is pulsed and may reach values of many kiloamperes. Schörling (1962, 64, 65) reports studies of the K series lines in iron, nickel, copper, cobalt, dysprosium, zinc, chromium, and vanadium. For details of this work the reader is referred to the original papers. General conclusions drawn from this work are that the K_{β_1} and K_{β_3} lines are broadened on the high energy side, the K_{α_2} on the low energy side, and the K_{α_1} on both sides; the total broadening tends to be greater for the K_β lines than for the K_α lines. In addition, energy shifts of a few tenths of an electron volt are observed.

Consider the potentials at 2 radii r_1 and r_2, in an atom with nuclear charge Z (fig. 3.36). In the absence of screening the potential difference between the 2 shells is given by

$$\frac{1}{4\pi\varepsilon_0} eZ\left(\frac{1}{r_1} - \frac{1}{r_2}\right) = \phi_{r_1} - \phi_{r_2} = \Phi$$

Fig. 3.36. Screening effect, for the case where $r_1 < R < r_2$.

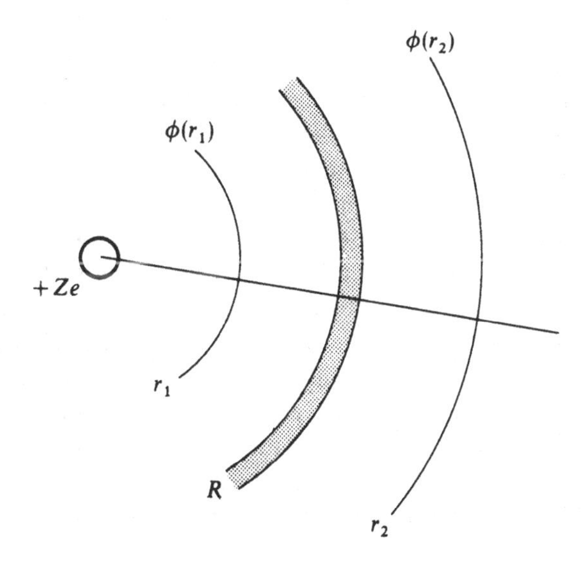

If we now introduce additional negative charge q spherically distributed at radius R we can distinguish three cases:

$$1.\ R<r_1:\ \phi_{r_1}=\frac{1}{4\pi\varepsilon_0}\frac{eZ-q}{r_1};$$

$$\phi_{r_2}=\frac{1}{4\pi\varepsilon_0}\frac{eZ-q}{r_2};\ \therefore\ \phi_{r_1}-\phi_{r_2}=\Phi-\frac{q}{4\pi\varepsilon_0}\left(\frac{1}{r_1}-\frac{1}{r_2}\right)$$

$$2.\ r_1<R<r_2:\ \phi_{r_1}=\frac{1}{4\pi\varepsilon_0}\left(\frac{eZ}{r_1}-\frac{q}{r}\right);$$

$$\phi_{r_2}=\frac{1}{4\pi\varepsilon_0}\frac{eZ-q}{r^2};\ \therefore\ \phi_{r_1}-\phi_{r_2}=\Phi-\frac{q}{4\pi\varepsilon_0}\left(\frac{1}{R}-\frac{1}{r_2}\right) \tag{3.68}$$

$$3.\ r_1<r_2<R=\phi_{r_1}=\frac{1}{4\pi\varepsilon_0}\left(\frac{eZ}{r_1}-\frac{q}{r}\right);$$

$$\phi_{r_2}=\frac{1}{4\pi\varepsilon_0}\left(\frac{eZ}{r_2}-\frac{q}{R}\right);\ \therefore\ \phi_{r_1}-\phi_{r_2}=\Phi$$

We see that in all cases the removal of such screening charge (by, for example, a double ionization) would result in a slight increase in the energy levels of atoms ionized in any shell, and that in cases 1 and 2 the *transition energy* between two levels would also increase, causing the appearance of high energy satellites (qv. (3.65), which corresponds to case 2), or a broadening of the existing lines on the high energy side. The greatest shift will be expected when the two shells are separated by the greatest amount, in agreement with the experimental observation that the K_β lines are broadened more than the K_α lines.

Multiple ionizing events are normally the result of two electrons being removed simultaneously by a single electron in the primary beam, or alternatively by a radiationless transition taking place in a singly ionized atom. Such processes cannot be increased relatively to single ionization merely by increasing the beam current. The ionizations giving rise to these high current phenomena must therefore be caused by two or more independent events. The lifetime of atoms ionized in an inner-shell is much too small in the solid target for this to occur with measurable probability, even with the high current densities available in the flash tubes developed by Schörling, and it is unlikely that the lifetimes of atoms ionized in an *outer* shell would be long enough for such events to be observable. However, Schörling established that a single pulse of the magnitude available to him would be able to evaporate a layer of target material several hundred atoms thick, so that an appreciable proportion of the X-rays observed would in fact proceed from the vaporized material. The lifetime of ionizations in outer shells is much longer in the gaseous than in the solid state (10^{-8} s as opposed to 10^{-16} s) so that the production of a doubly ionized atom by two or more independent primary electrons becomes more feasible. Doppler and Stark effects in the vaporized target are considered to be small.

The broadening of lines on the low-energy side cannot be explained in such a simple manner. Some progress can be made by considering the detailed behaviour of the orbital electrons immediately following the ejection of an electron. In the calculation of electronic binding energies by, e.g., the 'self-consistent field' method, the assumption has often been made that, when a bound electron is removed from, for example, the K shell to a stationary position at infinity, all the remaining electrons remain unchanged in their orbits. This 'frozen-orbital' approach is consistent with the statement that the binding energy of an electron is equal to the negative of its eigenvalue:

$$E_B = -\varepsilon_K \tag{3.69}$$

This is known as Koopmans' theorem. However an alternative situation may be postulated, viz., that all the remaining electrons 'relax' into new orbitals, which will be somewhat more tightly-bound, because of reduced screening around the nucleus, so that we must write

$$E_B = -\varepsilon_K - \delta\varepsilon_K \tag{3.70}$$

where $\delta\varepsilon_K$ is the energy of relaxation. The binding energy is now greater. Calculations taking relaxation into account have generally shown better agreement with experiment than calculations based on the 'frozen-orbital' assumption. However, relaxation may not be complete, particularly for outer electrons, and this would have some influence upon all the energy levels in the atom (including inner shells) and upon transitions occurring between them. This approach has been developed by Parratt (1959). Schörling (1961) has considered it in relation to the line broadening observed by him, suggesting that the 4s electrons show incomplete relaxation. The K_{α_1} and K_{α_2} line are supposed to originate from those atoms where the 4s electrons have not had time to readjust to their new orbitals before the K–L transition takes place, whereas the low energy broadening of these lines occurs in that proportion of transitions in which the readjustment to new orbitals, of smaller radius, has taken place, involving a lowering of the transition energy in accordance with example 2 in (3.68).

4

Experimental techniques for the study of X-rays

4.1 Introduction

The experimental study of X-rays involves consideration of the methods of producing X-rays over a wide range of photon energies, methods of detecting the X-radiation and analysing it in terms of intensity, photon energy and polarization, and the effect of materials placed between source and detector. We have already seen that X-radiation is produced whenever a beam of charged particles encounters any target material, solid, liquid, or gaseous, and that it is also emitted by various processes during radioactive decay. The charged particles used in X-ray generators are normally electrons, but protons or alpha particles are readily available in accelerated beams, and cause X-rays to be produced when slowed down in a target. Accelerating voltages, for electrons or other particles, may range from a few hundred volts to many MV, and the photon energies of interest extend from this upper limit down almost to the ultra-violet region of the electromagnetic spectrum. X-ray detectors depend upon the ionizing qualities of the radiation (or, very occasionally, upon nuclear excitation), and include photographic emulsions, gas-filled devices (such as the ionization chamber, and Geiger and proportional counters), the scintillation counter, and the solid-state detector. Some of these detectors have an intrinsic ability to distinguish between photons of different energy, but often this energy resolution will need to be improved upon by the use of diffraction gratings, or by utilizing Bragg reflection from single crystals. The effect of intervening materials may be an intrinsic part of the experiment, for example in the determination of attenuation coefficients or in the study of X-ray fluorescence, or it may be an unavoidable effect which has to be determined and then corrected for, as in the case of absorption of radiation in the windows of X-ray tubes and detectors, or in the intervening air.

4.2 X-ray tubes and other generators

Very many studies of X-rays have used tubes consisting of a solid metallic target (or 'anticathode') housed in an evacuated envelope together with a heated filament, and with a steady potential difference of 10–100 kV between these electrodes, the filament being negative with respect to the anode. The development of X-ray technology has depended heavily on this type of X-ray tube, which is in very wide use in medical radiodiagnosis and radiotherapy, industrial radiography, and X-ray crystallography. In the early days of X-ray studies, the gas-filled tube was important, but this has given way to the modern evacuated tube with its high stability and reliability. For radiotherapy the linear accelerator has rapidly gained favour, but tubes of the kind to be described will undoubtedly remain in use, in all the fields cited, for many years to come. The nature of the target will depend on the studies for which the tube is designed – if a continuous spectrum is needed, a metallic target of high atomic number will normally be used, and tungsten fulfils many requirements well. Such a spectrum will be required for Laue diffraction work, for the intense beams required for radiotherapy or radiodiagnosis, and is often suitable for the excitation of fluorescence radiation in secondary targets, as in X-ray fluorescence analysis. If the tube is designed for studies of the continuous spectrum, a variety of targets may be required. Metallic targets are usually more convenient than non-metallic targets, mainly because of the need for high thermal conductivity. Fig. 4.1 shows schematically an arrangement in common use. If the tube is designed with a single purpose in mind, for example radiotherapy, it will normally be sealed off, and no provision will be made for changing the target or filament. If a choice of target is needed, for the study of characteristic radiation, or for the production of beams of monoenergetic radiation of selected energy as in X-ray diffractometry, the

Fig. 4.1. Conventional X-ray tube (schematic).

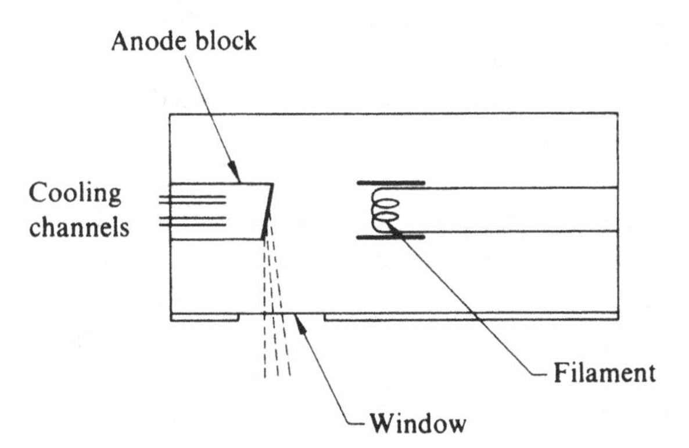

tube will often be of the continuously pumped type, which is demountable so that targets can be changed at will. This necessitates letting the tube down to air, and then re-pumping, but with modern equipment this can be achieved in a few minutes. Tubes have been designed in which the targets can be changed by external magnetic control applied to a sealed-off tube, but in any work in which radiation of known energy and purity is needed, a demountable tube is preferable – surface contamination of targets by tungsten evaporated from the filament is unavoidable in any tube, and access to the target for cleaning and repolishing is greatly to be desired. One convenient design has four targets mounted on the different sides of a square target holder. The target can then be changed by rotating the target assembly in a few seconds only, but the tube is demountable for cleaning purposes and for replacement when necessary. A design described by Nordfors (1956) has two targets mounted on the same block (fig. 4.2), and two filament assemblies which can be switched at will.

An important design problem in any X-ray tube is the removal of heat generated in the target. Currents incident on the target will often be in the region of 10–50 mA, which at the kilovoltages already quoted may require the dissipation of 2–3 kW of heat. This may be effected by circulating oil or water through the target assembly, or by directing a jet of oil or water on to the rear of the target, which then necessarily forms part of the vacuum wall. If the rate of heat production is not very great, the target may be extended out of the vacuum tube by means of fins, which will then become cooled by air convection. Water-cooling is most commonly used, unless the design of the tube is such that the target is at an elevated potential with respect to ground, in which case oil-cooling would be the appropriate method.

When the problem of removing the heat from the system has been solved, it is still necessary to ensure that the surface temperature of the target remains below an acceptable limit. Surface melting has to be avoided, and also the sputtering of target material, which can occur at temperatures appreciably below the melting point. To assist in this, the anode may be rotated at a rate sufficient to prevent undue temperature rise at any point of the target. The maximum temperature reached by any point will depend on the current density in the beam, the time for which the current is incident on any particular point (which in turn will depend on the rate of rotation of the target), and the thermal conductivity of the target material. High rates of rotation will be advantageous, and the life of the target will then be limited only by deterioration due to repeated thermal cycling. Rotating-anode tubes find an important application in diagnostic radiography, in which short exposure times are often needed in order to reduce blurring of the image due to movement of the patient. Currents up to the region of 500 mA are available in rotating-anode tubes designed for this purpose.

In diffractometry, it is often convenient for the X-ray source to be in the form of a 'line focus', several mm in length by 0.1–1.0 mm in width. This enables the electron beam to be spread out, thereby reducing the surface temperature reached by any point within the target surface. If an X-ray source of round or square shape is needed, a tube giving a line source can still be used, and the source viewed along its length, at a glancing angle in the region of 6 degrees. Line sources are in common use in diffractometry.

Secondary electron emission from the target may be appreciable, and this, in turn, can cause unwanted charge to appear on surrounding insulators, with consequent distortion of the accelerating field. To combat this, the target may be hooded to catch the secondary electrons, and this has

Fig. 4.2. Tube with two anodes: 1. Anodes; 2. Filaments; 3. Cooling; 4. Brass dividing wall; 5. Insulator (Nordfors, 1956).

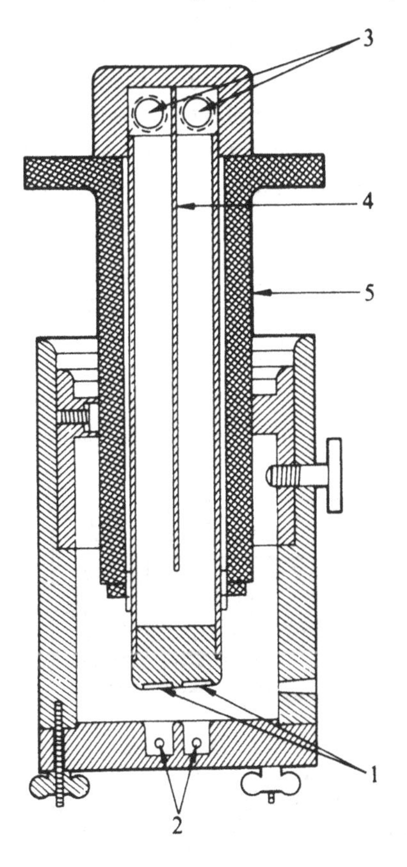

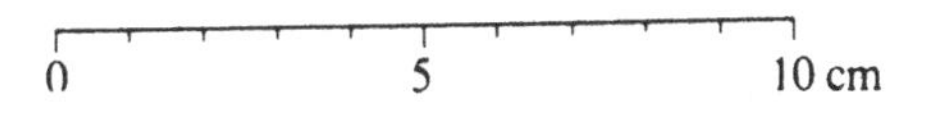

the further advantage of restricting to some extent the solid angle over which X-radiation leaves the tube, reducing the hazard from scattered radiation (fig. 4.3).

Non-metallic targets present problems because of their poor thermal conductivity, and current densities have in consequence to be limited because of this. Oxide targets have been frequently used in the study of characteristic spectra; their melting point is often high, and this offsets to some extent the disadvantage of low thermal conductivity. The oxide or other material would normally be in the form of a thin layer, pressed into the ribbed surface of a metallic support, to provide thermal and electrical conduction.

Chemical changes will often take place during prolonged bombardment of non-elemental targets, and the small shifts in energies which may occur because of this are of considerable interest. There is some discussion of this in section 8.2.

Although the target of an X-ray tube is normally solid, targets which are in the vapour phase have occasionally been used for special purposes. An early example is provided by the work of Nicholas (1929) whose studies of the polarization of X-radiation were carried out using a target of mercury vapour. A recent determination of h/e by the X-ray method has been

Fig. 4.3. Tube with hooded anode (Johns, 1964, after a design by Philips Laboratories) (from *The Physics of Radiology*, courtesy of Charles C. Thomas, publisher, Springfield, Illinois).

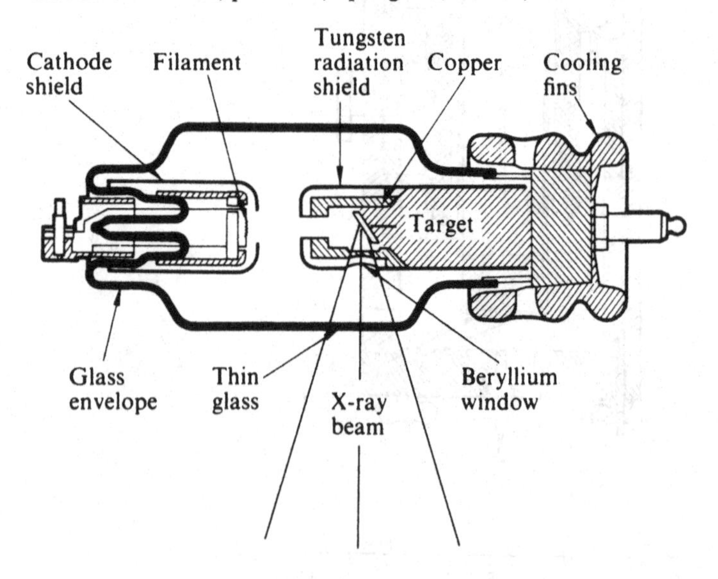

referred to in section 2.10. For this work a special X-ray tube, again using a mercury vapour target, is described by Bearden *et al.*, (1964). In experiments of this type, the minimum potential difference required to generate continuous X-rays of a particular photon energy is measured accurately. There is, however, some structure in the isochromat observed, which can be related to the band structure of the target material. Furthermore, it is important that the target should present only a relatively small number of atoms per unit area, so that an impinging electron has only one collision of any kind in the target material. This necessitates the use of a target which is either very thin or in a very attenuated form. Both these problems are eliminated if the target is a gas. The tube described by Bearden *et al.* had a mercury 'boiler' at its base, and a jet of mercury vapour was projected upwards. The vapour was ultimately condensed and the liquid mercury channelled back to the boiler. The electron gun (maintained at high vacuum) projected the electron beam horizontally and the electrons impinged on the mercury vapour jet, the electrons being subsequently stopped in a water-cooled anode at the opposite side of the tube. In this way a vapour target with an equivalent thickness of about 50 nm was produced, and the X-ray output in the vicinity of the high-energy limit was no less than that expected from a solid or liquid target of the same material.

A tube of the kind just described (i.e. with a gaseous target) would probably be adaptable to experiments with other target materials, for example the carbonyls of many metals. Slow decomposition of the molecules, with consequent deposition of the elemental metal, would however be expected in the course of use.

The cathode of a conventional X-ray tube usually consists of a tungsten wire in the form of a coil, and a simple focussing arrangement in the form of a 'focussing cup'. This produces a line focus at the target, which may then be viewed at the appropriate angle to present a circular or square source by foreshortening. Typical dimensions for the focal spot of a diffractometry tube are 10 mm × 1 mm. A finer focus is an advantage for accurate work, or for studies of small crystals, and a size of 6 mm × 0.1 mm would be typical for such a tube. In tubes for medical radiotherapy the requirements of spot size are somewhat less stringent, and a circular or near-circular spot of 1–5 mm diameter would be adequate.

The special requirements of microfocus tubes used in X-ray microscopy and X-ray emission microanalysis are described in section 8.1.

A thin window forms part of the vacuum wall of an X-ray tube in order to obtain maximum X-ray intensity in the emergent beam, and the most favoured material for this purpose is beryllium. The window has to support atmospheric pressure against the vacuum inside the tube, and to meet this requirement a thickness of 1 mm and a diameter in the region of 1 cm would be typical. Often two or more windows will be provided on opposite sides of the target assembly, to enable for example, several diffraction cameras to be set up and operated independently. Each window could have its own shutter (for independent use), and would often have provision for a β-filter, to remove the characteristic K_β radiation, immediately in front. For medical applications, it is usually desirable to attenuate, or to stop completely, radiation with a photon energy less than about 15 keV, to avoid delivering excessive radiation doses to the skin or to superficial tissues. The use of a beryllium window would then not be necessary, and aluminium is in fact commonly employed. An exception is of course made if the tube is designed specifically for the radiotherapy of superficial tissues or skin, when beryllium is used as the window material.

For the study of the K series of the heavier elements ($Z > 50$) and for most radiotherapeutic work, accelerating voltages in excess of 100 kV are normally needed to yield adequate intensities. For example, standard radiotherapeutic equipment operates at a kilovoltage of 240 kV (peak). The design of such tubes does not differ materially from those already considered except that the electrical insulation requirements are naturally more severe.

Several X-ray tubes for research and other purposes have used a transmission type of target, in which the target is thick enough to stop all the electrons within the target itself, but sufficiently thin to allow a substantial proportion of the X-rays to travel through. The target forms part of the vacuum wall, and a separate window is no longer necessary. An example of this is the X-ray tube of Thordarson (1939) for studies of the angular distribution of X-rays in the forward hemisphere. A target of tungsten 0.02 mm in thickness was able to transmit with little attenuation the more penetrating components of the radiation produced at accelerating voltages of 60–170 kV. For lower energy work, transmission targets forming part of the vacuum wall were not practicable, unless the diameter could be restricted to 1 mm or less, as in the early X-ray microscope of Cosslett and Nixon (1952), until beryllium sheet of adequate thickness and purity (and with no porosity) became available. In 1952, Botden *et al.* described a very compact sealed-off X-ray tube for operation at 10–25 kV in which the target consisted of a thin ($\sim$100 nm) layer of gold evaporated on to a beryllium window $1\frac{1}{2}$ mm in thickness (fig. 4.4). The tube produced relatively soft radiation of high intensity and was designed for the radiotherapy of superficial lesions, where the high surface dose-rate and rapid fall-off by absorption and inverse square law are desirable features. X-rays produced below 10 kV are used in microradiography, and tubes for this purpose frequently make use of the transmission type of target just described.

Power supplies for X-ray generators working up to 100 kV normally consist of a transformer operating at the mains frequency with a half-wave or full-wave thermionic valve or solid state rectifier system. In principle the rectifier system can be dispensed with altogether, but this is undesirable because under these conditions the peak inverse voltage appears across the tube once every cycle, and can cause heavy bombardment of the cathode (due to the thermionic electron emission from the target) with consequent deterioration, and spurious X-ray production. For medical purposes the pulsating DC produced from unsmoothed rectifier circuits is quite acceptable, but for much crystallographic work, and for X-ray spectroscopy, a smoothed and stabilised DC supply is essential. Such stabilisation may need to be 0.1% or better. In specialized tubes for microradiography, spot sizes in the region of 1 μm are in common use, and to achieve this, focussing by means of an electron-optical system using magnetic focussing lenses is often used. This imposes even stronger requirements on the stability of the accelerating voltage, and a stability approaching 1 part in 10^4 is required, to ensure that the focussing condition is accurately maintained over long periods. Practical X-ray generators, together with circuits, are described by e.g., Johns (1964), and Brown (1966).

For X-ray generators operating above 250 kV up to about 750 kV the Cockcroft-Walton generator is commonly employed. To produce higher potential, up to 2 MV, the 'resonant transformer' principle is often used.

Above 2 MV or so the devices for producing X-rays belong to the domain of nuclear particle accelerators, and the Van de Graaff accelerator is often

Fig. 4.4. X-ray tube for production of soft X-rays for radiotherapy of superficial tissues, showing anode can (A), beryllium window with gold layer, filament (G) and focussing cylinder (M) (Botden *et al.*, 1952).

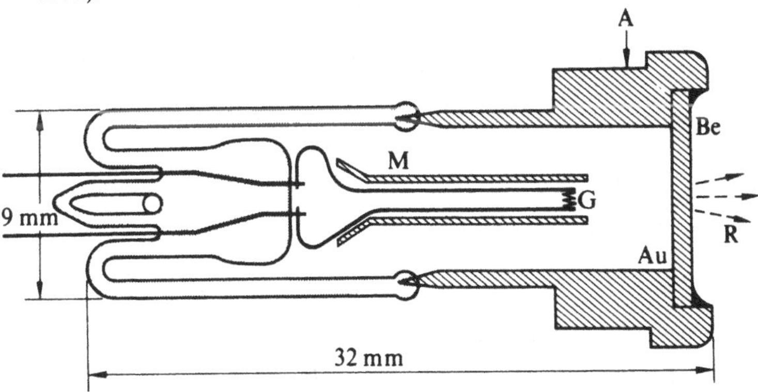

used for studies of X-ray production (fig. 4.5). In the Van de Graaff machine electric charge is 'sprayed' on to a moving belt which conveys the charge a distance of up to several feet to a region of low electric field within an isolated conducting electrode. The potential of this electrode rises to a limit set only by the qualities of the insulating system, and more particularly, by the onset of corona discharge at the high potential electrode. To assist in the attainment of a high potential the assembly is normally enclosed in a vessel filled with dry nitrogen to a pressure of several atmospheres. A detailed description of modern Van de Graaff machines has been given by England (1974).

For electron energies between 4 and 20 MeV the *electron linear accelerator* has been found to be an instrument of great reliability and versatility. In the linear accelerator electrons are produced by thermionic emission and are then accelerated by the electric field associated with a travelling wave generated at microwave frequencies and injected into a waveguide. Injection takes place at one end of the guide, normally circular, with the electric field parallel to the direction of propagation along the guide* (fig. 4.6).

* This is not at variance with the well-known requirement that the electric and magnetic fields must both be perpendicular to the direction of propagation – the field diagrams of fig. 4.7 show that this condition is in fact satisfied periodically along the guide.

Fig. 4.5. Van de Graaff generator (Van de Graaff *et al.*, 1946).

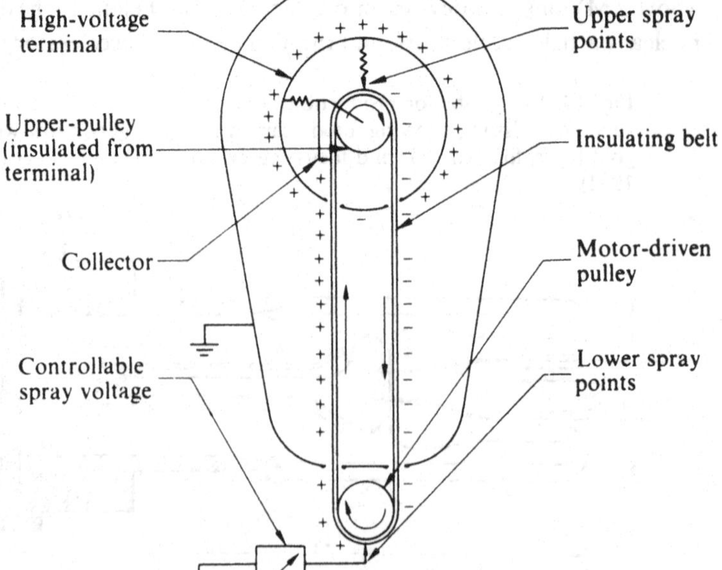

The electrons are accelerated in this field, and the phase velocity of the wave is adjusted by means of iris diaphragms (fig. 4.7) so as to equal the velocity of the electrons at each point. The injection velocity is in the region of $0.3c$ (where c is the velocity of electromagnetic radiation in free space), and this is achieved by an electron gun with a potential of about 20 kV between anode and cathode. The velocity increases gradually along the guide, and at the target the electron velocity will have risen to $0.999c$, corresponding to an energy of 10 MeV. A static longitudinal magnetic field serves to focus the electron beam, and the spot size at the target is in the region of 4 mm in diameter. A linear accelerator is normally pulsed in operation, at pulse rates which range between 50–600 pulses per second in

Fig. 4.6. Linear accelerator: Corrugated waveguide and microwave feedback arrangement (Miller, 1954).

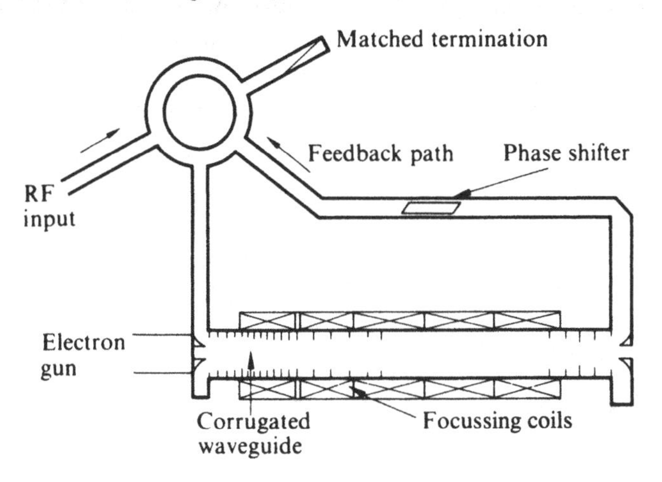

Fig. 4.7. Linear accelerator: field configuration in the corrugated waveguide (Miller, 1954).

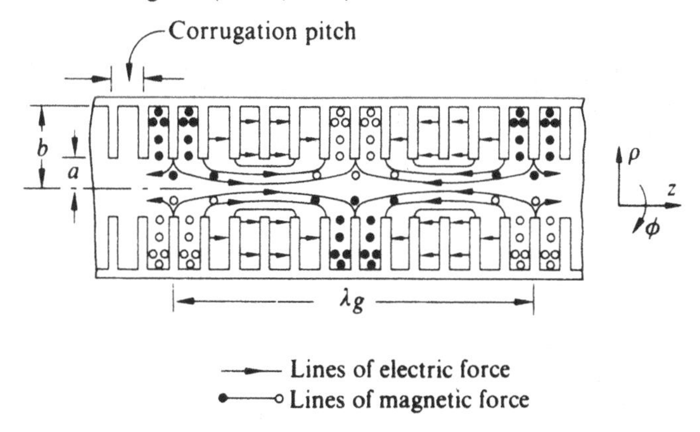

one commercially available linear accelerator*. In this equipment the pulses are 2 μm in duration, with a peak power of 5 MW, generated by a magnetron operating at 2998 MHz. An operating dose rate of 600 rad min^{-1} at 1 metre from the 'heavy metal' target is quoted.†

For energies above 15 MeV, the *betatron* provides a source of X-ray Bremsstrahlung which has been found to have considerable application in radiotherapy because of the highly penetrating nature of X-radiation at high energies. Although the advantages over the linear accelerator are marginal in this respect, the 'build-up' of dose with increasing depth below the surface referred to in section 4.8 in connection with X-ray dosimetry, takes place more gradually at the higher energies available from the betatron (maximum dose occurs at 3 cm below the surface in soft tissue for 18 MeV betatron radiation, compared with 1 cm at 8 MeV), so that the skin and superficial tissues receive doses which are relatively low.

In the betatron the electrons are accelerated by the EMF induced by a changing magnetic field. The field is longitudinal and the electrons are therefore accelerating in a circular orbit (fig. 4.8). In order to understand the magnetic field distribution required to accelerate the beam and at the same

* Medical Linear Accelerator SL 75 (MEL Equipment Company Ltd.)
† or 6 Gy min^{-1}. These radiation units are defined in section 4.8.

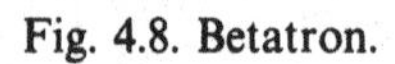
Fig. 4.8. Betatron.

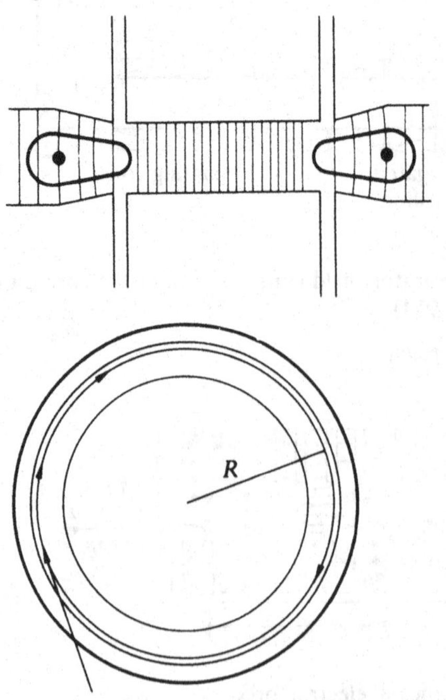

time maintain a circular equilibrium orbit,* we first write down the magnitude of the electric field produced by a changing magnetic flux,

$$E = \frac{1}{2\pi R}\phi \tag{4.1}$$

where R is the radius of the orbit.

We then use this to obtain the rate of change of linear momentum of the electron in terms of the Coulomb force acting on it,

$$\dot{p} = eE = \frac{e}{2\pi R}\dot{\phi} \tag{4.2}$$

Finally we obtain the magnitude of the magnetic field required to maintain an electron of momentum p on an orbit of radius R:

$$\frac{mv^2}{R} = Bev, \qquad \text{Hence } p = BeR \text{ and}$$

$$\dot{p} = \dot{B}eR \tag{4.3}$$

Equating (4.2) and (4.3) we see that

$$\dot{\phi} = 2\pi R^2 \dot{B}, \tag{4.4}$$

demonstrating that the field at the radius of the orbit must increase at the same relative rate as the field providing the accelerating flux, and that the flux needed is exactly twice that which would be obtained if the magnet produced a *uniform* field of magnitude B over the whole area (πR^2) enclosed by the orbit. The pole pieces are therefore designed to give increased flux near the centre of the machine.† The acceleration continues until the increasing flux reaches its limit, and then the electrons spiral inwards, losing some energy by radiation and finally striking a target near the centre of the polepieces. The accelerating field normally has a frequency in the region of 50–100 Hz. Kerst *et al.* (1950) describe a machine operating at 300 MeV (pulsed at a rate of 6 Hz) but for medical work lower energies, up to 45 MeV, are usual.

* The principle of the *cyclotron*, with its *spiral* orbit, cannot readily be applied to electron accelerators because of the very considerable relativistic increase in electron mass which occurs progressively as the energy is increased beyond about 500 keV.

† Improved designs separate these two fields, and achieve certain economies in construction.

4.3 The detection and analysis of X-radiation

(a) *Introduction*

Amongst the earliest properties of X-radiation to be discovered were its ability to discharge an electroscope and to blacken a photographic plate. These properties allowed X-rays to be detected, and enabled quantitative measurements to be made on the ionizing power of beams of X-radiation. Early work of this type laid the foundations of modern X-ray physics. The ionization in air caused by a beam of X-rays is in fact a more useful parameter than might at first sight be imagined, because the amount of energy required to produce an ion pair is nearly independent of the kinetic energy of the electrons causing it, and so the ionization per unit volume (or per unit mass) of air is closely proportional to the amount of energy absorbed in the process. This in turn is related to the energy flux, through the energy absorption coefficient. The ionization per unit mass of air forms the basis of the definition of the Roentgen, which is the amount of radiation producing 2.58×10^{-4} coulomb kg^{-1} of dry air. It will be shown later that, in the case of materials consisting of light elements ($Z < 8$), which includes much organic and biological material, the energy absorbed per unit mass can be inferred from measurements of ionization in air over a wide range of photon energies. This forms the basis of the method of ionization dosimetry which is used very widely for the determination of radiological dose delivered in radiodiagnosis and radiotherapy, in radiobiological research, and in occupational exposure to ionizing radiation. The theory of cavity ionization chambers is described in section 4.8, where we shall see that exact predictions can be made about the response of chambers with appropriate gas filling and wall materials. But from what has been said it is already clear that an air-filled ionization chamber used in conjunction with a device for measuring currents in the region of 10^{-9}–10^{-11} amperes would be very adequate for measuring relative intensities of radiation of fixed energy, e.g. the K or L characteristic radiation from an individual element, and its variation with accelerating voltage. It would not be suitable for studies of spectra of varying energy, on account of the variation of the absorption coefficient of air as a function of photon energy; for this type of measurement, and for studies of X-ray spectroscopy in general, some form of photon energy analysis is necessary.

It is customary to distinguish between *wavelength-dispersive* detectors, which depend upon the diffraction of X-rays (as in the crystal spectrometers described below) and *energy-dispersive* detectors, in which electrical signals, proportional in magnitude to the incident photon energy, are produced, as in the other devices described in this section.

(b) *Crystal spectrometers*

The discovery of X-ray diffraction by means of crystals soon led to the use of a single crystal to display an X-ray spectrum. W.H. Bragg used a crystal of rock-salt to give a spectrum of characteristic and continuous radiation from a target of platinum and thus initiated the era of X-ray crystal spectrometry. The Bragg reflection rule

$$n\lambda = 2a \sin \theta \qquad (4.5)$$

is the basis of measurements of this kind. In 4.5, θ is the angle indicated in fig. 4.9, λ the wavelength of the incident radiation, a the distance between the lattice planes, and n the order of the reflection. The quantity λ is the wavelength in the crystal, which will differ slightly from the wavelength in air because of refraction, but these two wavelengths differ by not more than a few parts in 10^5.

To measure the intensity distribution of an X-ray spectrum by this method, the incident beam has to be collimated, and the detector (which may be a photographic plate, ionization chamber, or Geiger counter) is placed at the corresponding position on the opposite side of the normal (fig. 4.10). The system effectively selects all rays which are incident at the

Fig. 4.9. Illustrating Bragg reflection: Definition of θ and a.

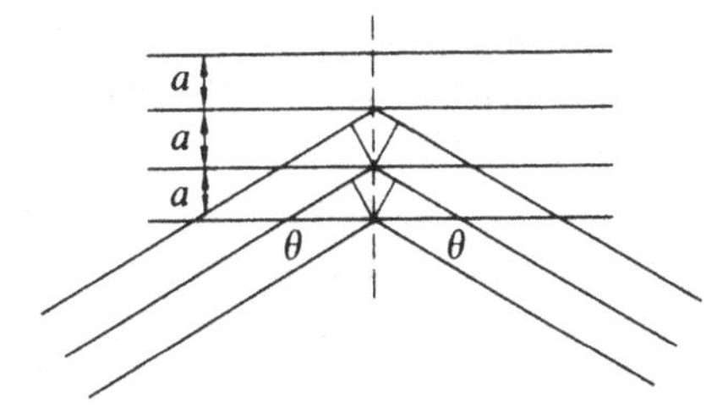

Fig. 4.10. Bragg spectrometer showing X-ray tube, crystal, detector, and the defining apertures.

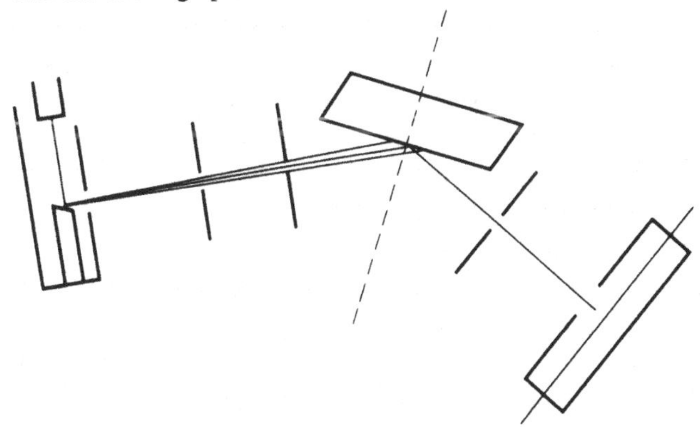

appropriate angle for Bragg reflection. If the angle of incidence is then varied, which will normally involve rotating the crystal through a controllable angle and the detector by twice this amount, the detector will receive radiation over a range of wavelengths and a spectrum will be obtained. If the system is such that the angular range $d\theta$ is constant over the whole range of wavelength investigated, as in the geometry illustrated in fig. 4.10, we can write

$$n\,d\lambda = 2a\,d(\sin\theta) = 2a\cos\theta\,d\theta \tag{4.6}$$

The intensity thus received will be proportional to $\cos\theta$ and to $d\theta$. After dividing by $\cos\theta$ and correcting for the variation with angle of the reflection coefficient of the crystal and the variation with wavelength of the detector efficiency, the recorded signal will be proportional to the intensity per unit wavelength interval, and it is in this form that continuous X-ray spectra are traditionally plotted.

The resolution of the spectrometer just described will depend, in part, on the geometrical definition provided by the slit. At any one setting, a small range of angles will be available, and any wavelength for which the appropriate Bragg angle lies within this range will be reflected and recorded. A further limit is imposed by the intrinsic resolution of the crystal. If the incident radiation were truly monoenergetic, the peak obtained in the spectrum would have a finite width determined by these two effects. This is known as the 'rocking curve' for the system. If the geometrical resolution is such that the crystal provides the only contribution to this finite width, the rocking curve is that for the crystal alone. In practice, characteristic X-ray lines have a finite width, which is related to the life-time of the state by the uncertainty principle, and the natural width of lines can be studied by spectrometers of the kind described in this section.

The natural resolution of crystals depends essentially upon the degree of alignment of the surface atomic planes, although it should be noted that a perfectly uniform crystal will show strong absorption of any radiation incident at a Bragg angle, and would thus not be ideal for spectrometry, even if such a crystal were available. In practice, crystals have a mosaic structure, in which the crystal may be considered to consist of separately identifiable regions, perfect within themselves, but slightly misorientated with respect to their neighbours. To be suitable for spectrometry, such a crystal should have a narrow rocking curve combined with a high integrated reflection coefficient, the latter being the reflection coefficient integrated over the angular width comprising the rocking curve. Surface grinding and polishing can detach small grains of material, which then lodge, at a different orientation, in small surface cavities. Carried to excess

Table 4.1 *Analysing crystals (Jenkins and de Vries, 1970)**

Crystal	Reflection Plane	2d Spacing (nm)	Lowest Atomic Number Detectable		Reflection Efficiency
			K Series	L Series	
Topaz	(303)	0.2712	V(23)	Ce (58)	Average
Lithium Fluoride	(220)	0.2848	V (23)	Ce (58)	High
Lithium Fluoride	(200)	0.4028	K (19)	In (49)	Intense
Sodium Chloride	(200)	0.5639	S (16)	Ru (44)	High
Quartz	(10$\bar{1}$1)	0.6686	P (15)	Zr (40)	High
Quartz	(10$\bar{1}$0)	0.850	Si (14)	Rb (37)	Average
Penta erythritol	(002)	0.8742	Al (13)	Rb (37)	High
Ethylenediamine Tartrate	(020)	0.8808	Al (13)	Br (35)	Average
Ammonium Dihydrogen Phosphate	(110)	1.065	Mg (12)	As (23)	Low
Gypsum	(020)	1.519	Na (11)	Cu (29)	Average
Mica	(002)	1.98	F (9)	Fe (26)	Low
Potassium Hydrogen Phthalate	(10$\bar{1}$1)	2.64	O (8)	V (23)	Average
Lead Stearate		10.0	B (5)	Ca (20)	Average

* Additions to this list have been given by Jenkins (1972).

this is clearly disadvantageous, and etching is then necessary in order to improve the surface reflecting properties of such a crystal; but it should be noted that a carefully controlled surface abrasion procedure can sometimes improve the performance of a too-perfect crystal (Jenkins and deVries, 1970).

Diffraction spectrometers have often used crystals of calcite, especially for highly accurate work (see fig. 3.33). Much spectroscopy today is however directed towards analysis of complex materials (e.g. metal alloys) into elementary constituents: for this work, high reflectivity is of primary importance and the angular resolution is not required to be so good. A list of analysing crystals is reproduced in table 4.1, showing their main characteristics together with the lowest atomic number detectable in each case.

For improved resolution at ordinary wavelengths the double spectrometer is in wide use for the examination of line profiles. The principle of this arrangement is illustrated in fig. 4.11. The dispersion available is twice that of a single crystal spectrometer, and the resolution is also improved. The first crystal acts, in effect, as a perfect collimator, from which parallel bundles of radiation emerge, each wavelength at its own Bragg angle. The second crystal can then be used to produce a rocking curve, the width of which will be determined only by diffraction at the second crystal and the intrinsic width and structure of the line radiation being investigated. Many spectrometers have been designed on this principle, and the modes of operation are also numerous. A full account, together with detailed drawings, is given by Sandström (1957).

To obtain increased intensity in a crystal spectrometer it is natural to devise a focussing arrangement using a crystal with its surface bent into an arc of a circle. If we consider an arrangement such as that shown in fig. 4.12(a), in which the crystal is bent to a radius of ON, it is clear that this will not be satisfactory because, although the angles PNQ and PMQ are the

Fig. 4.11. (a) The double spectrometer: 'Parallel' position. This position gives a non-dispersive arrangement if both crystals are reflecting in the same order, and provides a reference direction. (b) The double spectrometer: 'Anti-parallel' position.

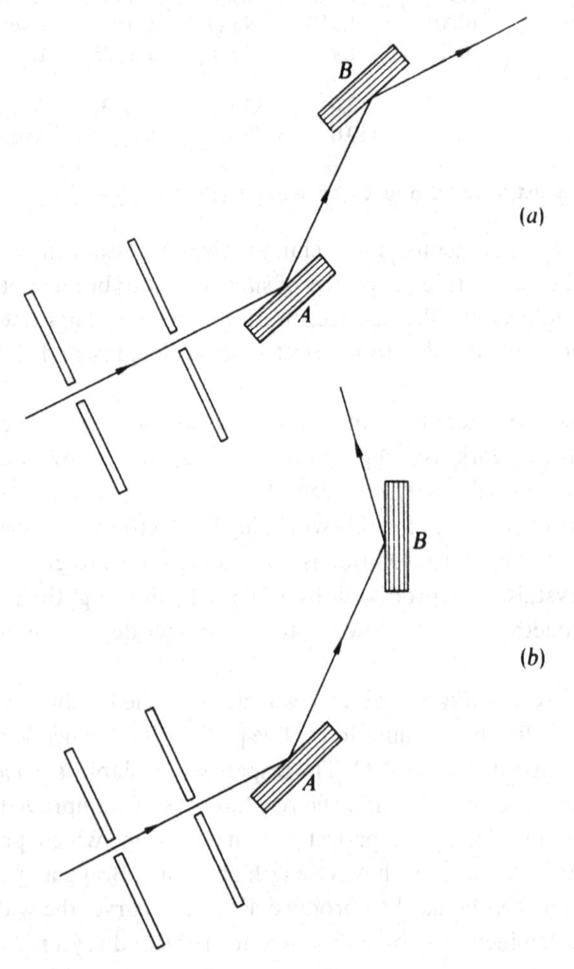

same, the reflecting planes at M are not perpendicular to the bisector MO' of PMQ, so Bragg reflection will not take place. If, however, the reflecting planes can be arranged to meet this condition, Bragg reflection will take place at all points around the concave surface, and the wavelength for which this is true will be the same at all points around the crystal because of the equality of PNO and PMO'. By altering the position of the source P, different wavelengths can be brought to a focus at the symmetrical position of Q, and a spectrum obtained. The required radius of curvature of the lattice planes is NO. Such a crystal can be prepared by grinding its surface to a radius NO' (which we shall call R), and then bending the crystal so that the lattice planes assume a radius R and the ground surface a radius $R/2$. The principle was enunciated by DuMond and Kirkpatrick (1930) and was achieved experimentally shortly afterwards by Johansson (1933) (fig. 4.12(*b*)).

In practice some degree of focussing can be obtained simply by bending an unground (flat) crystal to a radius R, but it will be appreciated from fig. 4.13(*a*) that the focussing is not exact. This method was introduced by Johann (1931) and is widely used. An alternative method of using the bent crystal, described originally by Cauchois (1932) is in the transmission arrangement shown in fig. 4.13(*b*). This method is complementary to the Johann method, as it is more suitable for rather small Bragg angles, that is,

Fig. 4.12. Focussing spectrograph: (*a*) This diagram illustrates the unsuitability of a crystal bent to radius ON ($=R/2$). The bisectors of PMQ and PNQ are normals neither to the surface of the crystal nor to its lattice planes. Therefore neither specular nor Bragg reflection can take place.
(*b*) Ground and bent crystal according to Dumond and Kirkpatrick.

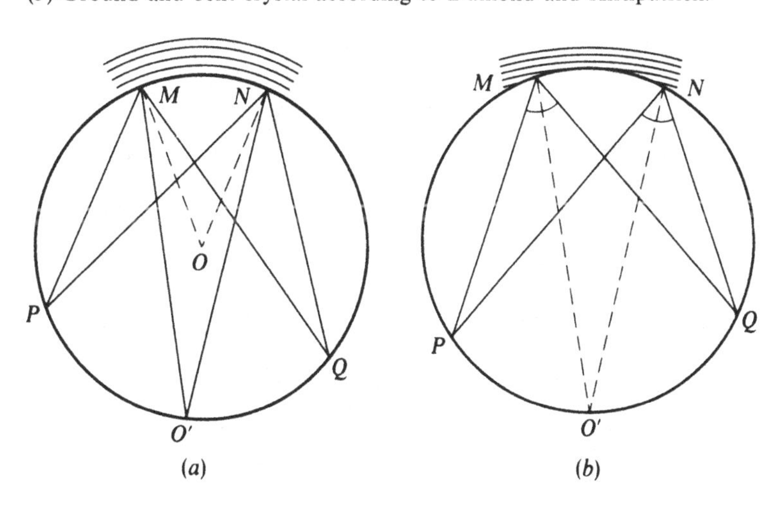

for shorter wavelengths. Absorption in the crystal in any case limits its use at longer wavelengths, for which the Johann arrangement is preferable.

The $R/2$ circle in fig. 4.12(*b*) is closely analogous to the 'Rowland circle' in a grating spectrometer and is often called by that name.

The diffraction grating has been developed for applications in the field of X-ray spectroscopy, by increasing the number of lines per cm and by working very close to grazing incidence. These requirements follow naturally from the very short wavelength of X-rays compared with optical wavelengths. Lens focussing is not possible, so the concave grating is used and has proved to be of importance in the absolute calibration of X-ray lines, and their intercomparison with lines in the optical part of the electromagnetic spectrum.

A detailed account of the grating spectrometer and of the many types of bent crystal spectrograph is given by Sandström (1957).

(c) *Proportional counters*

In the preceding discussion, the analysis of an X-ray spectrum has depended upon the dispersive property of a crystal or diffraction grating, and in consequence the radiation detector needed no ability to discriminate between different photon energies. We now turn to a group of detectors which are energy sensitive, and which yield considerable information about an X-ray spectrum without the use of a wavelength-dispersive device.

Fig. 4.13. Focussing spectrograph (*a*) Johann mode (*b*) Cauchois mode.

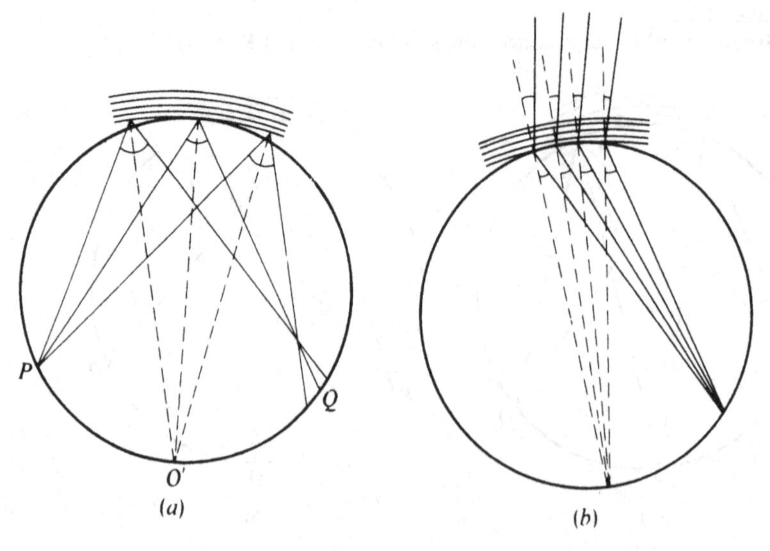

The earliest of these energy-dispersive detectors to find wide application in X-ray physics is the proportional counter, in which the magnitude of the electrical signal produced as a result of absorption of an X-ray photon is closely proportional to the photon energy. The charge liberated by the absorption of the photon is collected at an anode wire and appears across the self-capacitance of the counter. The resulting voltage pulse is amplified, shaped, and subjected to sorting and recording by the use of a single- or multi-channel pulse height analyser. Equally important is the scintillation counter, in which the photon energy is converted into optical scintillations which are then detected by a photomultiplier. Thirdly there is the solid-state detector which in recent years has become a powerful tool in X-ray work, with an energy resolution considerably better than that available from the proportional counter. The solid-state detector depends upon the separation of charge occurring within a semiconductor when ionizing radiation is absorbed. Standard electronic techniques can again be used for sorting and recording. We shall consider each of these detectors in turn.

The proportional counter is a member of the family of radiation detectors which depend on the ionization caused in a gas when photons are absorbed. Electrons liberated in this ionization process are accelerated by the electric field within the counter, and release further electrons by collision with gas molecules, a process known as 'gas multiplication'. The principal absorption mechanisms are the photoelectric effect, the Compton effect, and pair production, though we shall see in chapter 5 that the energy required for pair production is far above the useful upper limit of the proportional counter. These absorption processes are fully discussed in chapter 5, but we may note here that in the photoelectric effect the incident photon interacts with an inner electron causing it to be ejected, leaving the atom in an ionized state. In principle, an electron can be ejected from any shell by this process, or from the conduction band of a metal or semiconductor; but in the present context, ionization of the *K*, *L* or *M* shells is of greatest interest to us. The ionized atom subsequently deionizes with the emission of a fluorescent X-ray or an Auger electron. If an X-ray is emitted it will be reabsorbed in the gas, providing that the dimensions of the counter are large compared with the mean free path of the photon. The electrons released by these absorption processes are slowed down by further ionizing collisions with gas molecules (now usually involving *outer* electrons), and the electrons released in these secondary ionizing collisions will cause further ionization until the whole of the energy of the initial photon has been transferred to the counter gas. For this to take place one further requirement has to be met – all the released electrons must terminate their paths in the gas, and to ensure this, the counter must be sufficiently large. It is a property of gases that the energy required to produce one ion-pair is independent of the kinetic energy of the electron causing the ionization, so it follows that the number of ion-pairs produced as a result of the absorption of an incoming photon is independent of which of the many combinations of processes is actually responsible for the transfer of energy from photon to gas. It is this fact which causes a monoenergetic beam of X-radiation to produce a distribution of pulses of well-defined height, subject to certain statistical limitations to be discussed below.

We have seen that there are two limitations on the size of a proportional counter, both lower limits, one set by the need to reabsorb fluorescent radiation, and the other by the need to prevent the released electrons from losing any energy in the walls of the counter. The former restriction is the more difficult to meet, because the mean free path of X-ray photons is in general much greater than that of electrons of similar energy. If the first restriction is in fact not met, there will be an appreciable probability that some fluorescent X-ray photons will escape from the counter gas, causing the ionization produced in the gas to be deficient by an amount proportional to the energy of the K or L characteristic radiation of the counter gas. The principal peak in the pulse height distribution from a monoenergetic source would in these circumstances be accompanied by a smaller peak, the 'escape peak', with an intensity proportional to the product of the fluorescence yield of the counter gas and the escape probability. For counters with linear dimensions of a few centimetres filled with argon at atmospheric pressure, several per cent of the total counts may be located in the escape peak, if the K shell is excited, i.e., if the incident photons have an energy greater than 3.2 keV. In the case of a counter filled with, for example xenon, K excitation will occur for photons above 34.55 keV. The K escape peak will then be prominent, because of the high fluorescence yield of xenon and the relatively low absorption coefficient of xenon for its own characteristic radiation. For incoming photons below 30 keV, only the L shell will be excited, for which the fluorescent yield is lower and the absorption coefficient higher. The escape peak under these conditions is in fact much smaller than for an argon-filled counter of the same dimensions operated in the same energy region (3–30 keV). For this reason, and also for the higher absorption efficiency, a counter filled with xenon is more satisfactory for use in this region of X-ray energies.

The ion-pairs produced by the initial slowing-down processes normally consist of positive ions and free electrons, and if an electric field exists in the counter gas the charges may be separated and ultimately collected on electrodes. If the electrons remain unattached during their passage towards the positive electrodes they will arrive in a time of the order of a microsecond (assuming counter dimensions of a few centimetres and a gas pressure in the region of one atmosphere) and will produce a pulse with a rise time of a few tenths of a microsecond and a height proportional to the photon energy. To achieve this rapid mobility of electrons, electron-capturing gases (notably oxygen) must be rigorously excluded. It is possible to operate a counter in this way, without gas multiplication, but the pulses formed are small (40 μV for photons of 100 keV absorbed in a chamber of capacity 10 pF, assuming 30 eV per ion-pair), and it is difficult to reduce electronic noise to acceptable limits in the high gain amplifiers which are required. This mode of operation is therefore not much used.

The usual way of operating a proportional counter is to design the tube with a highly non-uniform field, and to make use of the 'avalanching' which occurs when electrons are accelerated in the high fields in the vicinity of a wire carrying a positive potential of 1–3 kV, in a gas at atmospheric pressure. In these circumstances a single electron may give rise to 10^3 or more electrons as a result of the successive accelerations and ionizing collisions experienced by electrons in such fields. If the X-ray photon being detected has an energy of 10 keV, the 300 or so ion-pairs produced will give rise to a final charge, after avalanching, of the order of 10^5 or 10^6 electrons. The pulse produced at the anode will now be several millivolts in magnitude and may be amplified without difficulty.

The proportional counter in practice, then, consists of a wire, commonly of tungsten, with a diameter in the region of 0.0025–0.005 cm mounted along the axis of a metal cylinder (dural, brass, stainless steel) of diameter 5 cm and length 10–15 cm, and filled with an inert gas to a pressure of 1 atmosphere (fig. 4.14). A positive potential of 1–3 kV is applied to the central wire, via an anode resistance of 10–100 MΩ. The field must be accurately uniform at all points along the wire, and to ensure this the wire must be smooth, and accurately circular in section. The counter gas must

Fig. 4.14. Proportional counter (schematic) showing end insulators, entrance and exit windows, centre wire and coupling capacitor.

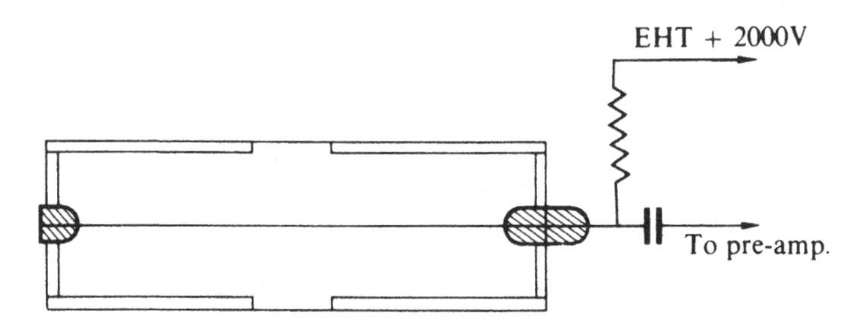

contain less than 1 part in 10^4 of oxygen or other electron-capturing gas, but it is usual to add 5–10% of a 'quenching' gas, normally methane or carbon dioxide. This serves two distinct functions: it absorbs ultra-violet photons produced in the avalanches which would otherwise release further, unwanted, electrons from the wire or the cathode cylinder. It also absorbs similar photons which are released when the positive ions reach the cathode, and which might otherwise initiate a second pulse.

The proportional counter is widely used in X-ray work as a sensitive detector of moderate energy resolution, and is capable of operation at counting rates of at least 10^4 per second without loss of resolution. For photons less than 50 keV the efficiency is sufficiently high to yield high sensitivity, and below 20 keV may approach 100% for a filling of krypton or xenon.

To examine the energy resolution of a proportional counter, let us suppose that the energy is completely deposited in the counter gas and that the number of ion-pairs produced is N. Let the gas multiplication averaged over the N electrons comprising one original event be $\bar{M}$. We shall denote the standard deviations of these quantities by σ_N and $\sigma_{\bar{M}}$. The output pulse height (in terms of the number of electronic charges arriving at the anode after the multiplication process), and its standard deviation, will be denoted by P and σ_P respectively.

We can write $P = N\bar{M}$ and

$$\left(\frac{\sigma_P}{P}\right)^2 = \left(\frac{\sigma_N}{N}\right)^2 + \left(\frac{\sigma_{\bar{M}}}{\bar{M}}\right)^2 \tag{4.7}$$

Remembering that $\bar{M}$ is averaged over N electrons released by the one event, it is clear that

$$\left(\frac{\sigma_{\bar{M}}}{\bar{M}}\right)^2 = \frac{1}{N}\left(\frac{\sigma_M}{M}\right)^2 \tag{4.8}$$

where σ_M is the standard deviation in the multiplication associated with *one initial electron.*

Hence

$$\left(\frac{\sigma_P}{P}\right)^2 = \left(\frac{\sigma_N}{N}\right)^2 + \frac{1}{N}\left(\frac{\sigma_M}{M}\right)^2 \tag{4.9}$$

$\left(\frac{\sigma_N}{N}\right)^2$ would be equal to $\frac{1}{N}$ if the production of ion-pairs were subject to statistical fluctuations of the order of $\sqrt{N}$ appropriate to a random process. In practice $\left(\frac{\sigma_N}{N}\right)^2$ is given by $\frac{F}{N}$ where F has often been taken to be in the region of 0.4, following Fano, (1947). $\left(\frac{\sigma_M}{M}\right)^2$ can be shown to be of the order of unity for a single initial electron.

Hence

$$\left(\frac{\sigma_P}{P}\right)^2 = \frac{F}{N} + \frac{1}{N} \tag{4.10}$$

In practice, therefore, the fluctuations in the number of ion-pairs contribute appreciably to the spread in pulse height, but the main contribution comes from the statistical spread in the multiplication process. If W is the energy (in keV) required to produce an ion-pair, and E_γ (in keV) the photon energy, we have

$$\left(\frac{\sigma_P}{P}\right)^2 = (1+F)\frac{W}{E_\gamma} \tag{4.11}$$

Values for W in argon have been given as 0.0264 keV (Valentine, 1952), and 0.0255 keV (Weiss and Bernstein, 1955). Hence

$$\frac{\sigma_P}{P} = 0.19\, E_\gamma^{-\frac{1}{2}}, \tag{4.12}$$

approximately, with E_γ in keV.

It is usual to express the resolution as the full width (ΔP) of the peak at one-half of the maximum height (FWHM). If the peak is assumed to be Gaussian in shape, this is given by $2.35\sigma_P$.

Hence, on the basis of (4.12),

$$\frac{\Delta P}{P} = 0.45\, E_\gamma^{-\frac{1}{2}}, \tag{4.13}$$

approximately.

A graph of (4.12) is shown in fig. 4.15 (Culhane *et al.*, 1966), together with experimental data by several authors. Resolutions are seen to be appreciably better than the values predicted from equation 4.12, probably because $\left(\frac{\sigma_M}{M}\right)^2$ for a single initial electron appears to be rather less than unity – Curran and co-workers (1949) obtained a value of 0.681 for this quantity, and this value has been closely confirmed by the measurements of Campbell and Ledingham (1966), who obtained a value of 0.651 for the variance in pulse height from single photoelectrons in an argon/propane mixture.

A further factor affecting resolution is the observation (Alkhazov *et al.* 1967) that the 'Fano factor' F varies markedly according to the nature of the counter gas. A value of 0.19 is obtained for argon and 0.8% methane, which is considerably lower than the value of approximately 0.4 estimated originally. For a mixture of argon and 0.5% acetylene the very low value of 0.09 is reported by these workers. Clearly these variations in F lead to the expectation that the resolution of proportional counters will vary considerably according to the details of construction, filling and operation.

We noted earlier that for best performance, i.e. to obtain maximum energy resolution, it is essential that the photoelectron which is initially released shall have a range which is small compared with the counter dimensions. A 10 keV electron has a range of about 1.2 mm in argon at atmospheric pressure. At 50 keV the range is 2.6 cm. If we suppose that the radius of the counter must be at least as great as the range of the photoelectrons, an argon-filled counter with a diameter of 5 cm would have a useful upper limit of 40 keV. At this energy, however, the detection efficiency would be very low (in the region of 1% only), so we see that for such a counter it is the photoelectric absorption coefficient which sets an upper limit to the energy at which the counter may usefully be operated, rather than the range of the electrons released in the absorption process. In the case of a counter filled with xenon at the same pressure, the absorption coefficient of the gas again sets the upper energy limit rather than the range of the photoelectrons, although the detection efficiency in the range 50–100 keV is several times higher than for an argon-filled counter of the same dimensions. If we regard an efficiency of 20% as the lowest acceptable

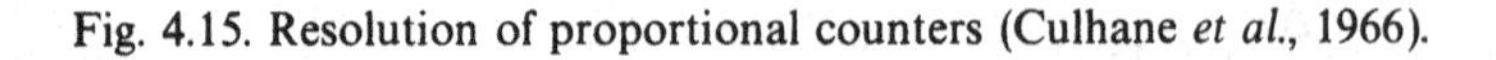

Fig. 4.15. Resolution of proportional counters (Culhane *et al.*, 1966).

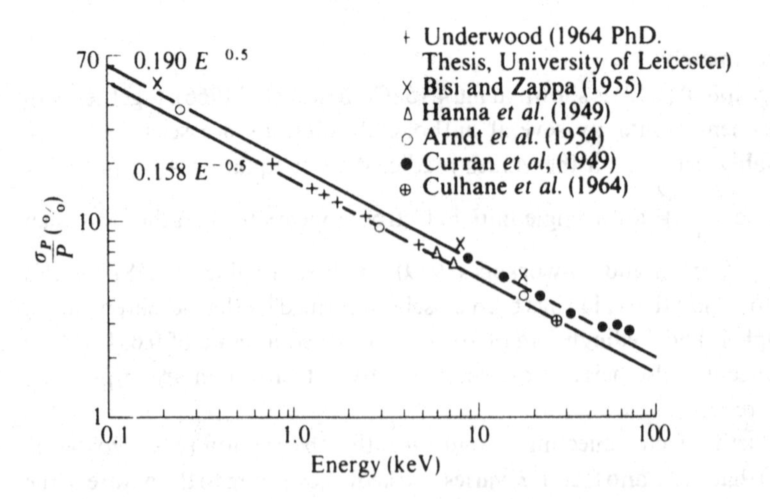

limit, this would restrict a 5 cm diameter counter to energies below 15 keV (argon-filled) and 60 keV (xenon-filled).

A further requirement for good resolution is that the counter gas should be free from electron-capturing gases, particularly oxygen and water vapour. If such molecules are present, attachment will take place with high probability, greatly reducing the mobility of the negative charges. The gas multiplication and energy resolution are much reduced because of this. A sealed-off counter has to be carefully outgassed before the gas is admitted, to ensure adequate purity, and even then a gradual deterioration in performance over a period of a year or two is not uncommon. Because of these somewhat stringent requirements, the 'flow' counter, in which the gas is passed continuously through the counter, has become popular. If the gas is of adequate purity and dryness, such counters can be made to give good resolution without the need for outgassing, and there is no deterioration in performance. Furthermore, the window no longer needs to be absolutely gas-tight or impervious to water, so foils of Melinex or other plastic materials can safely be used. One disadvantage of flow counters should be noted – they show a high temperature coefficient of gas multiplication, due to the variation of gas density, and consequent variation of mean free path, with temperature.

The great majority of electrons produced in the multiplication process are produced very close to the anode wire, because of the high field strength there. Once formed, the electrons move towards the anode and the positive ions move in the opposite direction. The pulse at the anode rises rapidly (in the negative sense) during the initial movement of charge, reaching 80% of its peak in about 0.3 microseconds. The rise then continues less rapidly, not attaining its limit until all motion has ceased with the positive ions finally arriving at the cathode. This may take several hundred microseconds. In practice, the pulse is passed through a differentiating circuit with a time constant in the region of 0.5 microseconds, to provide a short output pulse and to prevent 'pile-up' of pulses at high counting rates. The capacitance of the input circuit should be as small as possible, and will normally be in the region of 10–15 pF. Gas multiplications of 10^2–10^3 are used (and can be exceeded if necessary) and a subsequent electronic amplification of 10^3–10^4 will then yield pulses large enough to be processed by a pulse-height analyzer. Proportional counters can be operated down to energies of a few hundred eV (e.g. carbon K radiation, 282 eV) and have even been used to detect the single electrons released from metals by ultra-violet radiation.

It has been observed (Culhane *et al.*, 1966) that the rise time of pulses originating from X-rays is rather faster than that due to high-energy electrons or mesons. This appears to be due to the short range of photoelectrons with the consequence that all the ion-pairs produced by the absorption of an X-ray photon originate in a closely defined region of the counter, causing the electrons to have transit times which are closely similar. The difference in rise time has been used to discriminate against the cosmic-ray background, and pulse-shape discrimination is now well-established as a laboratory technique.

Although the proportional counter has been used effectively for the study of continuous radiation spectra, its main applications in X-ray physics have been in the study of characteristic radiation excited by electron bombardment or fluorescence. In this connection, one further deduction may be made regarding the available energy resolution. The energy of characteristic K is given approximately by the expression ($Z \leqslant 50$)

$$E_{K_\alpha} = 1.09 \times 10^{-2}(Z-2)^2 \; (E_{K_\alpha} \text{ in keV}) \tag{4.14}$$

The fractional difference in energy between adjacent elements is thus given by

$$\frac{\mathrm{d}E}{E} = \frac{2}{Z-2} = \frac{2\sqrt{1.09 \times 10^{-2}}}{\sqrt{E_{K_\alpha}}} = \frac{0.21}{\sqrt{E_{K_\alpha}}} \tag{4.15}$$

By comparing this with (4.12), and by supposing that elements may be distinguished if their energy separation is greater than one standard deviation of the pulse height spectrum, we see that adjacent elements may be distinguished with equal effectiveness throughout the whole series, providing that the elements are in a pure state. In practice, mixtures of elements differing by more than three units of atomic number can be analysed with ease; mixtures of elements closer than this give rise to spectra consisting of unresolved peaks, but procedures have been devised for 'unfolding' such distributions into their components. Dolby (1959) has discussed this problem in the context of X-ray emission microanalysis.

Proportional counters are widely used for element identification in X-ray emission and X-ray fluorescence microanalysis, and, as shown above, their resolution is adequate for many purposes. A particularly valuable application of proportional counters is when used as an adjunct to a Bragg crystal spectrometer, to eliminate higher order reflections. These reflections cause photons of 2, 3 etc., times the required photon energy to enter the detector, and a proportional counter can easily discriminate against these.

The literature of the proportional counter is extensive, and the references already cited provide a suitable starting point for more detailed studies. Curran and Wilson (1965) have given a comprehensive discussion of the proportional counter and its applications. Mulvey and Campbell (1958) have discussed the use of proportional counters in X-ray emission analysis.

(d) *Scintillation counters*

The scintillation counter depends for its operation on the luminescence of certain crystals, when excited by ionizing radiation. Photons are produced in the visible region, which are then allowed to impinge on the photocathode of a photomultiplier. Multiplication takes place at successive dynodes by secondary emission, causing a negative charge to appear at the anode. This yields a voltage pulse which may be differentiated, amplified, and analysed in a manner closely similar to that associated with the proportional counter. Scintillators are usually classified into organic and inorganic. In the former category the luminescence is an intrinsic process which is known to occur in many aromatic organic crystalline substances, whereas in the latter category the luminescent property arises from defects in ionic crystals. The defects may be due to imbalance in the stoichiometric proportions of cation and anion (in which case the scintillator is said to be 'self-activated') or may be due to interstitial or substitutional impurities deliberately introduced into an otherwise pure crystal ('impurity activated').

Organic crystals were widely used at one time for the detection and counting of β-particles in studies of radioactivity, and for charged particle detectors in studies of nuclear reactions and nuclear scattering. They have been largely superseded by polymerised plastic materials, which are cheaper and which can readily be machined into a variety of shapes. Both the organic and plastic scintillators have short luminescence decay times (3–20 ns) and are therefore suitable for fast counting work. But for X-ray work they are of limited use because of their low atomic number and low photoelectric absorption. Inorganic phosphors are preferred for X-ray studies, and sodium iodide activated with thallium is most often used. The decay consists of 2 components the principal luminescent decay time being 0.23 μs, along with a subsidiary component with a decay time of about 1.5 μs. This is short enough for many applications, and enables count rates up to the region of 10^4 cps to be used, with only a very small probability of 'random sum' pulses occurring.

The scintillator is normally surrounded by an optically reflecting substance (titanium dioxide and magnesium oxide have been used for this purpose on account of their high reflectivity) to ensure maximum channelling of light to the photomultiplier. Sodium iodide is deliquescent and crystals of this substance are therefore encapsulated in aluminium cans, to which may be fitted a beryllium window. One face of the crystal is in contact with a glass window which is then placed in optical contact with the end face of the photomultiplier, usually with a few drops of oil or transparent grease intervening to ensure maximum transmission of light. The whole assembly is then mounted in a light-tight container.

The photomultiplier consists of a photocathode, an anode and about 10 intervening dynodes at which the electron multiplication takes place. The photocathode is a thin layer of, for example, an antimony–caesium alloy deposited on the inside of the end face. Electrons released here drift across to the first dynode and the electron pulse then impinges on successive dynodes finally reaching the anode. The dynodes may be disposed in the 'Venetian blind' arrangement shown in fig. 4.16(*c*), or in the more complex arrangements of fig. 4.16(*a*), (*b*). The latter arrangements were devised to focus the electrons on to successive dynodes, in which case all the electrons take equal times to traverse the system, and the photomultiplier introduces only a negligible spread in the time of arrival of the electrons at the anode. The 'box and grid' structure of fig. 4.16(*d*) is sometimes used.

The photomultiplier requires potentials of about 100 volts between the photocathode and the first dynode, and between adjacent dynodes. These voltages are normally produced from a potential divider located within the

Fig. 4.16. Photomultiplier structures (Birks 1964, after Sharpe 1961). (a) focussed structure; (b) compact focussed structure; (c) Venetian blind structure; (d) box and grid structure.

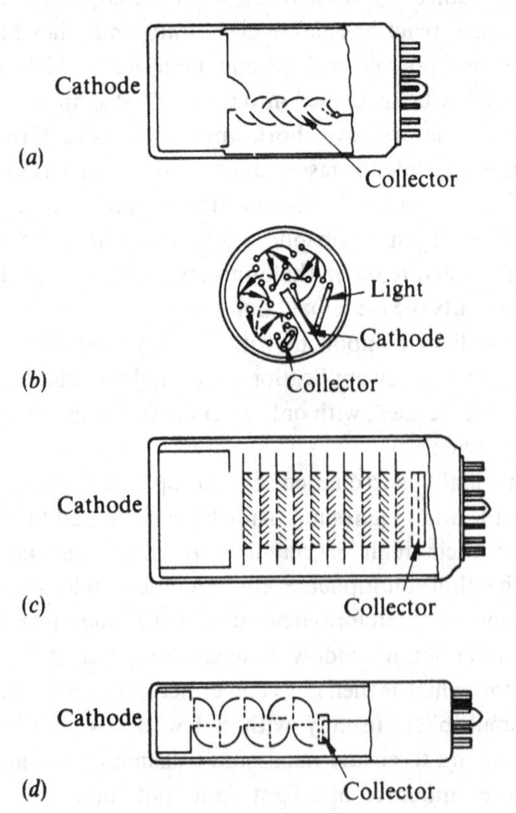

photomultiplier housing and connected to a positive DC supply in the region of 1000 volts. The anode is returned to the supply through an anode load, and is connected to the input of a cathode follower.

The number of photons released within the crystal is much smaller than expected from purely energetic considerations, because the majority of the excited centres decay by radiationless processes. As an approximate guide, the amount of energy in the initial ionizing particle (normally a photo- or Compton electron released in the crystal) required is about 70 eV per optical photon released in sodium iodide. Of the photons produced typically 10–20% will release photoelectrons at the photocathode. The multiplication taking place at each dynode is in the region of a factor of 4*, which, allowing for slightly less than 100% collection efficiency at the following dynode, yields an overall multiplication of 10^5–10^6 in a 10-dynode tube. A 40 kV X-ray photon may thus be expected to produce a charge of the order of 10^7–10^8 electrons at the anode. This is two orders of magnitude greater than available in a proportional counter, so the amplifier gain required is correspondingly less.

A scintillation counter will normally produce a large background of very small pulses, due to the release of single thermionic electrons from the photocathode. These single electrons yield pulses the rise times of which are rather shorter than the rise time of the main pulses, and so their pulse height can be greatly reduced by the use of a suitable integrating time constant, without affecting the size of the main pulses. A far more effective method of reducing this background is to cool the photomultiplier, and this is often done when photons of less than a few keV or when low disintegration rates at somewhat higher energies (e.g. from carbon-14) are being looked for.

The energy resolution of a scintillation counter is determined by the number of optical photons produced by the initial event, the efficiency of the photocathode, and the spread introduced by the multiplication process. The situation is thus very similar to that in the proportional counter, and the resolution (expressed as the relative width of the pulse height distribution due to a monoenergetic x- or γ-ray) varies as $E^{-\frac{1}{2}}$. In general, the resolution of the scintillation counter is inferior to that of the proportional counter, because of the smaller number of electrons initiating the avalanche process in each case. Comparison between the scintillation counter data of Birks (1964) and data for proportional counters reproduced in fig. 4.15 shows that the resolutions differ by a factor of about 4.

If the proportional counter, used alone, is only marginally adequate for X-ray spectral analysis, the scintillation counter is even less so. Its main uses

* Higher values are possible if the voltage between dynodes is increased.

in X-ray work have been in diffractometry and in the study of Bremsstrahlung spectra produced at $\frac{1}{2}$ MeV and above. Cylindrical crystals $1\frac{1}{2}''$ diameter and $1''$ in thickness are valuable for spectrometry up to 1 MeV, and larger crystals are available for higher energies. The pulse-height distributions from scintillation counters are somewhat more complex than from proportional counters because the absorption mechanism is less simple. Compton interactions are important over the whole range of usefulness of sodium iodide crystals, and above 200 keV the majority of the interactions in a crystal of the size quoted are by this process. If the scattered photon is reabsorbed in the crystal, the pulse will form part of the photoelectric peak; but if the scattered photon leaves the crystal, a pulse of reduced height will be produced and these pulses will form a continuous spectrum from zero up to a maximum energy of

$$E_{max} = \frac{2E_\gamma}{1+2E_\gamma} \tag{4.16}$$

where E_γ is the incident photon energy in units of m_0c^2. For $E_\gamma \gg 1$, this approaches $E_\gamma - \frac{1}{2}$, i.e. the separation between the 'Compton edge' and the photoelectric peak approaches $\frac{1}{2}m_0c^2$ or approximately 256 keV. A further feature of scintillation spectra is the 'back-scatter peak', caused by photons passing through the crystal, being scattered through 180° and absorbed on the way back through the crystal. The back-scattered photon is absorbed with considerably higher efficiency than the incident photon, so this sequence of events can take place with rather high probability. The energy of the back-scattered peak is given by $E_\gamma/(1+2E_\gamma)$ (in units of m_0c^2), and tends to a value of $\frac{1}{2}m_0c^2$ at high values of E_γ. The back-scattered peak may of course be enhanced by the presence of external shielding around the source and the crystal, but this would depend on the geometry of the system and would not be an intrinsic property of the detector.

A further factor which contributes additional features to scintillation spectra is the occurrence of pair-production in the crystal at energies above $2m_0c^2$. This effect becomes progressively more important at higher energies, and the positron produced in this interaction will normally come to rest and annihilate within the crystal. The escape of one or both of the annihilation photons which are thus produced gives rise to escape peaks at points m_0c^2 and $2m_0c^2$ below the main peak (fig. 4.17).

Finally we should mention the X-ray escape peak which is due to the escape of the iodine K_α and K_β radiation from the crystal. This escape peak is apparent only at rather low incident energies, for which the escape peak may be resolved from the main peak. It can be important for photon energies of 100 keV down to the K absorption edge of iodine at 33.16 keV.

All the effects referred to can be reduced as much as desired by sufficiently increasing the size of the crystal. Alternatively the crystal can be surrounded by a larger scintillator, and all coincident events in the two scintillators rejected by an anticoincidence arrangement. In this way any subsidiary peaks due to escape effects, and particularly the 'Compton continuum', can be greatly reduced. This method has the considerable advantage that the surrounding phosphor can be of poor resolution, or of plastic scintillator (available in large sizes at relatively low cost) with no appreciable photoelectric absorption at all. The inner sodium iodide crystal can then be of moderate size and high quality.

(e) *Solid-state detectors*

We turn now to a further class of radiation detector, which has become of great importance in X-ray studies, particularly in the fields of

Fig. 4.17. Scintillation spectrum of ^{24}Na using $1\frac{1}{2}'' \times 1''$ NaI(Tl) crystal (upper curve) and $3'' \times 3''$ NaI(Tl) crystal (lower curve). (Neiler and Bell, 1965).

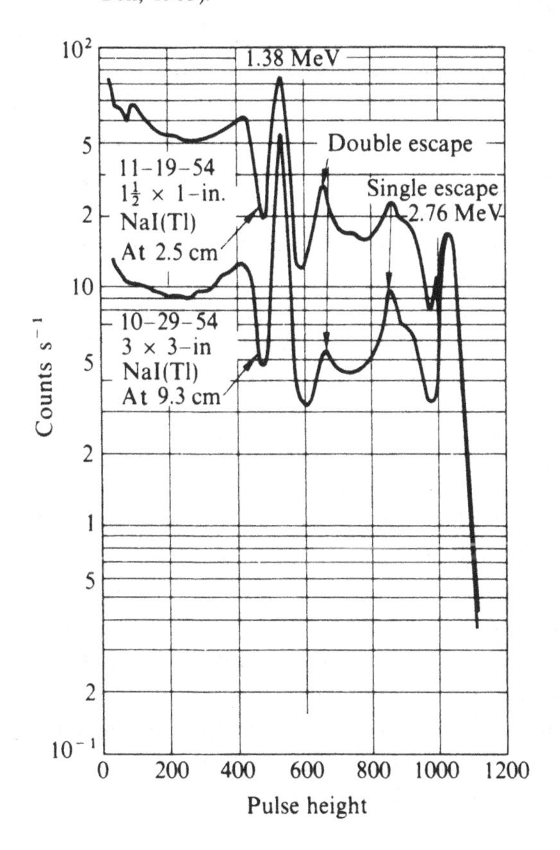

elemental analysis, and radioactivity. Semiconducting radiation detectors were developed originally for use as detectors of charged particles in the study of nuclear reactions, but are now finding increasing application as photon detectors down to energies in the region of 1 keV.

The *conduction counter* (or 'solid-state ionization chamber') consists of a wafer of semiconducting material across which an electric field is applied by means of metallic electrodes. The passage of a charged particle raises electrons into the conduction band, and the electron-hole pairs thus formed then separate in the field and migrate towards the electrodes. The energy needed for 'ion-pair' formation is in the region of 3–4 volts, and this gives the conduction counter a considerable advantage, in principle, in that the number of ion-pairs formed by the absorption of a given energy is much greater, and the statistical fluctuations much less, than in the gaseous ionization chamber. In practice, good performance is limited by noise and by the recombination of pairs. An account of these devices has been given by Gibbons and Northropp (1962). The diamond conduction counter can be used in special circumstances, e.g. elevated temperatures, when its high band gap (~ 5.6 eV) is an advantage.

The *semiconductor junction counter* is essentially a junction between n-type and p-type silicon or germanium which is operated under 'reverse-bias' conditions. In these circumstances the charge carriers are swept out by the applied field and a depletion layer is formed. In this way a device with very low current noise is constructed, and fields sufficiently strong to prevent recombination can be used. The entry of an ionizing particle into the depletion layer causes electron-hole pairs to be formed, which then move to their respective electrodes forming a signal.

One of the most widely used of the semiconductor radiation detectors is the *surface-barrier detector*. This consists essentially of an n-type silicon or germanium wafer with a thin gold film evaporated on to the front surface, and operated under reverse-bias conditions. A depletion layer is again formed, the thickness of which depends on the resistivity of the material and the reverse-bias voltage. The depth of the depletion layer can be chosen to suit the range of particles for which the detector is to be used, and longer range particles can be discriminated against to some extent. Depletion layers up to 5 mm can be achieved.

In the context of X-ray and γ-ray studies the *lithium-drifted detector* is of great importance. In this device the acceptors in a p-type semiconductor (silicon or germanium) are compensated by donor impurity atoms which are drifted into the material by means of raised temperature and an applied electric field. The material thus produced has a high resistivity, so that noise can be held at the low level, and thicknesses of several centimetres are at present available in lithium-drifted germanium detectors, ensuring good detection efficiency for gamma-radiation up to 1 MeV and above. These detectors have to be maintained continuously at liquid nitrogen temperatures to ensure that the lithium atoms occupy correct positions in the germanium lattice. The lithium-drifted germanium (Ge(Li)) detector is now used extensively for γ-ray spectroscopy; its high resolution and good efficiency make it the instrument of choice for this type of investigation. More recently the 'hyperpure' or 'intrinsic' germanium detector has been introduced, with comparable resolution. It has the advantage of not needing to be stored at liquid nitrogen temperature, but it must be cooled for operation to reduce noise to an acceptable level. The lithium-drifted silicon (Si(Li)) detector is well-suited for work with X-rays and low energy γ-radiation, particularly for X-ray fluorescence spectroscopy and X-ray emission analysis in connection with, for example, the electron microprobe (v. chapter 8).

A full account of these and other radiation detectors has been given by Knoll (1979).

The resolution potentially available may be deduced along the same line of argument as that which was developed for the proportional counter (q.v.) except that there is no charge multiplication in these solid-state devices. It is not difficult to establish that the fractional energy resolution is given by

$$\frac{\sigma_P}{P} = \left(\frac{FW}{E_\gamma}\right)^{\frac{1}{2}}$$

where P is the output pulse height and σ_P its standard deviation. As with other energy-dispersive detectors it is customary to express the resolution in terms of the full width at half maximum height (FWHM), which may be obtained by multiplying (4.17) by the factor 2.35. The mean energy, W, required for the formation of a charge (i.e. electron-hole) pair is given as 3.81 and 2.98 eV respectively for silicon and germanium (at a temperature of 77 K); Fano factors as low as 0.084 (Si) and 0.058 (Ge) have been reported. The resolution achievable in practice is limited by amplifier noise and other electronic factors, but, at the current time, resolutions (FWHM) of 0.45% for 122 keV γ-radiation and 2.7% for 5.9 keV X-rays are available commercially. These may be compared with equation (4.13) for proportional counters. A γ-ray spectrum for ^{24}Na is shown in fig. 4.18, which may be compared with the scintillation counter spectrum of the same radioisotope illustrated in fig. 4.17.

A detailed discussion of the values of the Fano factor is given by England (1974), together with a consideration of the several factors affecting the energy resolution obtainable under practical conditions.

4.4 X-ray attenuation – experimental aspects

The attenuation coefficients of the great majority of elements have been studied over at least some part of the X-ray spectrum. The principle behind all experiments to determine mass attenuation coefficients is simple – the attenuator is interposed between source and detector, and the reduction in intensity is measured. It will be appreciated that this reduction in intensity may be caused either by *absorption* of the photon within the intervening matter or by *scattering* out of the collimated beam. The attenuation coefficient measures the total contribution of both types of process, and the transmitted radiation is related to the incident radiation by the expression

$$I_t = I_0 e^{-\mu x}, \tag{4.18}$$

where μ is the *linear attenuation coefficient*. The amount of energy absorbed will not, in general, be given by $I_0(1-e^{-\mu x})$, because of the loss from the system caused by scattering, and a more detailed consideration of this is deferred until chapter 5. But in circumstances where the absorption processes (particularly the photoelectric process) strongly predominate, the

Fig. 4.18. Ge(Li) spectrum of the γ-radiation from ^{24}Na (Orphan and Rasmussen, 1967).

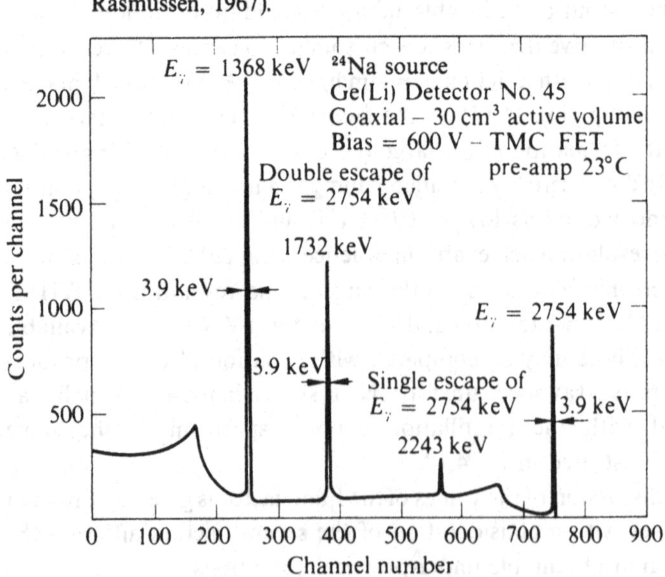

quantity μ may be referred to as the *linear absorption coefficient*. This usage is often encountered in X-ray work. It should be said, however, that, even in these circumstances, scattering processes do occur, and that a properly defined experiment will require a beam which is well-collimated, before and after the absorber/attenuator, so that the scattered radiation can be excluded from the detector when there is an absorber/attenuator in position.

The experimental methods used for attenuation studies are all necessarily similar in principle, the main differences being in the choice of detector, and the degree of collimation. In order to obtain reliable results, either the source or the detector must be mono-energetic, and the use of Bragg reflection in the incident beam has hitherto been the most satisfactory method of achieving this condition, especially if used in conjunction with an energy-sensitive detector to eliminate higher order reflections. A proportional counter or scintillation counter by itself does not have sufficiently good resolution to use with a radiation source which contains an appreciable continuous background spectrum (as is the case with the characteristic radiation obtained from conventional X-ray tubes) but a semiconductor detector is adequate in this respect. The use of a 'β-filter' to remove the K_β characteristic line, is often an advantage, in that it improves the composition of the incident beam to a considerable degree. Radioisotopes decaying by electron capture furnish very good sources for the determination of attenuation coefficients, but do not appear to have been used extensively for this work.

An important requirement in accurate determinations of attenuation coefficients is the exclusion from the detector of scattered and fluorescent radiation originating in the absorber. In chapter 5 we see that secondary radiation of this kind is associated with many of the interaction processes between X-radiation and matter – the photoelectric absorption process produces fluorescent radiation, some of which may escape from the absorber; Compton interactions necessarily produce scattered radiation, and pair production will inevitably cause annihilation radiation to appear in some part of the system. Moreover, all the elastic scattering mechanisms described in chapter 5 (of which the most important is Rayleigh scattering) yield some scattering events down to arbitrarily small angles, and this calls for clear definition of the quantities under investigation.

It is clear that an energy sensitive detector will exclude the secondary radiation from photoelectric and pair production interactions. The extent to which scattered Compton radiation is excluded will depend on the energy resolution of the detector. Photons scattered by elastic processes will enter the detector to an extent proportional to the solid angle subtended at the absorber by the aperture and we see that the 'narrow-angle' geometry shown in fig. 5.1 is essential if true values of the attenuation coefficient are to be obtained. One procedure is to measure the transmission as a function of aperture area and extrapolate the intensity per unit solid angle to zero angle. In order to prevent 'scattering-in' of radiation from 'off-axis' regions of the absorber to the detector, the radiation of the incident side of the absorber should be restricted by means of a collimator. The amount of scattered radiation entering the detector will clearly depend to some extent on the position of the absorber relative to source and detector. For this reason, a further useful precaution is to position the absorber in such a way as to minimize the transmitted intensity, thereby ensuring that the contribution caused by scattered radiation is as low as possible.

As an example of a study of attenuation coefficients in the range 25–130 keV, the work of McCrary *et al.* (1967) may be cited. These authors used an X-ray tube with a tungsten target which could be operated up to 150 kVp, and used a quartz crystal as monochromator. The system was such that the energy was defined to better than 1%; for about half their readings the energy definition was in fact better than 0.3%. The collimators were arranged so that the angular resolution at the absorber and the crystal was 5 min. of arc. This ensured that almost the whole of any radiation elastically scattered at the absorber would be rejected from the system, and it also defined the energy of the Bragg reflected radiation to adequate accuracy. One minute of arc corresponded to 1.1 keV at 130 keV and 0.04 keV at 25 keV. Their quoted precisions are in fact somewhat better than this, and are based on setting uncertainties which are rather better than 1 min. of arc.

The detector was a scintillation counter with a single channel analyser to reject higher order reflections. The samples used as absorbers were checked for purity and small corrections made when necessary. A diagram of their apparatus is shown in fig. 4.19, and their results are tabulated in appendix 2.

We shall see in chapter 5 that the photoelectric attenuation coefficient falls with increasing gamma-ray energy until the binding energy of an electron shell or sub-shell is reached: it then rises sharply as these electrons become able to interact with the incident photon. We now refer to two practical applications of this well-known phenomenon. The first is the so-called 'β-filter' used in X-ray diffractometry. The spectrum of characteristic radiation from targets of all except the lightest elements consists principally of the K_{α_1} and K_{α_2} lines (transitions $K \rightarrow L_3$ and $K \rightarrow L_2$) and the K_{β_1} and K_{β_2} lines (transitions $K \rightarrow M_3$ and $K \rightarrow N_{2,3}$). The lines of the K_β group are of higher energy than the K_α lines, and for vanadium and all elements above this it is possible to find another element (or, higher up the series, several elements) of which the K absorption edge lies between the K_α and K_β lines of the target element. The magnitude of the 'K-jump', expressed as the ratio of mass attenuation coefficients on either side of the K edge, is approximately 9 for titanium (the element immediately below vanadium) and falls to 5.5 for uranium. Over the whole range of interest it is thus sufficiently great to allow considerable attenuation of K_β lines without undue attenuation of the K_α lines. The transmission of nickel for the K characteristic radiation of copper is such that a 10-fold reduction in the K_β to K_α ratio can be achieved with a reduction of only about 36% in K_α intensity. The removal of K_β radiation is clearly advantageous in that it simplifies X-ray diffraction photographs and helps to reduce ambiguity in their interpretation.

The second application is the system of 'balanced filters' devised originally by Ross, and subsequently studied by Kirkpatrick (1939, 1944) and others. This system is used when it is desired to isolate for study a relatively narrow region of the continuous spectrum, and this is achieved by selecting two filters of adjacent atomic numbers, the K absorption edges of which lie on either side of the spectral region under investigation, and by taking measurements using each filter in turn. If the thicknesses are chosen such that their transmissions are the same at all energies except in the 'pass-band' lying between their respective absorption edges (usually K edges) subtraction of the two readings will give information relating solely to the radiation in the pass-band. The separation between the K edges of elements of adjacent atomic number is sufficiently narrow for many types of

Fig. 4.19. Apparatus of McCrary *et al.* (1967).

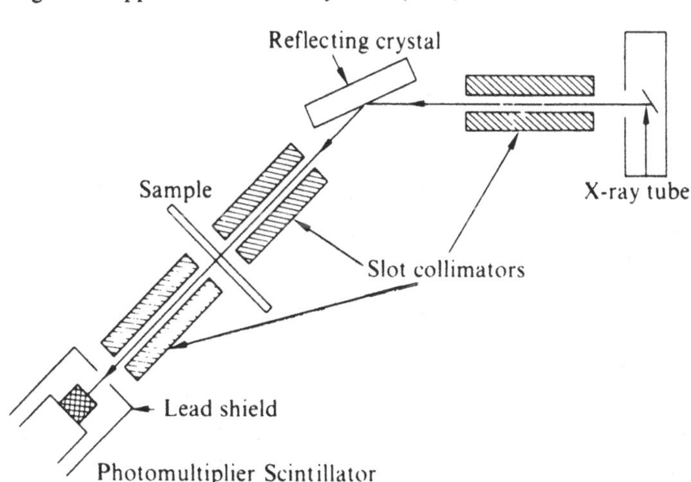

investigation, for example, the study of the angular distribution of the continuous spectrum within defined energy bands, and the method has the considerable advantage that the intensities available are high and that the boundaries of the selected energy band are sharp and well-defined. In this latter respect the method is superior to the use of the proportional counter.

Balance is achieved by equalizing the responses using a source of monoenergetic radiation with an energy chosen so as to be far from the absorption edges of the filters. Although fluorescent radiation consists of several lines, this may be used for balancing purposes providing that none of the lines fall within the pass-band of the filters. One filter may be mounted in a rotatable frame, so that balance may be achieved by adjustment of the degree of obliquity. In special cases, solutions may be used, and balance achieved by adjustment of the concentrations. Kirkpatrick has used aqueous solutions of thorium and uranyl nitrates to achieve a balanced Th/U filter.

In practice, balance is not quite perfect except over a rather narrow region of the spectrum on either side of the pass-band. Kirkpatrick has pointed out that in some situations it may be better to minimize the ratio of 'error power' to 'pass-band power' rather than simply to maximize the latter. These two distinct criteria require different thicknesses of absorber (leading to rather lower transmitted intensities in the former case) and the optimum thicknesses can be decided only in the context of a particular situation. As an example of relatively modern study using balanced filters we may cite the measurements of angular distribution of the continuous spectrum from thin targets carried out by Doffin and Kuhlenkampff (chapter 2) and the determination of fluorescent yields by Bailey and Swedlund referred to in section 4.6(c).

4.5 Soft X-ray techniques

In recent years the development of experimental methods using soft X-rays has proceeded very rapidly. In this section we shall consider those aspects of experimental technique which distinguish soft X-ray studies from work at higher energies.

Soft X-rays may be defined, rather loosely, as those which will not penetrate more than a few mm of air at ordinary pressures, and which therefore require that the spectrometer be either evacuated, or filled with helium. This definition places the dividing line at about 0.5 nm (or 2500 eV approximately), and the range of soft X-rays extends from this limit out to the region of 50 nm, where the far ultra-violet may be said to begin. The fields of soft X-ray study may be grouped under three headings. First, the study of characteristic K X-ray and band spectra of the lighter elements

($Z = 13$ or less) has been pursued for many years and has shed valuable light on the electronic band structure of these elements. The use of L-spectra extends this technique into the region of copper and iron. The second important field of application lies in the microanalysis of the very light elements, including carbon ($K_\alpha = 282\,\text{eV}$), nitrogen ($K_\alpha = 392\,\text{eV}$) and oxygen ($K_\alpha = 523\,\text{eV}$). The third main field of study is in stellar and plasma physics. All these fields of study are outlined in chapter 8.

X-rays in the region of 0.5–20 nm are produced by essentially the same means as X-rays of shorter wavelength. Because the production of continuous X-ray spectra takes place with rather low efficiency at low accelerating voltages, the main requirement is for high current and a high power dissipation at the anode. The anode is normally water- or oil-cooled and accelerating voltages extend from a few hundred volts up to 4 or 5 kV, depending upon the required radiation. There is no point in increasing the accelerating voltage beyond this range because the X-rays in the soft region will emerge only from the superficial layers of the target and any radiation produced at depths greater than about 100 nm will fail to emerge. Furthermore, in order to obtain the maximum current for a specified power loading, it is clearly desirable to keep the accelerating voltage as low as is reasonably possible. When characteristic radiation is required, similar considerations apply in regard to target penetration. The efficiency of ionization is relatively high, but against this has to be offset the very low fluorescence yields of the light elements, so that the X-ray yield tends to be low, and once again, high currents become of prime importance. Accelerating voltages of 3–5 times the critical excitation potential are usually adequate.

A major problem in tubes for work in this region is that the accumulation of even very thin layers (~ 1 nm) of tungsten contamination (from the filament) at the anode can greatly reduce the intensity of emitted soft characteristic radiation from the underlying light element target material. One way to avoid this is to place a shield between filament and anode to prevent evaporation of tungsten across the tube. In such an arrangement the electrons have to be emitted in the backward direction, and are then brought round to the anode by the focussing effect of the electric field. Another way of avoiding the difficulty is to use a cold-cathode discharge-tube type of X-ray generator, and such tubes are in fact found to be of value in soft X-ray work for this reason.

A major cause of severe attenuation of emergent X-rays is to be found in the window of the tube. This window is best dispensed with altogether and the evacuated spectrometer allowed to communicate directly with the X-ray tube. A sliding window can be fitted if desired so that the two systems can be let down to air independently if necessary.

If the X-ray tube is fitted with a thin window of, for example, mylar a few μm in thickness, it is necessary to avoid mechanical stresses and large pressure differences. An additional hazard to the window may be caused by its being struck by electrons scattered from the target which can set up strong electrostatic forces and can also cause heating and subsequent deterioration of the window. To avoid this, some tubes are operated with the cathode at earth potential and the anode raised to the accelerating voltage. In this way secondary electrons are all attracted back to the anode, and the window thereby protected.

X-ray tubes for soft X-rays may be of metal or glass construction. Glass is preferred by some, on account of its greater ease of baking and out-gassing. All inner surfaces must of course be maintained to a high degree of cleanliness and a high vacuum maintained. Carbon contamination causes the appearance of the carbon K line and can be troublesome. An X-ray tube for soft X-ray work has been described by Henke (1963, 1966).

The tubes so far described have been appropriate for the study of line spectra generated by electron bombardment, or for analytical procedures based on fluorescence. An alternative method of analysis is based on the study of absorption bands in the region of 10–40 nm in the spectrum transmitted by the sample, and for this an intense source of continuous radiation is essential. X-ray tubes produce only very weak continuous radiation at these wavelengths, but the spark discharge tube is found to emit intense X-rays in this region, when used with metallic electrodes. For more detailed descriptions of soft X-ray sources (and for details regarding other aspects of soft X-ray techniques) the reader is referred to the comprehensive article by Tomboulian (1957) which is directed particularly towards techniques for spectroscopy and band structure analysis. The articles by Henke (1957, 1960, 1963, 1966) provide a similarly authoritative account of techniques for microanalysis.

The proportional counter is widely used as a detector of soft X-rays. Resolutions become rather poor below 10 keV, but clearly defined peaks have been obtained down to the K_α radiation of carbon (282 eV; 4.4 nm); adjacent elements can be distinguished from each other quite clearly in this region of atomic number when in the pure state. A gas filling of argon, with a quenching agent, is often used in counters for this work, although Henke (1966) has used a filling of pure methane. The counter window has to be designed with some care – a window of thin nitrocellulose supported on a perforated brass disc with a thin aluminium layer to prevent electrostatic charging has been used by Campbell (1963). From absorption measurements it was deduced that the nitrocellulose was 80–100 nm thick and the aluminium layer 5–8 nm. Transmission of 50% for beryllium K radiation was achieved. Henke has used windows of Formvar (a plastic material used as a mount for electron microscope specimens) 300 nm and 600 nm in thickness for similar work.

A counter of unusual geometry (fig. 4.20) has been described by Duncumb (1960) for soft X-ray microanalysis, which gives good resolution in the wavelength range 0.8–4.4 nm. The needle has a spherical tip, and therefore gives acceptably uniform gas multiplication for these photons, few of which pass beyond the region of the tip because of their high attenuation in the argon filling.

The particular problems encountered in soft X-ray proportional counters for satellite and rocket observations have been discussed by Culhane *et al.* (1966). Thin Melinex or Mylar windows (3.75 μm) have transmission bands at 2.4–2.9 nm and 4.4–5.8 nm (2.4 nm and 4.4 nm being approximately the K absorption edges of oxygen and carbon respectively) and counters using these windows have been used to examine the continuous X-ray spectrum from the Sun and from X-ray stars (chapter 8). Flow counters are not convenient for this type of work, but sealed-off counters with these windows have been found to have useful lives of one month or more especially when used *in vacuo* or in a dry atmosphere. The 'lifetime' is adequate for much work of this kind.

The limited resolution available from proportional counters in the soft X-ray region has stimulated the use of crystals with large inter-atomic spacing. A list of such crystals is given in table 4.1. Lead stearate is an example of a 'multi-layer analyser' fabricated by depositing successive monolayers of metallic salts of long-chain fatty acids on to a glass slide. In the technique described by Henke, it is possible to deposit up to 150 such layers by repeated immersion and withdrawal of a microscope slide into and from a tank of water which carries a surface film of the compound in

Fig. 4.20. A proportional counter for soft X-rays (Duncumb, 1960).

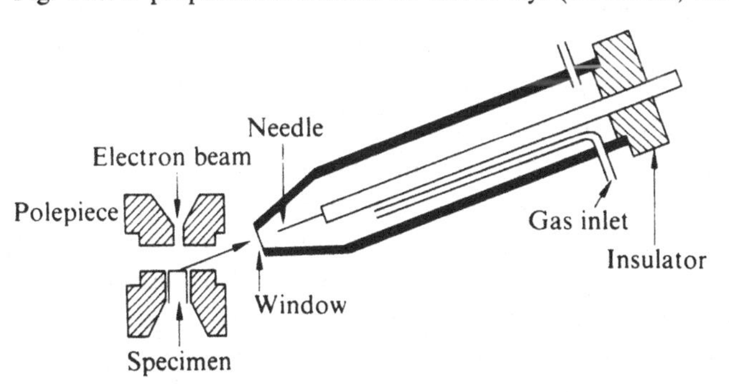

question. Resolving powers approaching the theoretical limit have been obtained from analysers prepared in this way.

When crystals or other wavelength dispersive devices are used, we have seen that the detector need possess no energy discriminating qualities of its own. A detector frequently used is the copper–beryllium photomultiplier (Jacob *et al.*, 1960, Bede and Tomboulian, 1961, Franks, 1964). The device is used without envelope, in a vacuum spectrometer and the electrons released from the first dynode are multiplied in the usual way. Beryllium–copper has the advantage of a high work function, so that the system is insensitive to visible light.

4.6 Study of secondary factors

(a) *Electron penetration*

We have already seen the importance in X-ray physics of knowing the range and energy-loss of the electrons within the target. This has entered into the discussion in two ways: when calculating the production of characteristic or continuous X-rays in a thick target it is necessary to know the rate of energy loss as a function of *path length* dE/dx, and we have already made use of theoretical expressions for this quantity in considering the production of X-rays in thick targets. (continuous radiation: (2.30)–(2.33); characteristic radiation (3.29)ff.). Secondly, when calculating the attenuation of the radiation as it emerges from the target (usually from the incident surface) some knowledge is required of the depth of production. For approximate calculations the mean depth is all that is needed, but in general the intensity of production per unit depth increment is needed, which can then be multiplied by the attenuation of the radiation on its way out and an integration or summation then performed. The *depth of penetration* as a function of energy, and the mean energy at various depths thus become relevant, and in general these cannot be obtained simply by integration of the energy-loss expressions. This is because the electrons undergo scattering during the slowing down process, and the obliquity of their paths causes the *path length* to be appreciably greater than the *penetration*. In appendix 1 we consider some of the experimental and theoretical data which bear on this, where it is shown that the difference between these two quantities is considerable.

In the case of light elements, the nuclear scattering cross-sections (responsible for deviation but not energy loss) are small compared with the electronic scattering cross-sections (responsible primarily for energy-loss), and the scattering angles are in any case small; in these circumstances the distinction between path length and penetration is not great, but for heavier elements the elastic scattering cross-section is relatively large: the condition of *full diffusion* is reached before the electrons have lost more than a small fraction of their initial energy. The difference will therefore be considerable. In this section we outline several methods used for determining electron penetration and energy-loss, but defer discussion of the results until appendix 1.

As an example of a study of the penetration of homogeneous beams of electrons into a solid medium, we refer to the work of Marshall and Ward (1937), who examined the penetration into aluminium for electrons in the range 421–1696 keV. Their data are reproduced in fig. 4.21.

If the linear portion of these curves is extrapolated to zero intensity, the intercept with the range axis is known as the *extrapolated range** and is a reasonably precise index of electron penetration. It is appreciably less than the *maximum range* which is the point at which the intensity curves merge with the background radiation. Clearly this is less precise than the extrapolated range but is a necessary parameter if it is desired e.g. to calculate the thickness of foil necessary to prevent *all* the β-particles emitted by a radioactive source from entering a detector. Such a situation may arise, for example, in the study of radioactive decay schemes, or the use of X-rays from radioactive sources for determinations of X-ray attenuation coefficients.

A second, closely related type of experiment is one in which 'electron transmission curves' are plotted. These are graphs of electron flux as a function of energy for a given foil thickness. As the electron energy is increased from a low value some of the electrons will eventually emerge from the far side of the foil, and after a further energy increase, all the electrons will emerge from the foil. The emergent flux may be recorded as a current (as in the work of Young, 1956b) or may be counted using a Geiger counter, providing due account is taken of the window thickness. If the

* or *practical range.*

Fig. 4.21. Data of Marshall and Ward (1937).

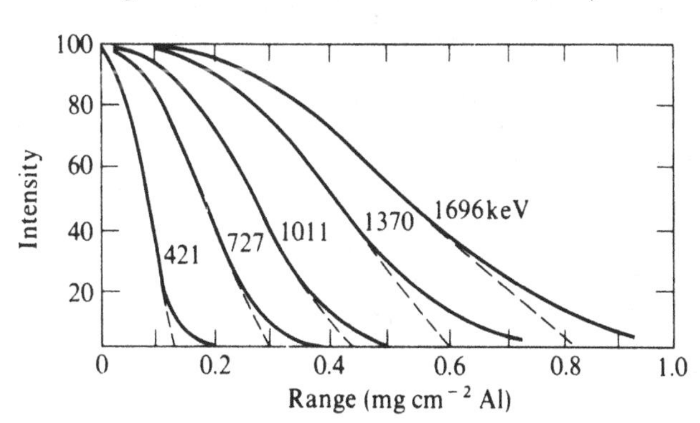

current is recorded, precautions must be taken to prevent the escape of secondary electrons. This may be achieved by placing a 'suppressor' grid, connected to a negative potential just before the collector; or the collector may be surrounded by a shield, (and electrically connected to it), the assembly being raised to a *positive* potential to ensure collection of the secondary particles.

The 'counter' method can be extended to higher energies. Hereford and Swann (1950) worked with an electron beam of 3–12 MeV and used three detectors, (two on the 'beam' side of the absorber, and one on the remote side) in a triple coincidence arrangement. By comparing the triple coincidence rate with the double coincidence rate (from the two detectors nearest the source of electrons), the fraction of transmitted electrons was obtained.

An alternative way of detecting the electrons is to deposit a layer of phosphorescent material on the far side of the absorber, and to record with a photomultiplier the light produced by the emergent electrons. The data of Young (1956a) and Feldman (1960) were obtained by this method. The main precaution to be observed here is to make a correction, by means of a subsidiary experiment, for the 'dead' thickness of the phosphor, which is a thin layer (adjacent to the absorber) insensitive to electrons. A set of electron transmission curves is shown in fig. 4.22. The linear part of the curves may be extrapolated to give an 'extrapolated energy' corresponding to the onset of emergence from the far side. A moment's reflection will lead to the conclusion that the relation between foil thickness and extrapolated energy will be closely similar to the relation between the incident energy and extrapolated range determined from data of the type illustrated in fig. 4.21. This is borne out by an examination of the work of Lane and

Fig. 4.22. Electron transmission curves (films of aluminium oxide) (Young, 1956b).

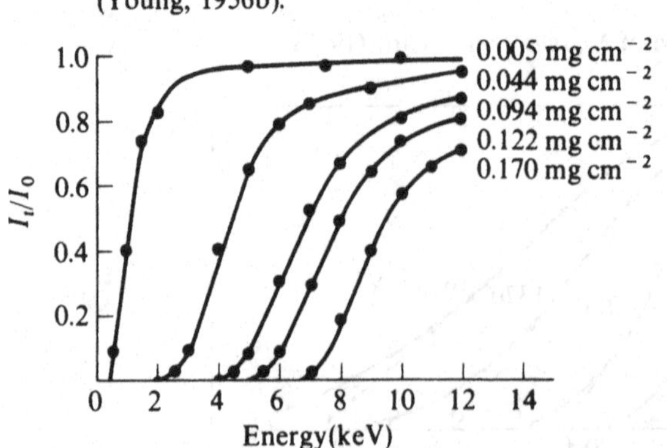

Zaffarano (1954) who obtained a set of electron transmission curves as a function of energy and then replotted the data to give absorption curves as a function of foil thickness. The extrapolated range/energy relation which they obtain is very similar to the curves obtained by plotting the extrapolated energy of their original transmission curves against foil thickness.

A third method uses transparent luminescent materials as the slowing-down medium. Koller and Alden (1951) deposited a thin layer of zinc sulphide on glass and used this as the luminescent screen in a demountable cathode-ray tube. As the voltage was increased the brightness of the luminescence rose approximately linearly at first and the voltage at which a flattening-off became detectable was taken as the voltage at which electrons began to emerge from the phosphor into the glass. This method clearly determines the *maximum range* of electrons in the phosphor.

Ehrenberg and co-workers (1953, 1963) have photographed the luminous cloud produced when a beam of electrons was allowed to slow down and diffuse in a luminescent medium. The results of experiments of this kind cannot be compared directly with data obtained by other methods, but are of interest, especially in view of the wide range of atomic numbers in the materials investigated by them.

A further class of studies consists in measuring the energy distribution of electrons emerging from a foil. This type of experiment provides information on electron straggling, and from an energy distribution of this type (fig. 4.23) the *mean energy loss* and the *most probable energy loss* can be determined as a function of initial energy and absorber thickness. Some

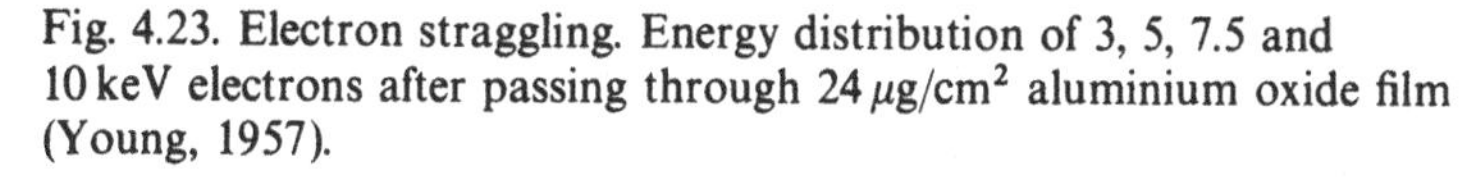
Fig. 4.23. Electron straggling. Energy distribution of 3, 5, 7.5 and 10 keV electrons after passing through 24 μg/cm^2 aluminium oxide film (Young, 1957).

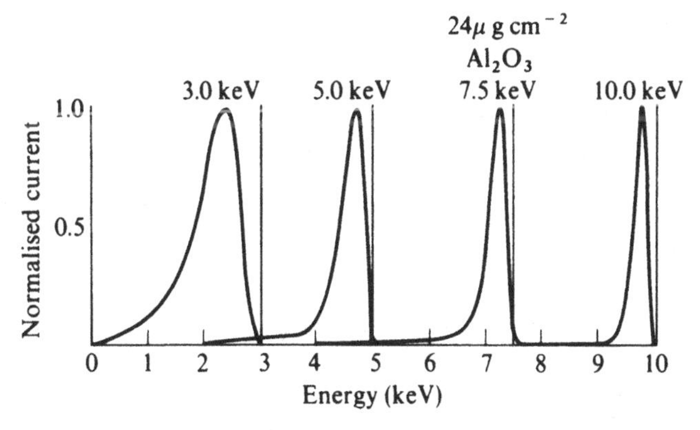

form of energy analyser is required for this type of experiment: it may be either a magnetic spectrograph (Hanson *et al.*, 1952), or an electrostatic analyser (Young, 1957). The distinction between the mean and the most probable energy-loss is important. In general the electrons lose energy in small amounts at a time, as a result of collisions which are generally of rather large impact parameter. Occasionally, an electron can lose a major part of its energy in one nearly head-on collision, which causes a long 'tail' to appear on the straggling curves.

The technique which yields the greatest amount of information is that using the cloud-chamber, because in this case complete trajectories of electrons are obtained from which the path length and the penetration in chosen direction can be simultaneously obtained. An experiment of this kind was carried out by Williams in 1931, in which electrons with energies up to 20 keV entered a cloud chamber filled with air or argon at atmospheric pressure. This experiment enabled a comparison to be made between path length and penetration, or 'range'. The graph of fig. 4.24 shows how considerable this difference can be. A cloud-chamber study has been carried out by O'Neill and Scott (1950) using a high pressure cloud-chamber (filled with hydrogen or helium) operated at 136 atmospheres. A magnetic field was applied and the electrons therefore produced curved tracks, spiralling inwards as they slowed down. The curvature at any point is a measure of the momentum at that point along the track, so it follows that each track can provide a complete range/momentum (or range/energy) relation-

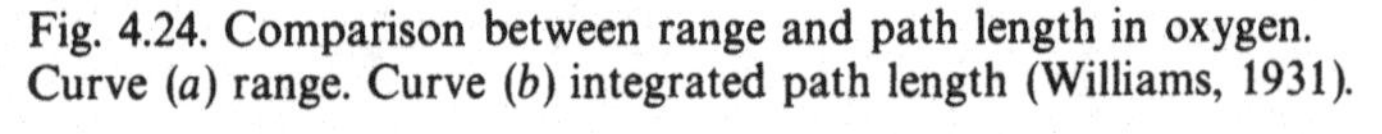
Fig. 4.24. Comparison between range and path length in oxygen. Curve (*a*) range. Curve (*b*) integrated path length (Williams, 1931).

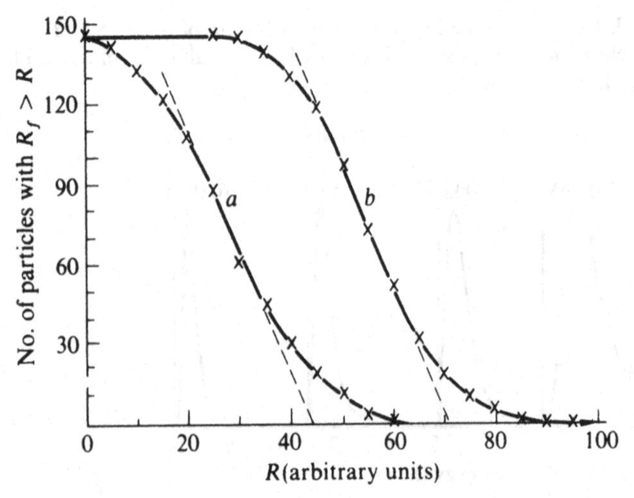

ship. In this way a range/energy relation for electrons in these two gases was determined between 81 keV and 700 keV.

One of the reasons why a knowledge of electron penetration is necessary in X-ray work is to facilitate calculation of the fraction of emitted X-rays which emerge from the surface of the target; a direct method of determining the mean depth of production of X-rays is therefore of particular interest to us, and is conveniently discussed here. Characteristic X-rays are emitted isotropically from the point of production, but the path length of the X-rays within the target depends on the take-off angle, θ, and this modifies the isotropy to an extent which may be considerable at small angles to the surface of the target. Green (1964) has measured the angular distribution of characteristic X-rays, and analyses his results as follows:

If all the radiation were produced at a depth x_m we could write

$$I(\theta)=I_0 e^{-\mu x_m \operatorname{cosec}\theta} \tag{4.19}$$

and

$$\ln I(\theta)=-\mu x_m \operatorname{cosec}\theta+\ln I_0 \tag{4.20}$$

If we plot $\ln I$ against $\operatorname{cosec}\theta$, the extrapolation to $\operatorname{cosec}\theta=0$ provides the value of I_0, and the slope of the graph (ideally a straight line) gives a value for μx_m from which the mean depth of production can be determined if the X-ray attenuation coefficient is known. This mean depth is very relevant to the production of X-rays and their attenuation on emerging, but cannot readily be related to other depth parameters. We have already illustrated mean depth as a function of energy in fig. 3.28.

(b) *Electron backscattering and secondary emission*

Although the phenomena of backscattering and secondary emission are quite distinct in origin, they both give rise to somewhat similar effects at the target, and the experimental study of the two phenomena follow somewhat similar lines. In a chapter devoted to experimental methods it is therefore not inappropriate to treat them together.

The backscattering of electrons is a parameter which we introduced in section 3.5(b) to calculate the intensity of characteristic X-rays produced by electron bombardment. To measure this quantity the incident electron beam must be measured carefully using a Faraday cup, and the sample then placed in the beam and the new (lower) current measured. The main problem here is to distinguish between the backscattered electrons and electrons released from the surface of the sample by secondary emission. Backscattered electrons are electrons which carry an appreciable (sometimes a large) amount of kinetic energy ultimately derived from the incident beam, and extend in energy down to indefinitely low values. Secondary electrons carry very little energy and are released from the top surface layer of the order of 2 nm in thickness. The dividing line between backscattered electrons and secondary electrons is thus rather arbitrary, and is conventionally taken at 50 eV. Fig. 4.25 shows a spectrum of electrons of both groups for electrons of 150 eV incident on a gold surface (Rudberg, 1936). The group of secondary electrons is seen to constitute a large proportion of electrons leaving the surface.

When measuring the backscattered electrons it is necessary to prevent the secondary electrons from leaving the target. To achieve this a 'suppressor' grid may be fitted around the target, at a potential of 50 volts negative with respect to the target, the latter being earthed. Alternatively, the target may be raised to 50 volts above earth, in order to attract secondary electrons back to the target. This method is preferable, and its use is described by Bishop (1966) who obtained the data of table 4.2. Bishop has also made measurements of the energy distribution of these backscattered electrons, and his paper may be consulted for details of this work.

The study of secondary emission makes use of techniques which are rather similar to those just outlined. Indeed, many studies of secondary emission fail to distinguish between these two phenomena, and the experiments consist of measuring the incident current, and then interposing the sample (usually in the form of a metallic foil) and measuring the current to it. The difference is the amount of current backscattered, reflected, or emitted as secondary electrons. Very many investigations of secondary emission have been reported in the literature, much of this work being

Fig. 4.25. Secondary emission and backscattering from gold - energy distribution (McKay 1948, after Rudberg 1936).

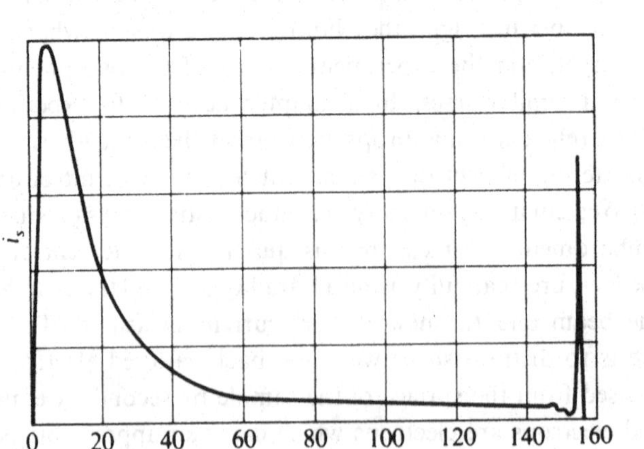

Table 4.2 *Backscattering coefficient of elements* (× *100*). E_0 *is the incident electron energy in* keV (*Bishop, 1966*)

Elements	Z	E_0		
		30	10	5
Carbon	6	6.0	7.2	8.5
Aluminium	13	15.5	17.7	18.6
Silicon	14	16.2	18.6	19.7
Titanium	22	25.4	26.8	27.0
Chromium	24	27.0	28.3	28.5
Iron	26	28.8	29.6	30.0
Nickel	28	30.8	32.3	33.3
Copper	29	31.9	33.9	35.2
Zinc	30	33.0	34.2	35.2
Germanium	32	33.4	34.9	36.2
Molybdenum	42	38.5	38.1	36.7
Silver	47	42.0	42.0	41.8
Tungsten	74	50.1	48.3	47.2
Platinum	78	51.6	50.3	48.6
Gold	79	52.1	50.1	48.9
Uranium	92	53.4	51.3	49.5

designed to elucidate the properties of thermionic valves and photomultipliers. For this reason, the majority of the work has been confined to incident electrons of low energy – a few hundred electron-volts or less. Many of the materials investigated are of a nature not normally required for use as the targets of X-ray tubes, for example, metal oxides. But studies of metallic emitters, at somewhat higher voltages, have been carried out and are directly relevant to our present purpose.

Secondary emission is normally characterised by the *secondary emission coefficient* γ, defined as the ratio of the number of electrons released from the surface divided by the number of incident particles. It rises as the energy of the incident electrons is increased, and then falls again for incident energies greater than a few hundred volts. The maximum value of γ is often in the region of 0.5–1.0, and for certain materials may considerably exceed this. Secondary emission can thus greatly modify the net recorded electron current flowing in the anode circuit of an X-ray tube, unless steps are taken to allow for this. The energy of secondary electrons is a few electron volts only, and the surface layer from which they emerge is only a few nanometres in thickness. Secondary emission is thus very sensitive to the surface state; if a pure metal surface is to be studied, oxide layers have to be avoided by

Fig. 4.26. Secondary emission coefficients for silver and copper (Bruining and de Boer, 1938).

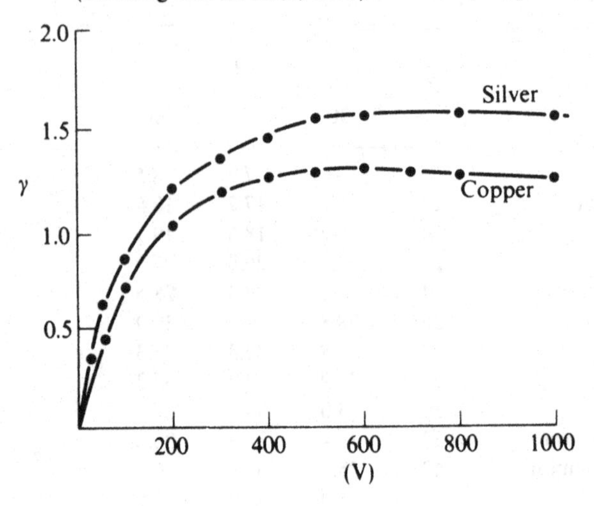

Fig. 4.27. Secondary emission at higher energies (Trump and Van de Graaff, 1947).

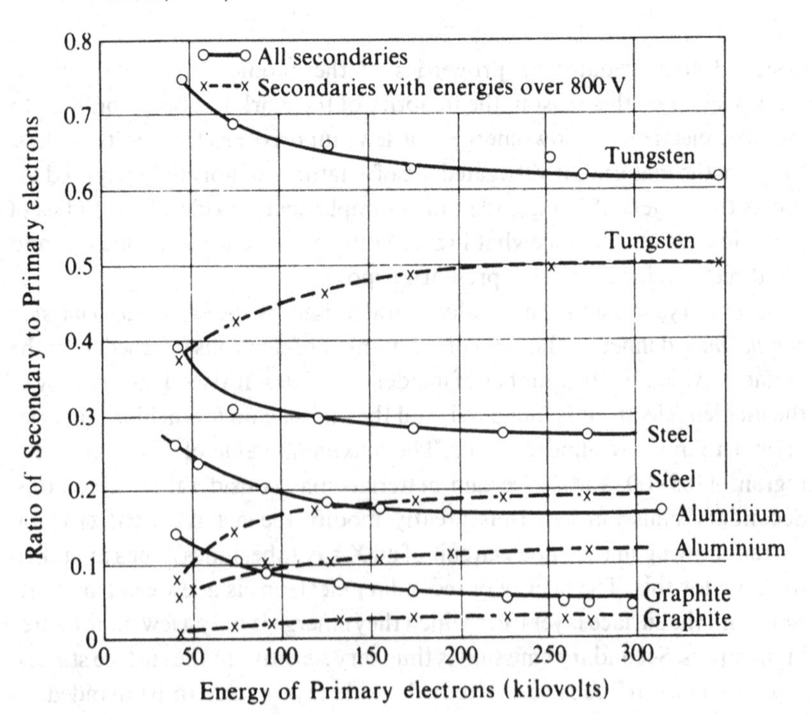

careful cleaning and outgassing of the evacuated vessel in which the measurements are carried out. The secondary emission coefficient for silver and copper is shown in fig. 4.26 as a function of incident electron energy, (Bruining and de Boer, 1938). The coefficient can be seen to exceed unity for incident electrons with energies above 200 eV. At higher energies we may quote the work (fig. 4.27) of Trump and Van de Graaff. The secondary emission falls gradually but the yield of inelastically scattered electrons (in this case with energies in excess of 800 eV) is seen to rise steadily as the incident energy is increased.

If the incident particles are protons or positive ions, the phenomena associated with secondary emission become more varied, we can distinguish between the 'reflected particle yield', the 'sputtering yield' (this being the proportion of *target* atoms which are released from the surface), and the 'secondary electron yield'. The yield of secondary electrons under proton bombardment is of interest in the present context, in view of the growing interest in the production of Bremsstrahlung-free characteristic radiation under these conditions (chapter 6). Fig. 4.28 shows data relating to secondary electron emission from a molybdenum target bombarded by protons.

The subject of secondary emission has been reviewed in detail by McKay (1948) and by Medved and Strausser (1965). The latter authors review particularly the production under ion bombardment.

Fig. 4.28. Secondary electron emission yield from proton bombardment of a molybdenum target (Large and Whitlock, 1962).

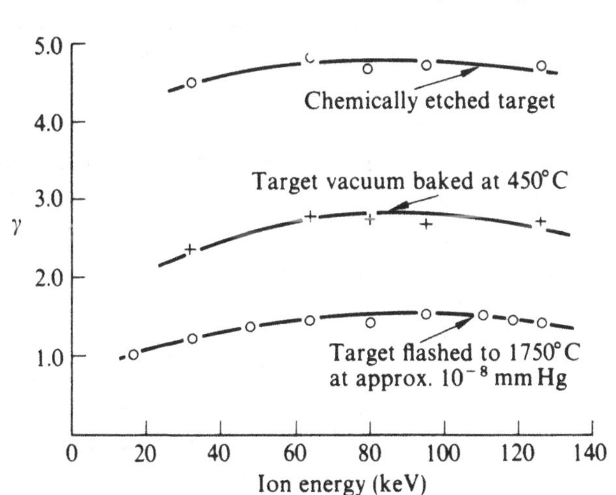

(c) *The fluorescence yield and the Auger effect*

Experimental methods for the study of these phenomena are numerous. First we refer to the cloud-chamber method. A beam of X-rays is allowed to enter a cloud-chamber containing the gas under investigation, and the tracks which are produced following photoelectric absorption are examined. If an ionized atom deionizes by X-ray emission, no further track will be produced at this point, but if decay occurs by Auger emission a second track will start from the same point. By comparing the number of pairs with the total number (pairs + individuals) the proportion of atoms decaying by Auger transitions can be determined. In the case of light elements up to the region of $Z = 18$, Auger tracks are very short, and the gas is often diluted with hydrogen to increase the path length of electrons. At higher atomic numbers the tracks are longer, because of the higher binding energy of the K electrons, but the situation becomes more complex because of the occurrence of multiple tracks caused by 2 or more Auger processes taking place in the same atom. The cloud-chamber method is one of the most accurate methods of determining the fluorescence yield, providing that the difficulties associated with the identification of tracks can be overcome.

Photographic emulsions provide another method of direct visualization of tracks. This technique has been used in the determination of fluorescence yield of radioactive elements. Astatine-211 decays by orbital electron capture to polonium-211, which in turn decays almost immediately (half-life 5×10^{-3} s), by α-particle emission. The number of α-tracks not associated with 59 keV Auger tracks divided by the total number of α-tracks is equal to the K-shell fluorescence yield.

The proportional counter provides a useful method of determining the fluorescence yield of gases. The absorption of X-rays in such a counter is primarily by the photoelectric effect, which results in either fluorescence or Auger emission. If the fluorescent X-ray escapes from the system, the pulse produced will be of reduced height, and if the fluorescence yield is high, the number of pulses in the escape peak may be comparable with those in the main peak. The size of the escape peak is proportional to the product of the fluorescence yield and the escape probability. The latter factor is essentially geometrical and can be calculated from the counter dimensions and the gas pressure. This method has provided values of the fluorescence yield for argon and other rare gases.

Direct methods of measuring the fluorescence yield involve measuring the intensity of the incident X-ray beam, introducing the material under examination into the beam, and measuring the intensity of the fluorescent radiation emerging under defined geometrical conditions. The radiation is emitted isotropically but the degree of attenuation on emerging will depend on the angle of emergence relative to the surface. The incident and fluorescent radiation are of different energies so it is essential to know the efficiency of the detector as a function of energy. The incident monoenergetic beam may be obtained either by Bragg reflection, by fluorescence from an intense primary source, or by the use of Ross filters. A major problem in measurements of this type is that the fluorescent beam from the material under investigation will be many times weaker than the direct beam, perhaps by a factor of 100. One method of overcoming this difficulty is to compare the fluorescent beam with radiation scattered from a standard sample which emits no fluorescent radiation of its own. As an example of this, we quote the work of Bailey and Swedlund (1967) in which the fluorescence yields of several elements were determined (fig. 4.29). The 'Ross filter' method was used to produce a monoenergetic primary beam which then impinged on the material under investigation, usually in the form of a metallic foil. The fluorescent radiation was detected by a proportional counter and was compared with the Compton scattered radiation from a helium-filled scattering chamber. It was necessary to assume a theoretical value for the scattering cross-section of helium, but this was thought to introduce an acceptably small error (less than 5%) into the results. A detailed account of the available methods of determining fluorescence yields has been given by Burhop (1952).

Fig. 4.29. Determination of fluorescence yield (Bailey and Swedlund, 1967).

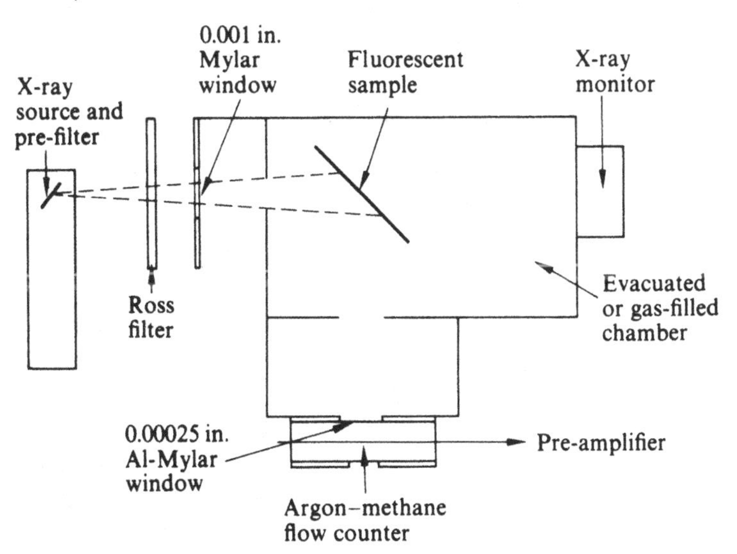

4.7 Polarization measurements

Several γ-ray interactions are polarization-sensitive and in principle any of these can be used as the basis of a γ-ray polarimeter. The photoelectric process is polarization-sensitive in that the photoelectrons are emitted preferentially along the direction of the polarization vector. The photodisintegration of the deuteron yields a distribution of the emitted neutron and proton which is polarization-sensitive. But the process used for polarimetry has almost invariably been the Compton Effect. For polarized radiation the Compton scattering cross-section depends on the direction of the electric vector of the incident radiation, the full expression being

$$d\sigma = \tfrac{1}{2} r_0^2 \left(\frac{\nu'}{\nu}\right) \left\{ \frac{\nu'}{\nu} + \frac{\nu}{\nu'} - 2\sin^2\delta\cos^2\Phi \right\} \delta\Omega \qquad (4.21)$$

where δ is the scattering angle and Φ the angle between the scattering plane and the plane containing the momentum of the incident photon and its electric vector* (fig. 4.30). The ratio R, of cross-sections for scattering with $\Phi = 90°$ and $\Phi = 0°$, known as the asymmetry ratio, will be greatest for scattering angles δ in the region of 90°, although a variation in δ causes a change in ν'/ν (5.9) so the optimum value can be obtained only by detailed examination of (4.21). Optimum values of δ are in practice somewhat less than 90°, and have been calculated by Metzger and Deusch (1950). For example, for incident photon energies of 0.511 MeV and 1.0 MeV, the optimum angles are 82° and 78° respectively.

For linearly polarized incident radiation (either fully or partially) the intensity of the scattered radiation N_Φ will vary as Φ is varied, and will be maximal when the scattering plane is perpendicular to the direction in which the average value of the electric field associated with the incident

* If $\cos^2\Phi$ is averaged over all directions, $\overline{\cos^2\Phi} = \tfrac{1}{2}$ and (4.21) reduces to (5.11b) for unpolarized radiation.

Fig. 4.30. Angles used in polarization work.

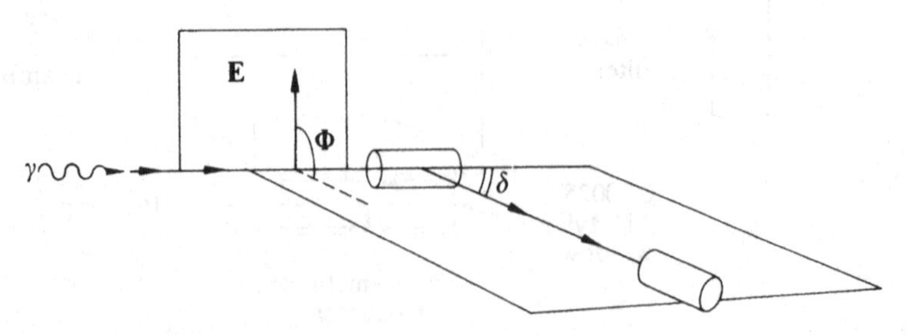

radiation is greatest. If the incident radiation be decomposed into intensities $J_{\parallel}$ and $J_{\perp}$ polarized respectively parallel and perpendicular to the direction of maximum electric field E, the ratio of scattered intensities for $\Phi=90$ to $\Phi=0$ will be given by

$$\frac{N_{90}}{N_0}=\frac{J_{\parallel}+J_{\perp}/R}{J_{\perp}+J_{\parallel}/R}$$

where R is the asymmetry ratio, from which we obtain

$$P=\frac{J_{\parallel}-J_{\perp}}{J_{\parallel}+J_{\perp}}=\frac{R+1}{R-1}\,\frac{N_{90}-N_0}{N_{90}+N_0} \tag{4.22}$$

where P is the polarization.

P and the direction of linear polarization may thus be obtained from measurements of counting rate as a function of Φ. Often the Compton scatterer is a sodium iodide crystal, so that the scattering events may be counted as coincidences between counters 1 and 2.

Most of the studies of the polarization of Bremsstrahlung have used this type of polarimeter, and it has also been used in the study of Rayleigh scattering (Sood, 1958).

Circular, or longitudinal, polarization of Bremsstrahlung has been observed when a target is bombarded with longitudinally polarized β-particles from a radioactive source. In order to detect this, use is made of the spin-dependent part of the Compton scattering cross-section – if a thick absorber of iron is magnetized longitudinally to technical saturation, alignment of the unpaired spins takes place, and this modifies the Compton scattering. If the field is then reversed, any change in transmitted γ-ray intensity will be indicative of a circularly polarized component in the incident radiation. Circularly polarized radiation produced under these circumstances was first observed by Goldhaber, Grodzins and Sunyar (1957) and an account of their method will be found in their paper. The work of Bisi and Zappa (1955) already referred to in section 2.9 used a similar method.

4.8 Radiation dosimetry

A discussion of experimental techniques in X-ray physics would be incomplete without a brief description of X-ray and γ-ray dosimetry. The purpose of measurements of this kind is to determine the amount of energy deposited in a unit mass of irradiated material. The determination of the so-called *radiological dose* is of paramount importance in medical radiotherapy, because the biological effects of X-radiation depend principally on the amount of energy deposited in the tissue by the ionizing radiation.* It is also important in radiobiological work and in radiation chemistry.

The S.I. unit of measurement of absorbed dose from a field of ionizing radiation is the Gray (Gy), which is defined as 1 Joule per kilogram of any specified material. For many years the *rad* ('radiation absorbed dose') was in extensive use, which is equal to one-hundredth of this quantity.

The measurement of absorbed dose may be achieved by measuring the current flowing in a suitably designed ionization chamber, and in order to understand this we need to consider the principles underlying the quantitative behaviour of such chambers.

Consider a sample of homogeneous solid material exposed to a beam of X-radiation. Transfer of energy will take place from the beam to the irradiated matter by means of the electrons released by the various interaction processes already considered. It will be seen that the electrons passing through (and depositing energy in) any small element of mass will originate not in that mass but in the material surrounding it. The intensity and energy distribution of the electrons depositing energy in the selected element of mass will thus rise to an equilibrium value as the amount of surrounding matter is increased, and any meaningful discussion of energy deposition will require that a state of equilibrium has been reached. This requirement is equivalent to a statement that the linear dimensions of the absorbing material shall be greater than the range of any of the electrons participating in the energy deposition.

If we introduce a cavity into the medium, it is our basic assertion that the energy distribution and angular distribution of the electrons will remain unchanged, if the dimensions of the cavity are sufficiently small. Let us now introduce a gas into the cavity. This will be ionized by electrons released in the surrounding medium, and the ionization may be measured by introducing electrodes and measuring the current flowing in an external circuit under the influence of a suitable applied potential difference.

Let the absorbed energy per unit volume in the gas and in the solid be denoted respectively by ${}_{\mathrm{v}}E_{\mathrm{g}}$ and ${}_{\mathrm{v}}E_{\mathrm{s}}$, and let the mean linear stopping powers of the electrons (i.e. averaged over all the electron energies during the slowing down process) be denoted by $(\mathrm{d}T/\mathrm{d}x)_{\mathrm{g}}$ and $(\mathrm{d}T/\mathrm{d}x)_{\mathrm{s}}$, respectively.

* It should be added that the other important parameter in determining the biological or chemical effect of a given energy absorption is the ionization per unit path length, or Linear Energy Transfer (LET), of secondary particles. Clearly this cannot be discussed without detailed consideration of the biological or chemical effect being studied, and so falls outside the scope of this book.

We may then write

$$\frac{{}_{\mathrm{v}}E_{\mathrm{s}}}{{}_{\mathrm{v}}E_{\mathrm{g}}}=\frac{(\mathrm{d}T/\mathrm{d}x)_{\mathrm{s}}}{(\mathrm{d}T/\mathrm{d}x)_{\mathrm{g}}}$$

If W is the mean energy required to produce an ion-pair and ${}_{\mathrm{v}}J$ the number of ion-pairs released per unit volume of gas

$${}_{\mathrm{v}}E_{\mathrm{g}}={}_{\mathrm{v}}JW$$

Hence

$${}_{\mathrm{v}}E_{\mathrm{s}}={}_{\mathrm{v}}JW\frac{(\mathrm{d}T/\mathrm{d}x)_{\mathrm{s}}}{(\mathrm{d}T/\mathrm{d}x)_{\mathrm{g}}}$$

Introducing the densities ρ_{g} and ρ_{s}, and the energies absorbed per unit mass ${}_{\mathrm{m}}E_{\mathrm{g}}$ and ${}_{\mathrm{m}}E_{\mathrm{s}}$, we obtain

$${}_{\mathrm{m}}E_{\mathrm{s}}={}_{\mathrm{m}}JW\frac{(\mathrm{d}T/\mathrm{d}(\rho x))_{\mathrm{s}}}{(\mathrm{d}T/\mathrm{d}(\rho x))_{\mathrm{g}}} \tag{4.23}$$

${}_{\mathrm{m}}E_{\mathrm{s}}$ is the absorbed radiation dose in the medium, and (4.23) is the formal statement of the *Bragg–Gray theorem*. ${}_{\mathrm{m}}J$ can be equated to $\frac{Q}{eM}$, where Q is the total charge collected, and M is the total mass of gas in the cavity. We are thus able to deduce the absorbed dose *in the medium* from measurements made of ionization current *in the gas*.

The fundamental assumptions made in the course of the above argument were as follows:

(a) the electron spectrum is unaffected by the presence of the cavity, i.e., the energy-loss experienced by any electron crossing the cavity is small, and the dimensions of the cavity are small compared with the range of the electrons in the gas
(b) photon interactions generating electrons *in the gas* are negligible
(c) the spatial photon flux around the cavity is uniform.

The Bragg–Gray equation therefore enables the absorbed dose to be experimentally determined in any material from which the walls of an ionization chamber can be made. By an extension of this treatment, a chamber may be used to determine the absorbed dose in any other material (for example water, biological tissue or chemical systems) provided that the ratio of the *mass energy-absorption coefficients* for the chamber walls and the material being investigated is known.

We may note here that the mass energy-absorption coefficient of a medium is of paramount importance in the field of radiation dosimetry, expressing, as it does, the energy absorbed from the radiation field. This quantity is of course quite different from the *mass attenuation coefficient* of the medium. The distinction is brought out in chapter 5.

Cavity ionization theory was developed originally for the accurate measurement and control of the dose delivered to the tissues in radiotherapy. To achieve this, the *Roentgen* was introduced which is a unit of *ionization in air.* (It is not defined for any other gas.) Its definition has been reworded from time to time, and for many years was the amount of radiation flux necessary to produce 1 electrostatic unit of charge of either sign in 0.001293 gram of dry air at STP. In terms of S.I. units this is approximately 2.58×10^{-4} C Kg^{-1} and the Roentgen has now been redefined as 2.5800×10^{-4} C Kg^{-1} exactly, and is a unit of the physical quantity known as *exposure.*

To assist in the measurement of exposure, small 'air-wall' ionization chambers are used in which the walls are made of bakelite or some other material with the same mass stopping power as air, i.e., the ratio in (4.23) is unity. Such a chamber is termed a *homogeneous* chamber, and has the advantage that the restriction on small size can be lifted.

If W is given in electron-volts per ion-pair, as is customary, an exposure of 1 Roentgen corresponds to an absorbed radiation dose of

$$\frac{2.5800 \times 10^{-4}\, W}{1.602 \times 10^{-19}}\,\text{eV kg}^{-1}$$

or

$$2.5800 \times 10^{-4}\, W\,\text{J kg}^{-1}$$

W for air may be taken as 33.7 eV per ion-pair, leading to the conversion factor

$$1\,\text{R} \equiv 0.869 \times 10^{-2}\,\text{Gy in air.} \tag{4.24}$$

To determine the dose delivered to biological tissue, this quantity must be multiplied by the ratio of the mass energy-absorption coefficients of tissue and air. Over a wide range of photon energies (100 keV–5 MeV) this is nearly constant, and is given by the ratio of Z/A for the respective media, because in these circumstances the absorption of energy is almost entirely due to the Compton process. For muscle tissue for example, this ratio may be taken as 1.1, giving a conversion factor

$$1\,\text{R} \equiv 0.956 \times 10^{-2}\,\text{Gy in tissue.*} \tag{4.25}$$

Outside this energy region other absorption processes become important and the energy absorption coefficients cannot be calculated so simply. The determination of absorbed dose (from, for example, a beam of high energy Bremsstrahlung) then presents special problems and for a detailed discussion of these problems standard reference works on radiation dosimetry should be consulted, for example, Attix, Roesch, and Tochilin (1966–9). This work includes reviews of ionization chamber principles (Boag, 1966) and cavity chamber theory (Burlin, 1968).

* More accurately (after Attix *et al.*, 1966–9):

1 R of X- or γ-radiation produces in soft tissue (muscle) under charged particle equilibrium conditions:

0.951×10^{-2} Gy at 0.1 MeV
0.960×10^{-2} Gy at 0.3 MeV
0.956×10^{-2} Gy at 1.0 MeV
0.953×10^{-2} Gy at 3.0 MeV
($W_{air} = 33.7$ eV/ion-pair)

Fig. 4.31. Dose rate as a function of photon energy. The units are Roentgens hour^{-1} mCi^{-1} at 1 cm in air, and the graph is based on 1 (unconverted) photon per disintegration.

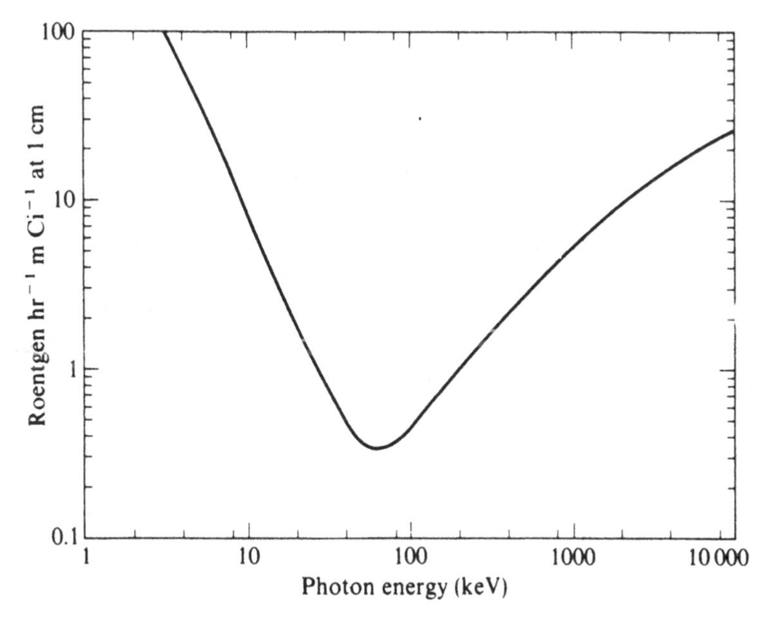

The ionization current recorded in a cavity chamber will be reduced by the attenuation of X-ray in the chamber walls, and this conflicts with the desire to have the walls sufficiently thick for equilibrium to be achieved. In these circumstances the true dose can be determined by extrapolating to zero wall thickness. Finally, we should remark that the dose rate as a function of depth within the tissue or other material being irradiated will depend on the attenuation experienced by the beam, and also on the scattering-in of radiation. The latter is not readily calculable under the varying degrees of 'broad-beam' geometry used in radiotherapeutic practice, and is determined empirically by means of an ionization chamber immersed in a suitable medium (often water). Comprehensive sets of 'depth-dose curves' for many field sizes and radiation qualities are available (see figs. 5.15 and 5.16).

We have yet to consider, in this context, the relation between ionization in air and total energy flux. This will clearly depend on the energy absorption coefficient of air, and will be a function of photon energy. This is of considerable practical importance in that it enables the ionization associated with a given disintegration rate of a radioactive source to be determined.

The *specific gamma emission*, or '*k-factor*', (defined as the number of Roentgens per hour per mCi at 1 cm in air for a source yielding one photon per disintegration) can be calculated to be

$$1.95 \times 10^{-4} E_\gamma \left(\frac{\mu}{\rho}\right)_{\text{air}} \tag{4.26}$$

where $\left(\frac{\mu}{\rho}\right)$ is the mass energy-absorption coefficient in $\text{cm}^2\,\text{g}^{-1}$, and E_γ is in eV.

A graph of this quantity is given in fig. 4.31. The ionization at any distance from a radioactive source of known activity can be calculated from this graph if the decay scheme is known. A simple example may be given – ^{60}Co emits gamma radiation at 1.17 and 1.33 MeV, yielding one photon of each kind per disintegration. Its *k*-factor may be seen from the graph to be 13.3. If the radionuclide decays by electron capture or internal conversion, the contribution from characteristic X-rays may be considerable, and can be calculated in a similar way.

The specific gamma emission of a radionuclide is often quoted in milliroentgens $\text{hr}^{-1}\,\text{mCi}^{-1}$ at 1 metre, in which case the numerical values are one-tenth of the values appropriate to the units of fig. 4.31. It is also sometimes expressed in terms of absorbed dose in air, water, or biological tissue, rather than as ionization in air, this being useful in the assessment of radiotherapeutically administered doses. It is likely that the use of the roentgen (the unit of 'exposure') will become progressively less, in favour of units of absorbed dose, and in this connection it may again be noted that the rad can be expressed very simply in SI units (1 rad $= 10^{-2}$ J kg^{-1}) whereas the roentgen (1 R $= 2.58 \times 10^{-4}$ C kg^{-1}) does not lend itself to conversion so readily.

5

The absorption and scattering of X-rays

5.1 Absorption and scattering cross-sections

The interaction between electromagnetic radiation and matter represents one of the most varied classes of phenomena in the whole of experimental physics. Even within the range of energies normally associated with X-rays (itself covering several orders of magnitude of the electromagnetic spectrum) many different processes occur, all of which possess their own individual characteristics.

The nature of the matter with which the radiation interacts offers almost as wide a range of phenomena as does the nature of the radiation. This is true even within the relatively restricted domain of X-ray physics. For example, the subject of X-ray crystallography is essentially a study of the interactions between ordered matter and a radiation field, and any discussion of the absorption and scattering processes in crystals must have as its basis the collective behaviour of a large number of atoms bound by chemical bonds or other interatomic forces into a recognisable structure.

However, in the present work we are concerned mainly with situations in which the overall behaviour of an absorber or scatterer can be deduced by regarding it as a collection of individual atoms each absorbing or scattering independently of its surroundings. In such cases we can assert that interactions between X- or γ-ray photons and matter are single, identifiable, processes, each associated with an individual atom, and can therefore be characterised by a *cross-section.* Such an interaction may be primarily a scattering event, in which little or no energy is imparted to the atom in question or to any of the electrons associated with it; or it may be essentially an absorption process, in which the great majority of the energy of the photon will be transferred to the atom or to one of its electrons. We shall see that the removal of photons from the incident beam and the absorption of energy by the irradiated atoms are two distinct consequences of X-ray interactions, requiring careful distinctions to be made in the theoretical treatment. We shall speak mainly of the *removal cross-section* σ, from which the *mass attenuation coefficient* of the interacting medium can be defined as follows:

Suppose that a beam of the incident radiation of intensity I is reduced by an amount δI on passing through a thin homogeneous layer of thickness δx, consisting of one kind of atom only (fig. 5.1(a)). We may write $-\dfrac{\delta I}{I} = \sigma n \delta x$, where n is the number of atoms per unit volume and σ the removal cross-section per atom.

After integration, and assuming an incident intensity I_0, we obtain

$$I = I_0 e^{-\sigma n x}, \tag{5.1}$$

demonstrating the well-known experimental attenuation of a beam of X-rays passing through matter. The quantity σn is known as the *linear attenuation coefficient* μ, and may be written also as $\dfrac{\sigma N_A \rho}{A}$ when N_A is Avogadro's number, A the atomic weight, and ρ the density.

Fig. 5.1(a). To illustrate cross-sections. Thickness of slab $= \delta x$. No. of atoms in slab $= n\delta x$. Fraction of incident photons removed $= n\sigma\delta x$. (b) Geometry in an attenuation experiment.

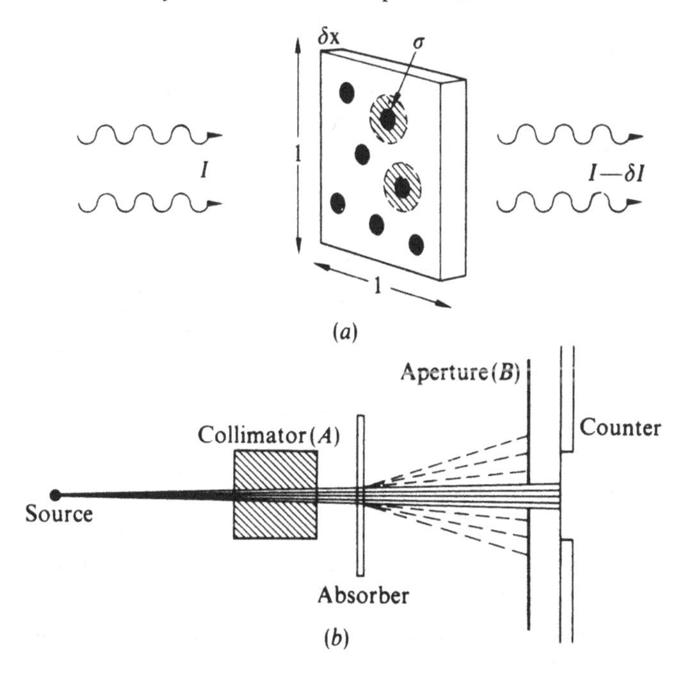

Because the interaction cross-section is unaffected by the density of the interacting medium, it is convenient to define the mass attenuation coefficient as $\frac{\sigma N_A}{A}$, or $\frac{\mu}{\rho}$. This is the form in which attenuation coefficients are commonly quoted. Because the interactions are also unaffected by chemical combination (except as discussed in section 8.2), we can write, for the removal cross-section of a molecule, $\sigma = \sum_i \sigma_i$, from which it follows that the mass attenuation coefficient of a compound may be obtained from its constituent elements by taking μ/ρ for each element, multiplying by the 'fractional mass' appropriate to each element, and adding:

$$\left(\frac{\mu}{\rho}\right)_{\text{cpd}} = \left(\frac{\mu}{\rho}\right)_A p\frac{A}{M} + \left(\frac{\mu}{\rho}\right)_B q\frac{B}{M} + \ldots \tag{5.2}$$

where M is the molecular weight of compound $A_pB_q \ldots$ and $A, B \ldots$ are the atomic weights of the respective elements. If only cross-sections are available we can write

$$\left(\frac{\mu}{\rho}\right)_{\text{cpd}} = \frac{N}{M}(p\sigma_A + q\sigma_B \ldots) \tag{5.3}$$

Similar arguments can of course be applied to mixtures or to alloys.

In this chapter we shall speak of the attenuation coefficient when referring to the fall of intensity as radiation passes through a material sample, and we link this with the 'removal cross-section' as already indicated. We shall often separate the removal cross-section into two 'partial' cross-sections σ_a and σ_s, which correspond to the proportions of energy absorbed and scattered respectively.

$$\sigma = \sigma_a + \sigma_s \tag{5.4}$$

We shall speak of the *energy absorption coefficient* when referring to the absorption of energy in the irradiated matter. In this way we preserve the necessary distinctions, although we may loosely speak of an 'absorber', even though a portion of the energy is known to be removed from the beam by a scattering process.

5.2 The photoelectric effect

This is one of the most important interactions in the energy region of 1–100 keV, and consists of the removal of a bound electron from an atom in the absorber. In general the inner electrons are the predominant contributors to the photoelectric interaction, so we speak of interaction (or 'absorption') in the K, L, M shell etc. This is subject to the overriding consideration that the incident photon energy must be greater than the binding energy of the electron in question, if photoelectric absorption is to take place.

Photoelectric absorption occurs most readily if the binding energy is comparable with the photon energy, and has certain of the qualities of a resonance process. For example, it decreases with increasing photon energy, corresponding to the physical notion that there is a decreasing probability of the electron being found in a field sufficiently strong for the resonance to take place (Grodstein, 1957). If we suppose that the Coulomb force on the electron must be of order C or greater, where $\omega = \sqrt{\frac{C}{m}}$ (angular frequency for resonance of a classical harmonic oscillator of mass m and restoring force C), the inverse square law for the Coulomb force leads to the conclusion that the maximum distance from the nucleus for photoelectric interaction to occur for photon energy $h\omega$ must be equal to $(Ze^2)^{\frac{1}{2}}(m\omega^2)^{-\frac{1}{2}}$; the *volume* surrounding the nucleus must be proportional to the cube of this, or ω^{-3}. This inverse variation of the photoelectric cross-section with the cube of the photon energy is observed to be followed in practice. A simple extension of this argument, by comparing this volume with the volume enclosed by the K shell leads to a $Z^{\frac{9}{2}}$ dependence which is also observed to be approximately true in practice.

For photon energies which are large compared with the binding energy, Heitler (1954) obtains

$$\tau_K = \sigma_R 4\sqrt{2}\frac{Z^5}{137^4}\left(\frac{mc^2}{h\nu}\right)^{7/2} \tag{5.5}$$

where $\sigma_R = \frac{8\pi}{3} r_e^2$, where r_e is the classical electron radius and the other symbols have their usual meanings. We shall see that σ_R, which has the dimensions of a cross-section, is important in the theory of the elastic scattering of γ-radiation.

At energies such that $h\nu \gg mc^2$, the photoelectric cross-section varies much less rapidly than $h\nu^{-\frac{7}{2}}$, and

$$\tau_K = \tfrac{3}{2}\sigma_R \frac{Z^5}{137^4}\frac{mc^2}{h\nu} \tag{5.6}$$

The variation of photoelectric cross-section with photon energy is illustrated in fig. 5.2.

The K shell makes the greatest contribution to the photoelectric absorption, followed by the L, M, N shells etc., in decreasing order of importance. If the photon energy is less than the binding energy of the K shell, ejection of the K electrons is no longer possible, and only the L, M . . . shells can interact in this way. The photoelectric absorption cross-section therefore falls sharply as the photon energy decreases through this critical value, the magnitude of the K-jump, or K-absorption discontinuity, being in the region of 7–8 for elements of medium atomic number. Similar discontinuities occur, though smaller in magnitude, at the binding energies of the L and subsequent electrons. Within the L and M shells it would be expected that the L_1 and M_1 electrons would make a much greater contribution than any of the other L or M sub-shells, because of the small

Fig. 5.2. Photoelectric cross-section as a function of energy. The dotted lines have slopes corresponding to E^{-1} and $E^{-\frac{7}{2}}$ laws.

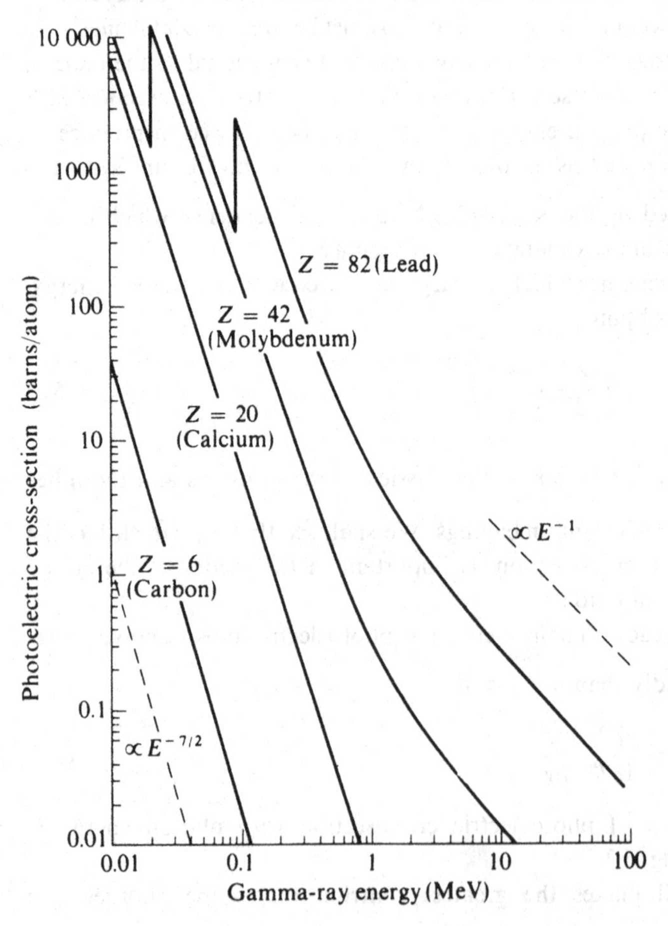

probability of the other electrons being found in the vicinity of the nucleus. This is found to be true in regions far from the L and M edges, but in the proximity of the L absorption edges the resonant nature of the absorption process becomes apparent, and the L_2 and L_3 sub-shells assume a correspondingly greater importance. The same is true for the M_{2-5} sub-shells in the region of the M absorption edges, though these have not been studied in detail.

In practice, simple expressions such as those quoted above do not lead to predictions sufficiently accurate to be used in the calculation of photoelectric absorption cross-sections for practical use. The variation with Z is in fact slightly less than Z^5 even at high energies. At lower energies the variation is nearer Z^4. The $(h\nu)^{-3}$ variation at relatively low energies becomes progressively less rapid at high energies. We illustrate the variation with Z in fig. 5.3(a), from which it may be seen that the variation with Z is approximately as Z^4.

The exact law of variation, Z^n, can be found by measuring the slope of the data of fig. 5.3(b) and noting that $n = g + 4$ for this method of plotting, where g is the slope of the curves in this diagram.

The theory of the photoelectric effect has been reviewed by Hall (1936), and calculations have been given by Grodstein (1957). These calculations have been reproduced in a conveniently accessible form by Davisson (1965). When photoelectric absorption occurs, an energy $h\nu - E_K$ appears as kinetic energy of the ejected electron, the remainder appearing as fluorescent X-rays or Auger electrons. A commonly occurring situation is one in which the Auger electrons are brought to rest in the absorbing medium (because of their short range) and the fluorescent X-rays escape. The average absorbed energy per incident photon is then $h\nu - E_{K_\alpha}\omega_K$. In terms of the partial cross-sections for scattering and absorption we may write

$$\tau = \tau_s + \tau_a$$

$$\frac{\tau_s}{\tau_a} = \frac{E_{K\alpha}\omega_K}{h\nu - E_{K\alpha}\omega_K} \tag{5.7}$$

Clearly, if the absorber is sufficiently thick, the fluorescent radiation will be reabsorbed. In practical situations where knowledge is required of the energy absorption in organic or biological material, the quantity $E_{K\alpha}\omega_K$ is usually small compared with $h\nu$, in which case the energy removed from the beam may be assumed to be transferred entirely to the absorbing medium. However, practical situations do exist (for example in the design of shielding for radiation experiments) in which the fluorescent radiation from a heavy element (e.g. lead or tungsten) is produced in significant amounts and which would contribute undesirably to any X-ray or γ-ray spectrum being investigated.

The fluorescence yield has been discussed in chapter 3, and a graph of the yield in the *K* shell as a function of atomic number is given in fig. 3.7.

The greater part of the momentum of the incident photon is taken up by the interacting nucleus, and is not necessarily carried by the photoelectron. The direction of emission of the photoelectrons is in fact predominantly at right-angles to the incident photons, and is determined mainly by the

Fig. 5.3(*a*). Photoelectric cross-section ($K+L+M$) as a function of Z.

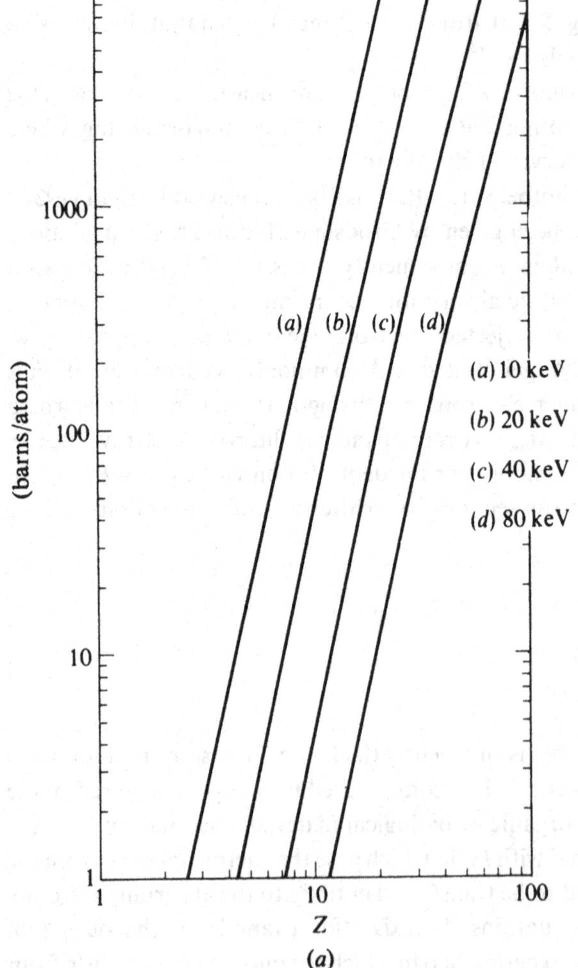

direction of the electric field in the incident radiation. That is, if the incident radiation is linearly polarized, the photoelectron is emitted with a large component of its velocity parallel to the direction of the electric field of the incident radiation. At higher energies, however, the photoelectron acquires a forward component of momentum as required by conservation of the momentum of the incident radiation. The angular distribution of photoelectrons for two values of incident photon energy is shown in fig. 5.4.

The detailed study of the intensities of photo and Auger electrons has become important in connection with electron spectroscopy for chemical

Fig. 5.3(*b*). τ/Z^4 at 10, 20, 40 and 80 keV.

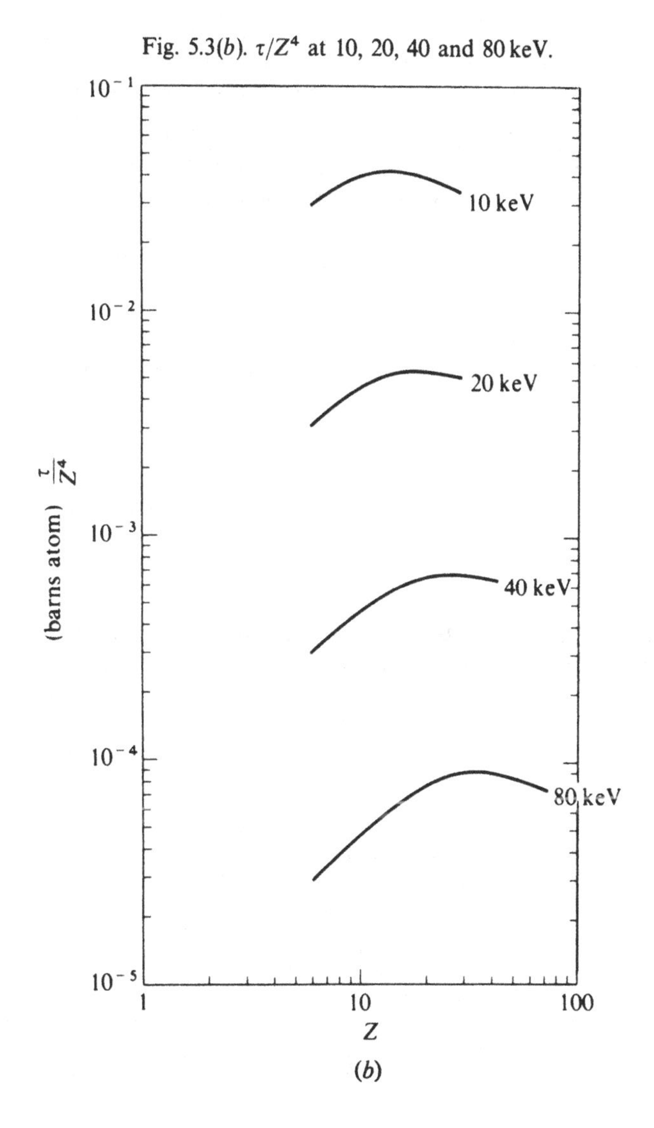

analysis (or ESCA). In this type of investigation the sample is irradiated with X-radiation of known energy and the intensities and energies of electrons emitted from the surface of the sample are measured. From the results, identification of elements and the determination of their concentrations is possible.

For a study of photo and Auger emission from the surface of absorbers under fluorescent X-ray excitation, the paper by Eliseenko *et al.* (1967) may be consulted.

5.3 The Compton effect

The second interaction which we shall consider is that in which the electron may be considered free, and not bound to an atom.

This situation is more amenable to exact theoretical treatment than the photoelectric effect, because we can now apply the principle of the conservation of momentum to the electron, and the incident and scattered photon in order to ascertain the division of energy between photon and electron.

From the conservation of energy we can write

$$h\nu - h\nu' = T$$

where T is the kinetic energy of the ejected electron and is given by the relativistic expression

$$T(T + 2m_0c^2) = (pc)^2$$

where m_0 is, as before, the rest mass of the electron and p its momentum.

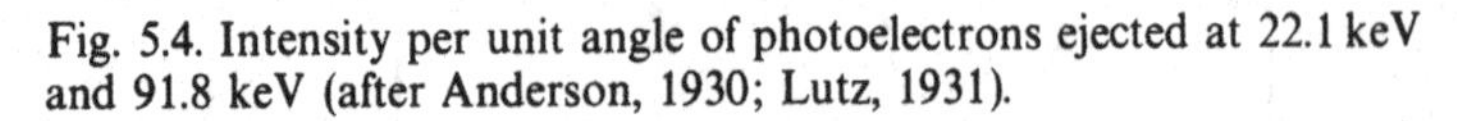

Fig. 5.4. Intensity per unit angle of photoelectrons ejected at 22.1 keV and 91.8 keV (after Anderson, 1930; Lutz, 1931).

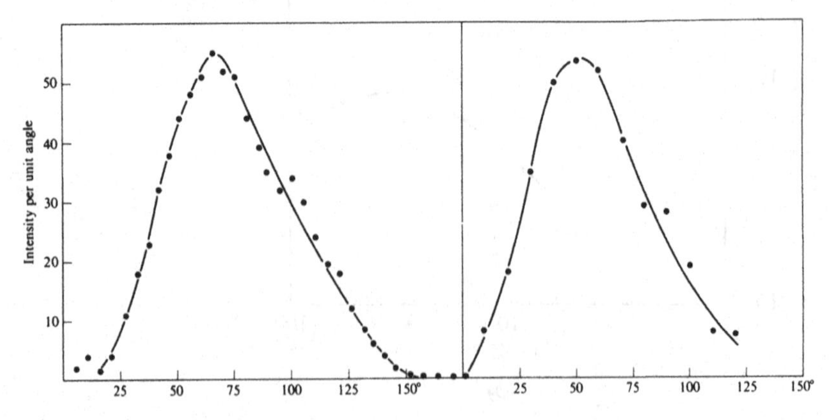

From the conservation of momentum we have (fig. 5.5)

$$\frac{h\nu}{c}=p\cos\phi+\frac{h\nu'}{c}\cos\theta$$

and

$$0=p\sin\phi-\frac{h\nu'}{c}\sin\theta$$

From these equations, by eliminating p and ϕ, we obtain the expression for the energy of the scattered photon

$$h\nu=\frac{h\nu}{1+\dfrac{h\nu}{m_0c^2}(1-\cos\theta)} \tag{5.8}$$

in terms of the photon scattering angle θ.

Clearly Compton scattering invariably results in a photon of reduced energy being produced. In Compton scattering, therefore, we must distinguish clearly between σ_a and σ_s, the partial cross-sections for absorption of energy and for scattering.

It should be noted, however, that these partial cross-sections cannot be used in quite the same way as total cross-sections, and that the energy absorption cross-section cannot be used without reference to the other cross-sections involved if the energy absorbed in an absorber of finite thickness is being calculated. The absorbed energy is in fact given by

$$I_{\text{abs}}=I_0(1-e^{-\sigma_t nx})\frac{\sigma_a}{\sigma_t}$$

not

$$I_{\text{abs}}=I_0(1-e^{-\sigma_a nx})$$

Fig. 5.5. Illustrating Compton scattering.

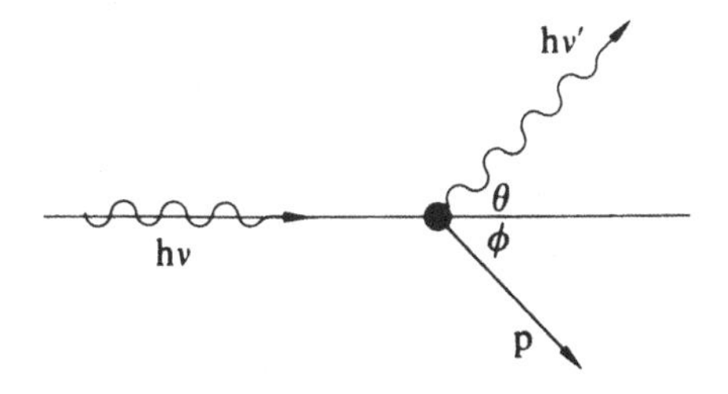

For thin absorbers (such that $\sigma_t nx \ll 1$) these expressions do however approach equality.

It is never possible for a free electron to absorb the whole of the energy of an incident photon, because to conserve momentum, the electron would always require more momentum than the photon can supply. In the limiting case of a 'head-on' collision, a scattered photon must emerge backwards in order to enable a balance of momentum to be achieved.

Equation (5.8) can be written conveniently in the form

$$\frac{c}{\nu'} - \frac{c}{\nu} = \lambda' - \lambda = \frac{h}{m_0 c}(1 - \cos\theta) \tag{5.9}$$

where $\dfrac{h}{m_0 c}$ is the 'Compton wavelength' and is equal to 2.426×10^{-12} m.

The change in *wavelength* for a given angle of scattering is thus independent of the incident photon energy, and this affords a ready method of calculating, for example, the energy of backscattered radiation, remembering that the wavelength and photon energy are related by the expression

$$E = h\nu = \frac{hc}{\lambda}$$

This becomes

$$E\,(\text{keV}) = \frac{1.2399}{\lambda}(\lambda \text{ in nm}) \tag{5.10}$$

It will be seen that for incident photon energies which are much less than m_0c^2 (511 keV), the change of energy on scattering through 180° is small, but at higher incident photon energies the change of energy becomes greater. For photon energies considerably greater than m_0c^2 the energy of backscattered radiation approaches a limiting value, which can be shown to be $\frac{1}{2}m_0c^2$, or approximately 256 keV.

The differential collision cross-section for unpolarized radiation is often given in the form

$$d\sigma = \frac{r_e^2}{2}\left\{\frac{1}{\{1+\alpha(1-\cos\theta)\}^2}\left[1+\cos^2\theta + \frac{\alpha^2(1-\cos\theta)^2}{1+\alpha(1-\cos\theta)}\right]\right\}d\Omega \tag{5.11a}$$

which may be reduced by means of (5.8) to

$$d\sigma = \frac{r_e^2}{2}\left(\frac{\nu'}{\nu}\right)^2\left(\frac{\nu}{\nu'} + \frac{\nu'}{\nu} - \sin^2\theta\right)d\Omega \tag{5.11b}$$

for scattering through an angle θ.

In (5.11a), α is $h\nu/mc^2$, and r_e is the classical electron radius $(1/4\pi\varepsilon_0)(e^2/mc^2)$. This equation is discussed by, e.g., Davisson and Evans (1952) and Evans (1955), and is commonly referred to as the Klein–Nishina formula, after its joint discoverers. $d\sigma/d\Omega$ is illustrated as a function of α and θ in fig. 5.6(*a*), from which it is seen that the cross-section falls with increasing angle θ, and that for high values of α the fall-off occurs rapidly: the scattered radiation is strongly peaked in the forward direction.

To obtain the total collision cross-section, (5.11) has to be integrated. This quantity is directly relevant to our discussion of the attenuation coefficient, because it represents the probability that the incident photon be removed from the beam by a Compton interaction. The expression is

$$\sigma_t = 2\pi r_e^2\left\{\frac{1+\alpha}{\alpha^3}\left[\frac{2\alpha(1+\alpha)}{1+2\alpha} - \ln(1+2\alpha)\right] + \frac{\ln(1+2\alpha)}{2\alpha} - \frac{1+3\alpha}{(1+2\alpha)^2}\right\} \tag{5.12a}$$

which for $\alpha \ll 1$ can be reduced to

$$\sigma_t = \tfrac{8}{3}\pi r_e^2(1 - 2\alpha + 5.2\alpha^2 - 13.3\alpha^3 + 32.7\alpha^4 \ldots) \tag{5.12b}$$

The exact expression is illustrated in fig. 5.6(*b*) in the form of a cross-section per *electron*, as a function of incident photon energy. When this cross-section has been multiplied by Z, the number of electrons in the atom, it may then be added (along with the atomic cross-sections for further processes yet to be discussed) to τ, the photoelectric cross-section, in order to calculate the attenuation of a narrow beam of X-rays passing through an absorber.

To calculate the fraction of *energy* scattered out of the beam it is necessary to refer back to (5.11b) and multiply by $\frac{\nu'}{\nu}$ before integrating. This yields the quantity of σ_s in fig. 5.7. $\sigma_t - \sigma_s$ is equal to σ_a. The difference between σ_a and the total σ_t becomes less as the incident photon energy is increased, because increasing amounts of the energy are transferred to the electron, but at lower energies σ_a and σ_t differ considerably.

The cross-sections already discussed relate to unpolarized radiation and are obtained by integrating the expressions for linearly polarized radiation

Fig. 5.6(*a*). Compton collision cross-section per unit solid angle as a function of θ and α (see text) (Davisson and Evans, 1952).

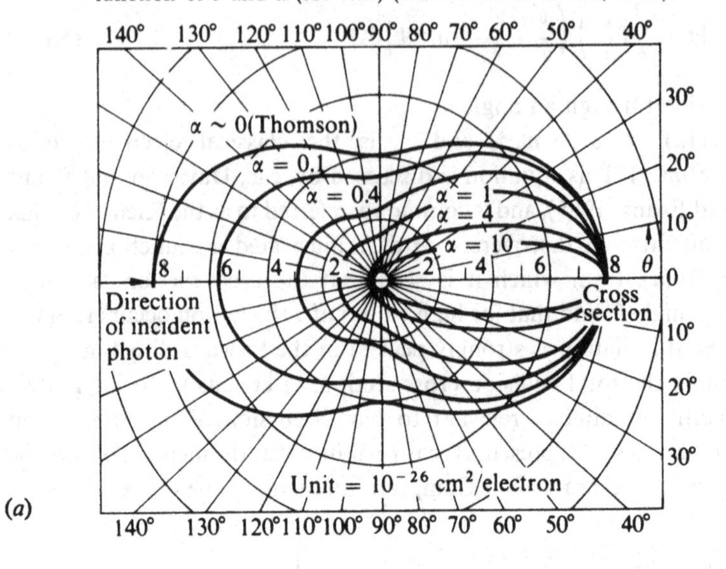

Fig. 5.6(*b*). Total Compton collision cross-section per electron as a function of energy.

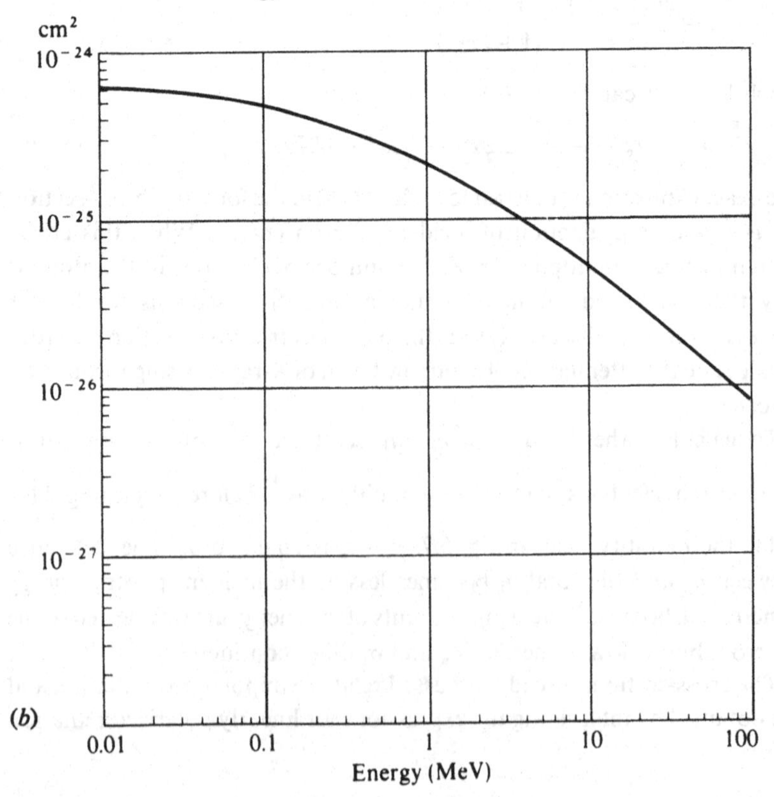

over all orientations of the electric vector of the incident wave. The cross-section contains a term dependent on the angle between the direction of the incident electric vector and the scattering plane, and the cross-section is a maximum for cases where the scattered photon and recoil electron are ejected approximately at right-angles to the electric vector of the incident photon. This 'polarization sensitivity' can be made the basis of a γ-ray polarimeter, and this application of the Compton effect was described in section 4.7.

We should note again that the theory of the Compton effect as set out here, and as calculated from expressions derived from the Klein–Nishina formula, relates to interactions with free electrons. Departures from these predictions will occur if the γ-ray energy is reduced to the point at which it becomes comparable with the binding energies of atomic electrons. In these circumstances the atomic cross-section for Compton scattering will fall, and this will be true particularly for small scattering angles, for which the transfer of momentum to the scattering atom is small.

This reduction in the cross-section for Compton scattering is expressed by a factor $S(\theta)$, the 'incoherent scattering function'. For K electrons, the effect is marked even at relatively high γ-ray energies. Fig. 5.8 illustrates the reduction of the electronic cross-section to values below the Klein–Nishina value at all angles below about 60°, and in this connection it is noteworthy that for larger angles, experimental values in excess of the Klein–Nishina cross-section are observed.

Fig. 5.7. Attenuation coefficients for various interactions in Al (Evans and Evans, 1948).

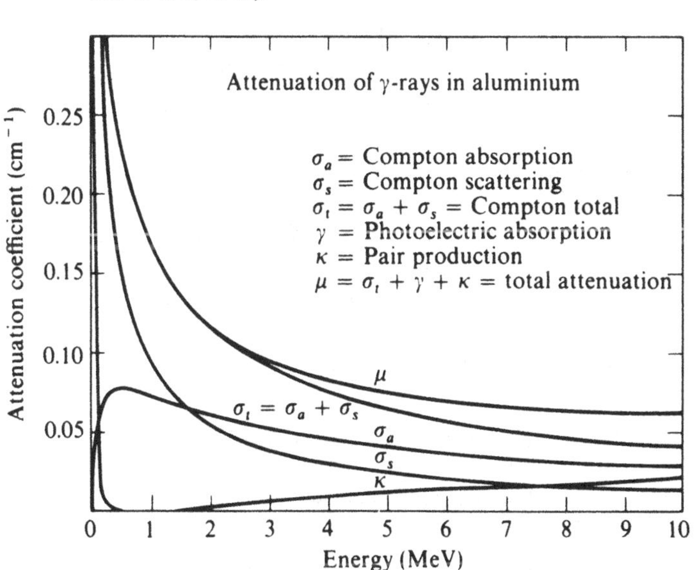

When the scattering cross-section is integrated over all angles, and over all electrons in the atom, the overall cross-section (i.e. 'atomic' bound-electron Compton scattering cross-section) is reduced. This has been discussed by Hubbell (1969) where it is shown that, for photon energies below 100 keV, the effect, at least in the heavier elements, can be considerable.

The theoretical treatment of electronic binding effects has been developed by James (1967) and cast into a form suitable for practical computation by Hubbell (1969). The function $S(\theta)$ may be written as $1-R$, and R has been shown, by these and other authors, to be related to the atomic form factor (or scattering factor), f, which will be introduced in our discussion of coherent (or Rayleigh) scattering. In this sense the two processes are complementary, but it should be said that the reduction in the Compton effect caused by electronic binding is more than offset by the onset of Rayleigh scattering, so that the total scattering by these two processes undergoes a steady rise as the photon energy is reduced to the level at which the effects of electronic binding become important.

Fig. 5.8. Ratio of observed Compton scattering from K-shell electrons (Motz and Missoni, 1961) to Klein-Nishina theory. (Hubbell, 1969).

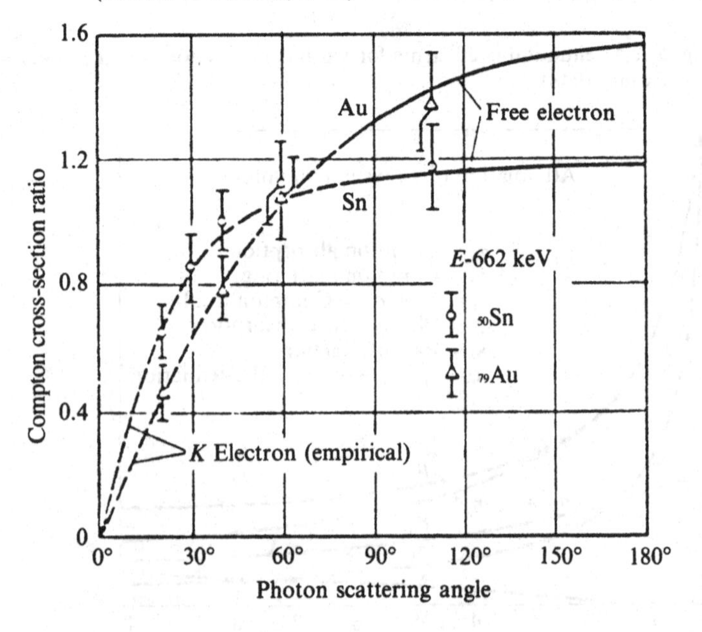

5.4 Pair production

The third interaction mechanism between γ-radiation and matter is known as pair production, and occurs if the photon energy exceeds $2mc^2$. The essential feature is the disappearance of the photon and its replacement by a positron–electron pair. This creation of particles is explained on the Dirac theory, in which it was proposed that in addition to the familiar electrons with positive energy $> mc^2$ there exists also a continuum of unobservable electrons with an energy $< -mc^2$. These negative energy states are illustrated in fig. 5.9. In order to cross the gap, an energy of at least $2mc^2$ has to be supplied. When this occurs, an electron appears, and the vacancy left in the negative continuum has the same properties as the electron except that it has a positive charge. This 'hole' is familiar in physics as the positron.

The momentum needed for this is less than that carried by the incident photon*, so forward momentum can only be conserved if a particle is already present to take up the excess. Pair production therefore takes place in the Coulomb field of a nucleus. It can also take place, though with a reduced probability, in the field of an electron.

The energy available is shared between the positron and the electron, and in principle this division may be in any proportion.

* It is never possible for a photon to pass on the *whole* of its energy to a free electron, because more momentum would be needed than the photon can supply; but as the photon has to *create* the particle as well as set it in motion, there is always more than sufficient momentum available.

Fig. 5.9. Illustrating pair production by a photon of energy E_γ.

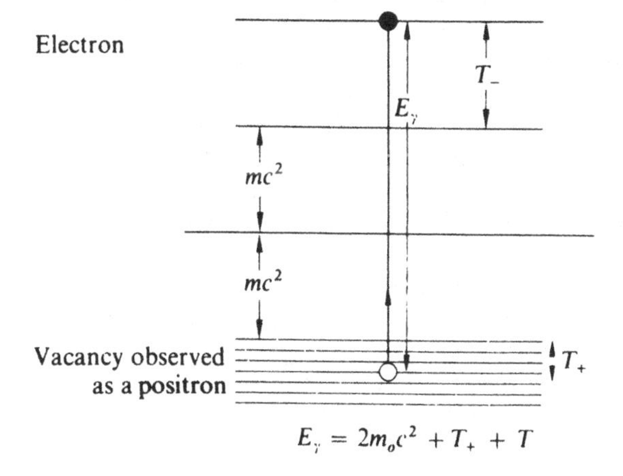

The differential cross-section (i.e. with respect to an infinitesimal range of positron kinetic energy) is given by

$$d\kappa = \frac{\kappa_0 Z^2 P}{h\nu - 2mc^2} dT_+ \tag{5.13}$$

where $\kappa_0 = \alpha r_e^2$ (where α is the fine structure constant and r_e the classical electron radius) and P is a function which rises to a rather flat maximum when the division of energy between the two particles is of equal order (fig. 5.10). By integrating $d\kappa$ over all positron energies the cross-section as a function of photon energy can be obtained. Following Evans (1955) we may write

$$\kappa = \kappa_0 Z^2 \bar{P}\,\text{cm}^2/\text{nucleus},$$

when $\bar{P}$ is the average value of P in fig. 5.10. $\bar{P}$ rises approximately logarithmically with incident photon energy.

Pair production thus increases with the square of the atomic number. Qualitatively this may be explained by noting that the region of high Coulomb field round the nucleus becomes more extensive as Z rises. In pair production a roughly equal division of energy is more likely than an unequal division, and it can be shown that this is because the former situation requires a smaller take-up of momentum by the nucleus.

Fig. 5.10. Division of energy between electron and positron in pair production (Davisson and Evans, 1952).

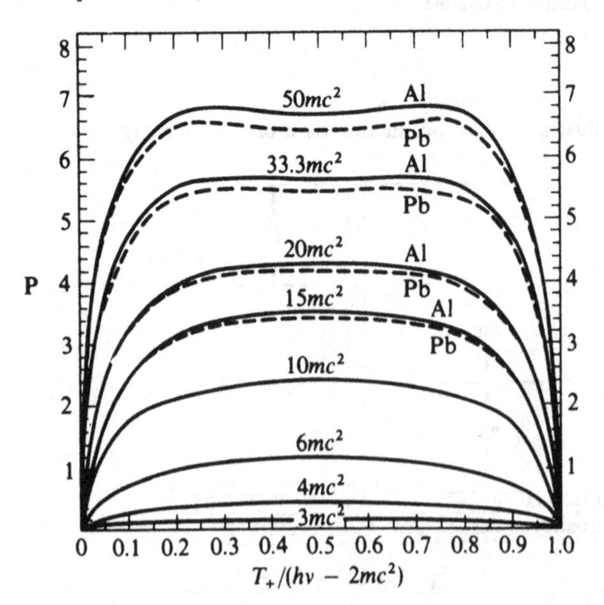

The positron and electron will, in practice, often come fully to rest within the absorber, and the positron will annihilate with an electron. Annihilation is usually a 2-quantum process* yielding 2 photons each with an energy of $h\nu = mc^2$.

Their escape probability can be calculated in principle, if desired. If escape occurs, the photon will behave like isotropically scattered radiation and we can write

$$\frac{\kappa_a}{\kappa_t} = \frac{h\nu - 2mc^2}{h\nu} \tag{5.14}$$

and

$$\frac{\kappa_s}{\kappa_a} = \frac{2mc^2}{h\nu - 2mc^2} \tag{5.15}$$

where κ_t, κ_a and κ_s are the total, absorption, and 'apparent scattering' cross-sections for pair-production.

We see that at high energies ($h\nu \gg mc^2$), the *absorption* cross-section approaches the *total* cross-section in value. At lower energies pair production will in any case be small compared with other processes. So in practice the distinction between κ_a and κ_t can often be disregarded.

5.5 Thomson and Rayleigh scattering

The remaining interactions are scattering processes, which deposit virtually no energy in the absorber, but which contribute to the attenuation or 'removal' cross-section.

Thomson scattering is well explained in terms of the classical theory of the electron. When an electromagnetic wave is incident on an electron, a periodic force is exerted which sets it into vibration at the forcing frequency. The electron then radiates electromagnetic energy, the intensity and angular distribution of the radiation being determined by its behaviour as an electric dipole. This radiation appears as scattered radiation, and is an important process at low energies. A truly free electron would of course be subject to momentum conservation as we have seen in the discussion of the Compton effect, but in the classical treatment of the Thomson effect, no account is taken of this. The classical approach can, however, be seen to be valid, by noting that the electrons involved in the scattering are in fact

* This occurs if the annihilation takes place from a *singlet* state. If a *triplet* state is formed, annihilation proceeds by a 3-quantum process. This usually happens with low probability, but if the positron–electron pairs form *bound* states (positronium), three-photon annihilation can become more important. In all cases, however, the total annihilation energy is $2mc^2$, and this is what concerns us in the present context.

loosely bound to their parent atoms which in turn may be bound to the crystal lattice. They are far from resonance (so that the quantitative aspects of the theory are not appreciably modified) but are sufficiently strongly bound to enable the momentum of the incident radiation to be taken up by the parent atom, or by the lattice as a whole.

Consider a free electron, irradiated by electromagnetic radiation which is plane-polarized with the electric vector in the plane of the diagram. Let the field strength of the incident radiation, and scattered radiation at a distance r from the electron, be E_y and E_s respectively. The energy flux per unit time associated with these fields is given by the Poynting vector for an electromagnetic wave,

$$\mathbf{N} = \mathbf{E} \times \mathbf{H}$$

which, averaged over a large number of cycles, leads to an energy flux

$$W = \langle E^2 \rangle \varepsilon_0 c \tag{5.16}$$

where ε_0 and c are the permittivity of free space and the velocity of light respectively.

The energy which is scattered in a time $\mathrm{d}t$ across an area $\mathrm{d}S$ may be written as

$$\langle E^2{}_s \rangle \varepsilon_0 c \, \mathrm{d}S \, \mathrm{d}t,$$

or, at a distance r from the electron, into a solid angle $\mathrm{d}\Omega$

$$\langle E^2{}_s \rangle \varepsilon_0 c r^2 \mathrm{d}\Omega \mathrm{d}t \tag{5.17}$$

If the scatterer is supposed to contain just one electron per unit area, the energy represented by (5.17) may also be written as the incident energy flux (in a time $\mathrm{d}t$) multiplied by the *differential* cross-section of the electron, i.e., as

$$\langle E^2{}_y \rangle \varepsilon_0 c \, \mathrm{d}\sigma \, \mathrm{d}t \tag{5.18}$$

By equating (5.17) and (5.18) we obtain

$$\mathrm{d}\sigma = \frac{\langle E^2{}_s \rangle}{\langle E^2{}_y \rangle} r^2 \, \mathrm{d}\Omega \tag{5.19}$$

To obtain $\langle E^2{}_s \rangle$ we use a result from classical electrodynamics which states that the field strength E_s at a distance r from a charge e with a position (displacement) f is given by

$$E_s = \frac{1}{4\pi\varepsilon_0} \frac{e\ddot{f}}{rc^2} \cos\theta, \tag{5.20}$$

where θ is the angle shown in fig. 5.11(*a*)*.

In the present circumstances

$$\ddot{f} = E_y \frac{e}{m}$$

Hence

$$E_s = \frac{1}{4\pi\varepsilon_0} \frac{e^2}{mrc^2} E_y \cos\theta \tag{5.21}$$

and $d\sigma$ becomes (from (5.19) and (5.21))

$$\frac{1}{(4\pi\varepsilon_0)^2}\left(\frac{e^2}{mc^2}\right)^2 \cos^2\theta \, d\Omega$$

i.e.

$$d\sigma = r_e{}^2 \cos^2\theta \, d\Omega \tag{5.22}$$

where r_e is the classical electron radius.

If the polarization of the incident radiation is perpendicular to the plane of fig. 5.11(*a*), all the rays scattered in that plane have their polarizations

* Compare with (2.3c), but note that in that equation the acceleration is *parallel* to the direction from which the angle θ is measured, whereas here the acceleration is *perpendicular* to the reference direction, necessitating $\cos\theta$ instead of $\sin\theta$.

Fig. 5.11(*a*). Illustrating Thomson scattering.

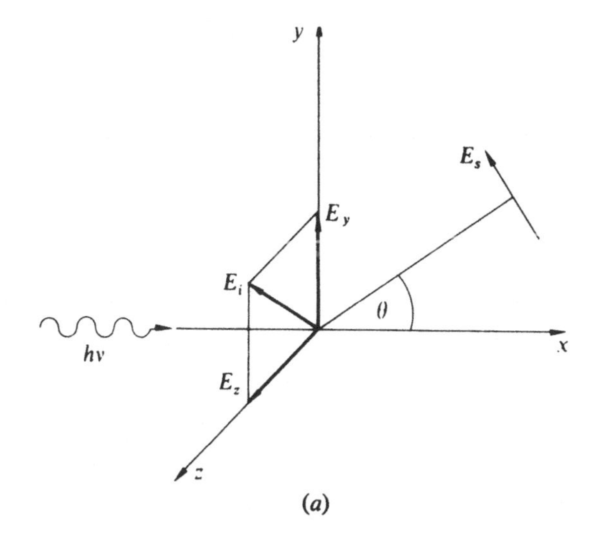

perpendicular to the plane, so we may replace $\cos\theta$ by 1. For unpolarized radiation we must use the average value $\frac{1}{2}(1+\cos^2\theta)$.

Hence

$$d\sigma=\frac{r_e{}^2}{2}(1+\cos^2\theta)\,d\Omega \tag{5.23}$$

Polar diagrams representing (5.22) and (5.23) are illustrated in fig. 5.11(*b*).

To obtain the total cross section σ_R, we integrate (5.23) over all values of θ, and obtain

$$\begin{aligned}\sigma_R&=\tfrac{1}{2}r_e{}^2\int_0^\pi 2\pi\sin\theta(1+\cos^2\theta)\,d\theta\\&=\tfrac{8}{3}\pi r_e{}^2\end{aligned} \tag{5.24}$$

Fig. 5.11(*b*). Polar diagrams for radiation polarized parallel to the *y*-direction and for unpolarized radiation.

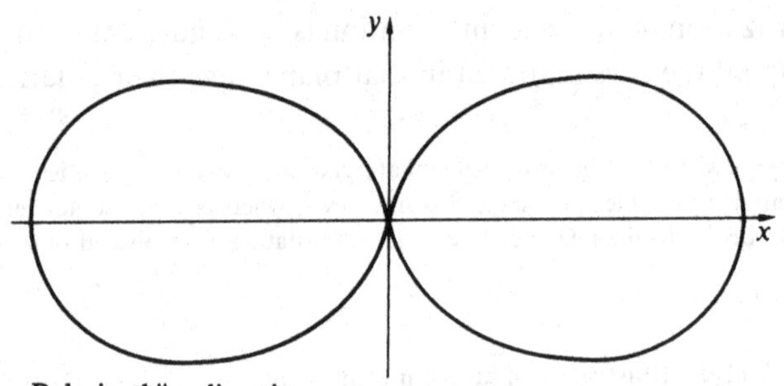

Polarized || *y* direction

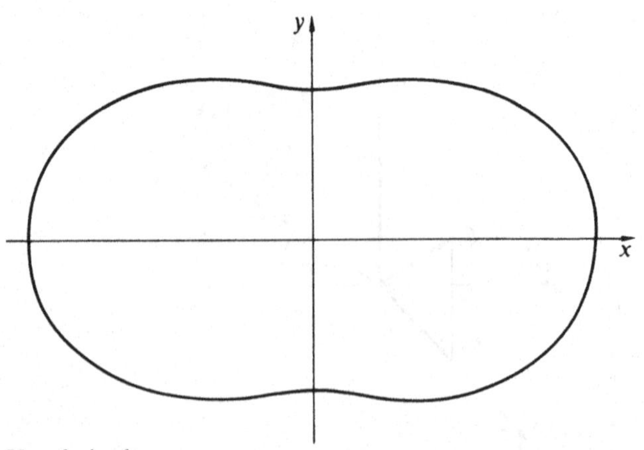

Unpolarized

for the cross-section per electron for Thomson scattering of unpolarized radiation. σ_R has already occurred as a constant in (5.5).

This simple result is independent of photon energy.

We next consider the co-operative effect of all the electrons in an atom, which gives rise to the process known as Rayleigh scattering. The electrons are all forced into vibration at the frequency of the incident radiation, and if this frequency is much greater than the natural frequencies of the atomic electrons, the scattered radiation from all the electrons will be in *antiphase* with the incident radiation.

The scattered contributions will therefore be coherent with each other, and in these circumstances the amplitudes of the scattered waves must be added *before* squaring to give the scattered intensity. The scattering cross-section per electron has little meaning in such a situation because of its non-additive nature, so we speak of an *atomic* differential scattering cross-section

$$d\sigma_R = r_e^2 f^2(\theta) \frac{1+\cos^2\theta}{2} d\Omega \tag{5.25}$$

where $f(\theta)$ is the atomic scattering factor and is a function of angle. It is defined with reference to the classical Thomson scattering for a free electron. For scattering in the forward direction, the interference between the scattered contributions will be constructive, the amplitude will be Z times, and the intensity Z^2 times, the value for a single electron. As the scattering angle increases, the interference will become destructive and the differential cross-section will be strongly peaked in the forward direction (fig. 5.12). The fall-off with increasing angle will become more marked as the incident photon energy is increased, because a gradual reduction in wavelength implies greater phase difference between contributions scattered from different parts of the electronic cloud. We would also expect the forward peaking to become *less* marked as the atomic number is increased because the charge distribution becomes somewhat more concentrated as Z increases. The variation of Rayleigh scattering with angle for 3 elements is shown in fig. 5.12, in which the slower fall-off for the heavier elements is clearly demonstrated.

To make progress in the calculation of a cross-section, the factor $f(\theta)$ has to be evaluated. This requires a knowledge of the distribution of electronic charge within the atom. The most reliable calculations are based on the Hartree self-consistent field model, which has been the basis of much work on the electronic structure of atoms. A detailed account of the calculation of atomic scattering factors by this and other methods, together with graphical examples, has been given by James (1967). An analytical approach, based on the Thomas–Fermi model has been given by Franz (1935), and his equations have been discussed by Moon (1950) and by Manninen *et al.* (1984). The analysis is expressed in terms of a parameter* u, given by

$$u = \frac{b}{\lambda}\sin\frac{\theta}{2} \tag{5.26}$$

where $b = 4\pi a_{\mathrm{TF}} Z^{-1/3}$. The quantity b is closely related to the atomic dimensions, because in the Thomas–Fermi theory of the atom, the 'atomic radius' is given by $a_{\mathrm{TF}} Z^{-1/3}$, where a_{TF} is the Thomas–Fermi constant, and is related to the more familiar Bohr constant ('Bohr Hydrogen radius') by the expression

$$a_{\mathrm{TF}} = (9\pi^2/2^7)^{\frac{1}{3}} a_{\mathrm{H}} \tag{5.27}$$

where the numerical coefficient is 0.885.

At large values of u (broadly speaking, when the wavelength is small

* u as used by Manninen does not include the factor $Z^{-\frac{1}{3}}$, which is incorporated elsewhere in his expressions.

Fig. 5.12. Rayleigh, Compton, and nuclear Thomson scattering as a function of angle. The upper, middle, and lower curves of each group of three refer to lead, copper, and aluminium respectively. (*a*) 2.8 MeV (*b*) 0.41 MeV. (after Moon, 1950).

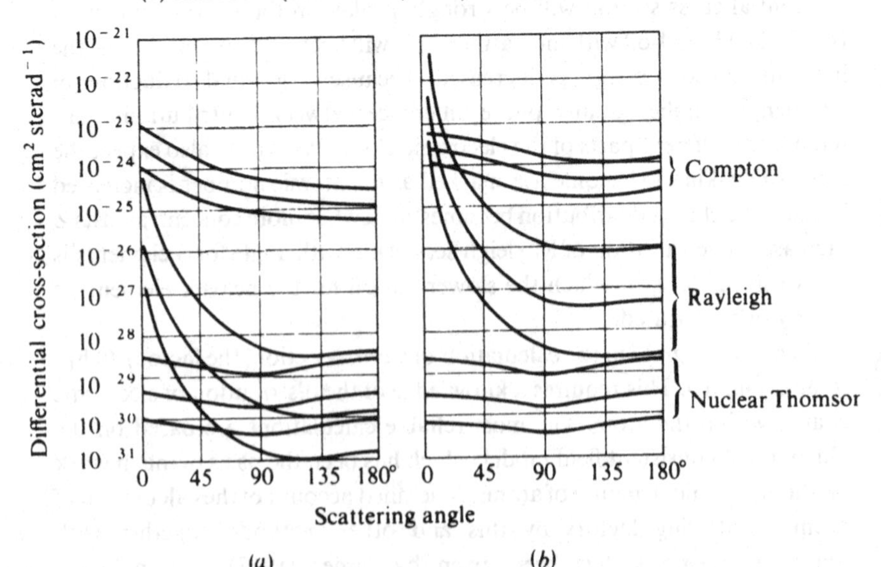

compared with atomic dimensions, and for scattering angles which are not small), the factor $f(\theta)$ has been approximated, by Franz, to

$$f(\theta)=Z\frac{1}{u}\sqrt{\left(\frac{\pi}{2u}\right)} \tag{5.28}$$

This leads to an expression for the *differential* Rayleigh cross-section

$$d\sigma=r_e{}^2\frac{1+\cos^2\theta}{2}\left(\frac{\pi}{2u^3}\right)Z^2\, d\Omega \tag{5.29}$$

Substituting for u, this leads to the expression

$$d\sigma=\text{const.}\times\frac{\frac{1}{2}(1+\cos^2\theta)}{\sin^3\theta/2}\left(\frac{mc^2}{E_\gamma}\right)^3 Z^3\, d\Omega \tag{5.30}$$

The constant may be evaluated to be $8.737\times10^{-33}\,\text{cm}^2$.

This expression illustrates the strong Z dependence of Rayleigh scattering, the strong inverse dependence on photon energy, and (in the circumstances imposed by $u\gg1$) a rapid fall-off with increasing scattering angle.

The parameter u may be understood by reference to fig. 5.13. It is equal to the phase difference (in radians) between a ray scattered from the centre of the atom and a ray scattered from the end of a diameter illustrated in that figure. An important parameter is the Rayleigh characteristic angle θ_0, corresponding to a phase difference of 2π between these rays. Franz's equations show that a fraction f_s of the scattering, where

$$f_s=\frac{0.6}{0.8-\lambda/8a_{TF}Z^{-\frac{1}{3}}} \tag{5.31}$$

Fig. 5.13. Relation between u, b and $\sin\frac{\theta}{2}$. (see (5.26)). (r is the atomic radius $=a_{TF}\,Z^{-\frac{1}{3}}$; $b=4\pi r$.)

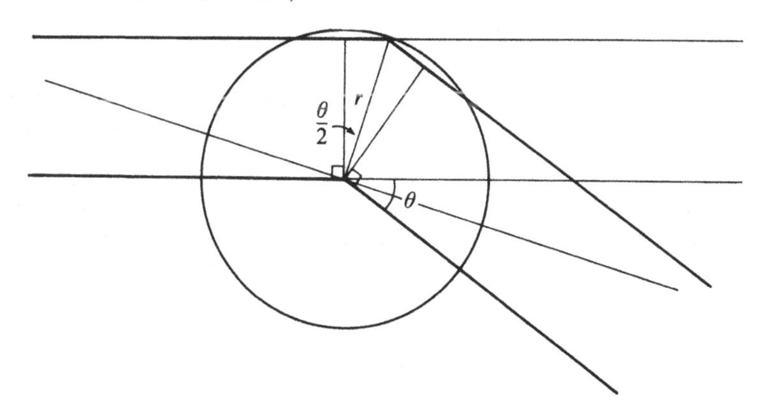

occurs within this characteristic angle. At high photon energies the scattering often takes place within small angles, and it follows that more than three-quarters of the radiation is scattered within the cone defined by θ_0.

Setting $u=2\pi$ in (5.26), we see that

$$\theta_0 = 2 \arcsin \frac{\lambda}{2a_{\mathrm{TF}}Z^{-\frac{1}{3}}} \tag{5.32}$$

which may be written as

$$\theta_0 = 2 \arcsin \left(\text{const.}\, Z^{\frac{1}{3}} \frac{mc^2}{E_\gamma} \right) \tag{5.33}$$

where the numerical constant is equal to 2.590×10^{-2}.

Turning to the total cross-section, i.e. integrated over all angles, we find that convenient analytical expressions do not exist. But the expression

$$\sigma_{\mathrm{coh}} = 8\pi r_e^{\,2} Z^2 \left(\frac{\lambda}{4\pi a_{\mathrm{TF}} Z^{-\frac{1}{3}}} \right)^2 \left(0.8 - \frac{\lambda}{8a_{\mathrm{TF}} Z^{-\frac{1}{3}}} \right) \tag{5.34}$$

may be deduced from the paper of Franz*, valid, as before, for wavelengths less than the diameter of the scattering atoms. This leads to an approximately $Z^{\frac{8}{3}}$ and E_γ^{-2} dependence, and is restricted, at least in the case of heavier elements, to γ-ray energies greater than about 60 keV. We have reproduced graphically (fig. 5.14) some calculations tabulated by Grodstein (1957). The variation with photon energy is approximately as E^{-2}, and the variation with Z lies between $Z^{2.5}$ and $Z^{2.7}$. Empirical rules of this kind are useful when comparing different modes of interaction, for example photoelectric absorption with Rayleigh scattering. The importance of Rayleigh scattering relative to the three processes already discussed will be taken up in section 5.9, and we shall see that there are circumstances in which it can contribute appreciably to the removal cross-section.

We have seen that in Rayleigh scattering the contributions from the electrons in a given atom all stand in a definite phase relationship to each other. The description 'coherent' is applied to a situation of this type. The question now arises as to the extent to which this description may be applied to the scattered radiation from different atoms. Rayleigh scattering is elastic in the sense that the internal state of the atom is the same after the scattering process as before, but it should be remembered that some

* Equation (12) of his paper and following equations, but restoring a factor of 2 lost in the course of the integration.

momentum has to be disposed of during the scattering process. This will amount to $2p \sin(\theta/2)$, where p is the momentum of the photon. The energy associated with this may reach a value of

$$\frac{2E_\gamma^2}{Mc^2} \tag{5.35}$$

where M is the mass of the recoiling atom.

In a scatterer such as a monatomic gas the interactions between atoms are small, and the scattering atom will therefore recoil freely. For example, a γ-ray of 10 keV scattered from an argon atom would impart a recoil energy of 5 millielectron volts. For a γ-ray of 100 keV this would become 0.5 eV. Rayleigh scattering between different atoms is thus not necessarily

Fig. 14(*a*). Rayleigh scattering as a function of photon energy.

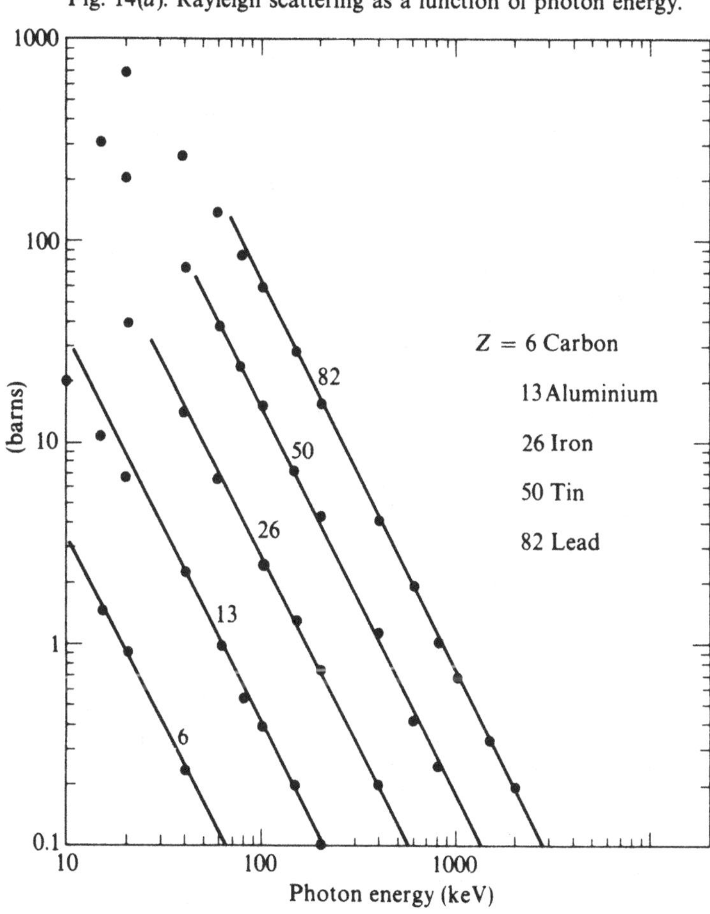

(*a*)

Fig. 5.14(*b*). Rayleigh scattering as a function of atomic number (the numbers on the curves refer to the photon energy in keV).

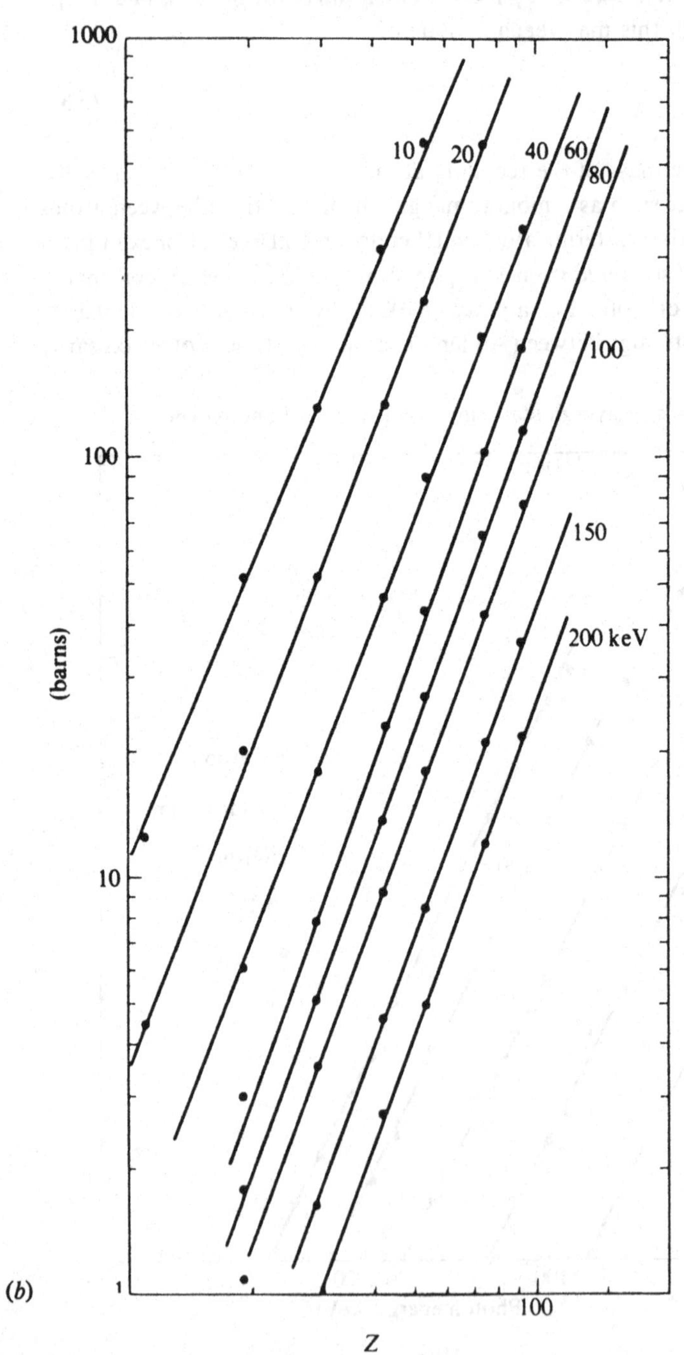

coherent, but if the atoms are bound in a solid, it becomes possible, in principle, for the whole crystal to recoil, in which case the quantity M in (5.35) would be larger than the atomic mass by many orders of magnitude. This process is sometimes referred to as 'recoilless' scattering. In these circumstances the scattering from the different atoms in the crystal will be coherent and therefore subject to interference effects. This is the process which gives rise to Bragg reflections from crystals, and on which the techniques of crystallography are based. This recoilless process will be the more likely if the atom is unable or unlikely to exchange a photon with the surrounding lattice, and so is favoured by a low ambient temperature and a high Debye temperature for the lattice. Further consideration of this extremely important aspect of X-ray scattering would lead us away from the main themes of this book, and to pursue these topics, standard texts on crystallography must be consulted. We note simply, that the distinction between 'recoilless' Rayleigh scattering and scattering with recoil of the quasi-free atom is of great importance in the interaction between X- or γ-rays and crystals. The reader should note that the scattering by a free atom is properly described as 'incoherent elastic' scattering and the recoilless scattering as 'coherent elastic' scattering. However, in contexts where the distinction is disregarded, Rayleigh scattering of both kinds is often referred to as 'coherent'.

5.6 Nuclear Thomson scattering

Much of the interest in the remaining scattering processes lies in the fact that they are all coherent with each other and also with Rayleigh scattering, and are thus able in principle to show interference effects. Nuclear Thomson scattering is the classical scattering of the incident electromagnetic radiation by the nucleus considered as a point charge. The scattering cross-section is given by (5.21–3), with the *nuclear* mass in the denominator. The total cross-section (i.e. integrated over all angles) is therefore very small compared with any of the processes so far examined, but its polar diagram suggests that it might become comparable with Rayleigh scattering at angles of 90° or more, and under these circumstances its effect has in fact been observed by Davey (1953) who examined the elastic scattering from lead at 2.76 MeV. The amount of scattering observed at 90° and above exceeds that which would be expected from Rayleigh scattering alone, and is consistent with the calculated data of fig. 5.12, within experimental error.

The graphs in fig. 5.12 show calculated values of the nuclear Thomson scattering for lead, copper, and aluminium. For 0.411 MeV radiation it can be seen that even at scattering angles approaching 180° the differential cross-section is at least an order of magnitude less than for Rayleigh scattering. At energies above 1 MeV they become more comparable in magnitude.

5.7 Delbrück scattering

This process is the least familiar of the γ-ray interactions treated in this chapter, and consists essentially of pair production in the nuclear field followed by annihilation *in the same field.* It therefore presents the same appearance as an elastic scattering process and would be coherent with other elastic processes. In principle it can occur for γ-ray energies below $2mc^2$ because the pairs can be virtual, i.e. can have negative energies and need not be observable. Moffat and Stringfellow (1960) have made observations of the scattering of 90 MeV photons in uranium through angles between 1.2 and 4.5 milliradians. In their experiment several processes may contribute to the observed scattered radiation: Compton scattering is relatively small (7.3 barns/sterad) and is essentially constant over the small angles involved. Bremsstrahlung produced by electrons and positrons released during pair-production is appreciable, but by using a thick scatterer to scatter the electrons and positrons over appreciable angles by multiple scattering, the Bremsstrahlung can be spread out over a

Fig. 5.15. Delbrück scattering (Moffat and Stringfellow, 1958).

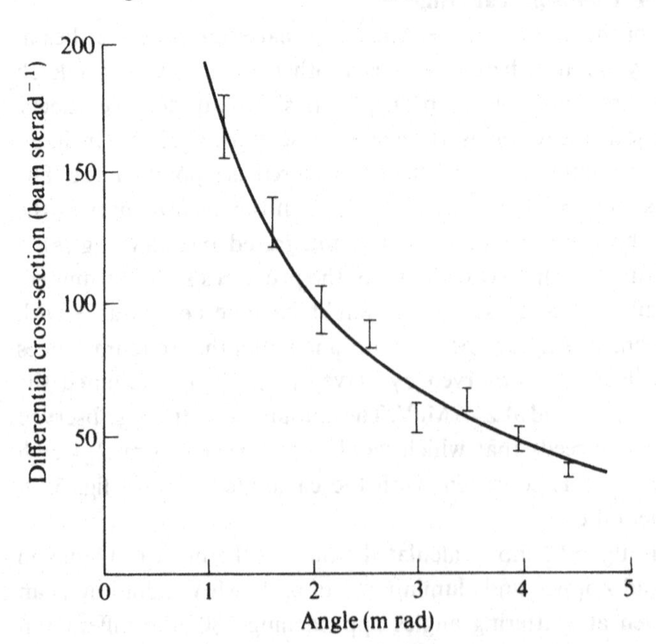

relatively wide angular range and becomes substantially constant over the angular range investigated. Nuclear Thomson scattering is negligible and the Rayleigh scattering is thought to be small ($\frac{1}{10}$) compared with the observed effects.

The scattering observed by Moffat and Stringfellow (reproduced in fig. 5.15) is therefore thought to be substantially due to Delbrück scattering, except for a constant background, caused by the effects referred to, requiring a correction of about 18 barns.

Bosch *et al.* (1962) have observed the scattering of 9 MeV γ-rays and find an intensity which exceeds the scattering predicted by the coherent superposition of Rayleigh and nuclear Thomson scattering by a factor of about 10. This is considered to be due to Delbrück scattering, and the work of Moreh and Kahana (1973) at 7.9 MeV, using targets of uranium and thorium, would appear to confirm this. The process may thus be regarded as being definitely established.

5.8 Nuclear resonant scattering

The remaining process to be discussed is nuclear resonant scattering and is unique amongst γ-ray interactions in that its characteristics depend upon the properties of nuclear excited states. When nuclei which possess a level lying not more than a few hundred keV above the ground state are irradiated with γ-radiation, it is possible, in principle, for the nuclei to be excited into the level in question. This, however, requires the establishment of a resonant condition between the absorbing nuclei and the incident radiation.

Suitable conditions by which this resonance may be demonstrated are not easy to achieve, mainly because of the very narrow widths of such excited states. In order to provide an incident flux with an adequate intensity within the very narrow band width of the resonance, the use of, e.g., a continuous X-ray spectrum would appear unprofitable, and characteristic lines would, on a basis of pure chance, also seem unlikely to be suitable; sources of radiation for this work have therefore consisted almost invariably of radionuclides decaying by β-emission or electron capture to the actual excited state in question (fig. 5.16(*a*)), providing that the subsequent decay to the ground state is not internally converted to a prohibitive degree. The natural widths of these states can be related to their lifetimes through the uncertainty principle; lifetimes in the range 10^{-8}–10^{-15} seconds correspond to natural widths which do not exceed 1 eV, and which are sometimes less than this by several orders of magnitude. The atoms of the source and absorber are in a state of thermal motion or vibration, and the emitted radiation will therefore be subject to a Doppler broadening because of this.

If the emitting atom is free, it will recoil with an energy of $\frac{E^2_\gamma}{2Mc^2}$, and the scattering atom will also recoil by this amount if a resonant photon is absorbed. There is therefore a discrepancy of $\frac{E^2_\gamma}{Mc^2}$ between the energy of the incident radiation and that required for resonant scattering (fig. 5.16(*b*)). This discrepancy normally amounts to the order of 1 eV, though in favourable circumstances, e.g. for a heavy nucleus and for low energy γ-rays, this may be reduced to a value within the range of thermal energies. Some overlap between the thermally-broadened lines in the emitter and scatterer then becomes possible.

This situation may be examined as follows: If a nucleus is moving with a velocity v towards the scatterer, the observed energy of its emitted γ-

Fig. 5.16. Nuclear resonant scattering. (*a*) Nuclear levels in source and scatterer. (*b*) Doppler broadened profiles of source and scatterer, showing overlap. (after Metzger, 1959).

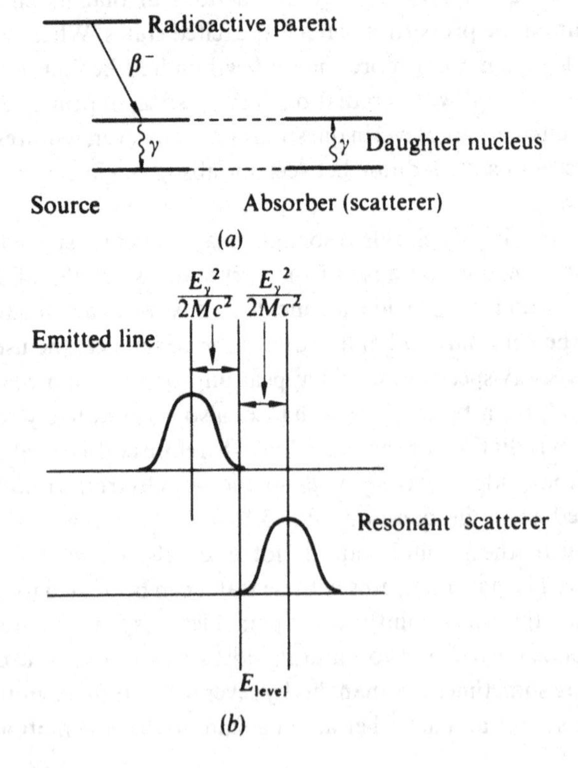

radiation is increased to $E\left(1+\frac{v}{c}\right)$ by the Doppler effect. If the velocities of the nuclei in the source are distributed according to a Maxwellian law, we can write

$$w(v)\,\mathrm{d}v=\left(\frac{M}{2\pi kT}\right)^{\frac{1}{2}}\mathrm{e}^{-\frac{Mv^2}{2kT}}\,\mathrm{d}v \tag{5.36}$$

for the probability of the nucleus having a velocity component v towards the observer.

The distribution of energies is obtained by substituting $\frac{(E'-E)c}{E}$ instead of v and $\frac{c\mathrm{d}E'}{E}$ instead of $\mathrm{d}v$ in (5.36). We obtain

$$w(E')\,\mathrm{d}E'=\frac{1}{\Delta\pi^{\frac{1}{2}}}\mathrm{e}^{-\left[\frac{E'-E}{\Delta}\right]^2}\mathrm{d}E' \tag{5.37}$$

Where $\Delta=E\left(\frac{2kT}{Mc^2}\right)^{\frac{1}{2}}$, and is known as the *Doppler width.*

The ratio of the Doppler width to the separation $\frac{E_\gamma^{\ 2}}{Mc^2}$ is thus given by

$$\frac{\Delta Mc^2}{E_\gamma^{\ 2}}=\frac{\sqrt{2}}{E_\gamma}(kTMc^2)^{\frac{1}{2}} \tag{5.38}$$

For $kT=0.025\,\mathrm{eV}$ and $M=200$ this ratio equals 1 for a γ-ray energy of 100 keV. Overlap is thus possible, although for observing the variation of overlap with temperature in a conveniently accessible temperature range, (300–1500 K), Metzger (1959) shows that somewhat higher γ-ray energies are preferable. He considers that this method of observing nuclear resonant scattering is best suited to γ-rays between 150 and 450 keV. This 'thermal' method has been used for the study of resonant scattering in several nuclides, following the work of Malmfors (1953).

Nuclear resonant scattering was first observed by utilising mechanical motion of source or scatterer, producing a Doppler shift which is thus superimposed on the thermally broadened emission profile. This shift is in the correct sense to restore resonance if source and scatterer are approaching each other, and it requires a magnitude of the order of

$$E_\gamma\frac{v}{c}\sim\frac{E_\gamma^{\ 2}}{\mathrm{M}c^2} \qquad \text{or} \qquad v\sim\frac{E_\gamma}{Mc} \tag{5.39}$$

For $E_\gamma\sim 500$ keV and $M=200$ this is of the order of $10^3\,\mathrm{m\,s^{-1}}$, which can be achieved by high speed rotors with linear dimensions of a few centimetres. These experiments used the 411 keV transition in ^{198}Hg, the source (^{198}Au) being mounted on the tip of a rotor. These observations are described by Moon (1951) and by Davey and Moon (1953).

Another method of restoring resonance is to make use of the recoil associated with a preceding particle or neutrino emission during β-decay. This method has been reported by Ilakovac (1954), Grodzins (1958) and others. It is suitable, however, only for those cases in which the resonant state has decayed before the recoiling atom suffers a collision. For solid and liquid sources this time is of the order of 10^{-12} seconds, but for gaseous sources lifetimes of up to 10^{-8} seconds can be used, the time between collisions being of this order.

Up to the present point in the discussion we have concerned ourselves with nuclear resonance from relatively low-lying states, accessible by radioactive β-decay. Gamma-rays of higher energy are available from nuclear reactions, and resonances can readily be excited in a scattering medium, because γ-rays produced in a charged particle reaction are considerably broader than is necessary to overcome recoil effects. Amongst the reactions used for these observations are $^{19}F(p,\alpha)^{16}O$, which produces γ-rays of 6.91 and 7.12 MeV, and $^{15}N(p,\alpha)^{12}C$, ($E_\gamma = 4.43$ MeV). It also becomes possible to excite resonance by means of radiation from a different nuclear species, as in the work reported by Rasmussen *et al.* (1958), in which 4.43 MeV radiation, from the second of the two reactions just referred to, was used to excite resonance in the 4.46 MeV excited state of ^{11}B. Nuclear resonant scattering at 7 MeV, using recoil-broadened radiation, has been examined by Reibel and Mann (1960). Cross-sections are of the order of millibarns.

Nuclear resonant processes take on a particularly interesting form when the nucleus is bound into a crystal lattice. In this circumstance the emission and resonant absorption of γ-radiation involves the whole lattice in recoil, and the energy difference of (5.33) becomes even less than the natural width of the line. The phenomenon of 'recoilless resonance' was observed in ^{192}Ir by Mössbauer in 1958. We recognise the similarity between this situation and the recoilless Rayleigh scattering discussed in section 5.5, and, for the reasons given there, it will be appreciated that a reduction in temperature causes a *rise* in the recoilless fraction, and hence in the magnitude of observed absorption or scattering. By contrast, in the absence of an appreciable recoilless fraction, a reduction of temperature would cause a *fall* in the observed effect because of the reduction in Doppler broadening and consequent reduction in overlap. Mössbauer observed a *rise* in scattering as the temperature was reduced, and also measured the width of the line by superimposing small Doppler shifts of a few cm per second by mechanical motion of the source.

The natural shape of the emitted line has the Lorenz form

$$I(E) \propto \frac{\Gamma^2}{(E-E_\gamma)^2+\left(\frac{\Gamma}{2}\right)^2}\frac{\Gamma_\gamma}{\Gamma} \tag{5.40}$$

where Γ is the natural width of the state from which the radiation is emitted, and E_γ is the most probable energy of emission. The absorption cross-section is given by a similar expression:

$$\sigma(E)=\frac{\lambda^2}{4\pi}g\frac{\Gamma^2}{(E-E_\gamma)^2+\left(\frac{\Gamma}{2}\right)^2}\frac{\Gamma_\gamma}{\Gamma} \tag{5.41}$$

This is the so-called Breit–Wigner relation (derived originally for the study of slow neutron resonances), in which λ is the wavelength at resonance, Γ_γ is the 'partial width' appropriate to the absorption of γ-radiation, (or its emission in competition with other processes such as internal conversion), and g is a statistical factor*.

The peak value of σ from (5.41) is given by

$$\sigma_{\text{peak}}=\frac{\lambda^2}{\pi}g\frac{1}{1+\alpha} \tag{5.42}$$

where α is the internal conversion coefficient.

This presupposes a truly monoenergetic incident radiation, but in fact the emitted line has the same width and profile as the absorption line. The effect of this is to reduce the peak absorption cross-section by a factor of two, and to increase the width by the same factor. The maximum absorption cross-section is therefore given by

$$\sigma_{\max}=\frac{\lambda^2}{2\pi}g\frac{1}{1+\alpha} \tag{5.43}$$

For $\lambda \sim 10^{-10}$ m and $\alpha \sim 10$ this cross-section is of the order of $10^{-18}\,\text{cm}^2$, and is thus considerably higher than the other elastic scattering processes. It can even dominate all other absorption processes, and is then readily observable in a simple attenuation experiment. A superimposed velocity of

* $g=\frac{1}{2}\frac{2I_e+1}{2I_g+1}$ where $2I_e+1$ and $2I_g+1$ are the multiplicities of the excited and ground states respectively, and the factor $\frac{1}{2}$ accounts for the fact that the incoming photon has 2 states of polarization.

order $\frac{c\Gamma}{E_\gamma}$ (which may be typically 10^{-2} cm s^{-1}) will be sufficient to destroy resonance, thereby enabling the resonant effect to be measured by subtraction. To obtain a large effect, the temperature of source and absorber should not exceed their Debye temperatures, the nuclide in question should be present in the absorber in reasonable abundance, E_γ should be rather low (< 50 keV), and the line should not be so narrow that a small shift due to, e.g., chemical effects would be sufficient to destroy resonance.

A nuclide which has been studied extensively is ^{57}Fe. In this nuclide there is a level at 14.4 keV above the ground state which has a mean lifetime of 1.4×10^{-7} s and a natural width of 4.5×10^{-9} eV. Equation (5.43) leads to an absorption cross-section at resonance of 1.19×10^{-18} cm^2, which for a paramagnetic sample such as stainless steel* (abundance of ^{57}Fe in natural iron 2.17%; recoilless fraction at room temperature 0.7) gives a nuclear resonant absorption coefficient of 196 cm^2 g^{-1}†, due to nuclear resonance alone. For comparison, the photoelectric absorption coefficient in pure iron is approximately 66 cm^2 g^{-1}. In this example, it is clear that the nuclear resonance process is predominant, although this is true only in the very small energy region ($\sim$1 in 10^{12}) covered by the resonance. Clearly resonance has only an infinitesimal chance of occurring except when radiation from the same transition is being used as the source.

For a general introduction to the theory and experimental techniques of the Mössbauer effect, texts by May (1971) and Gibb (1976) may be consulted. A detailed review of the scattering of recoilless radiation, both by the Rayleigh and nuclear resonant processes, has been given by Champeney (1979), who also gives a comprehensive bibliography.

In the course of early studies of the Mössbauer effect, several factors, such as temperature and pressure, which in normal circumstances have no measurable effect on nuclear transitions, have been shown to modify the energy of the recoilless line to a minute but observable extent. We shall discuss one of these effects – that of chemical combination – because of its close similarity to the 'X-ray chemical shift' described in chapter 8.

Nuclear energy *levels* are modified in practice by a small term which arises because certain orbital electrons, notably those in *s* states, exist for a fraction of time within the nuclear volume. The energies of nuclear

* For pure metallic iron, the internal magnetic field associated with the ferromagnetism causes a hyperfine splitting of the absorption and emission spectra which modifies the calculation.

† To be multiplied by a fraction representing the proportion of iron in stainless steel. Typically this would be in the region of 80%.

transitions are therefore modified if the nuclear radius *changes* during a transition. This modification in transition energy is given by

$$\Delta E_\gamma = \text{const.}\, \psi^2(0)(R_e^{\,2} - R_g^{\,2}) \tag{5.44}$$

$$= \text{const.}\, \psi^2(0)R\left(\frac{\delta R}{R}\right) \tag{5.45}$$

The magnitude of this term thus depends upon the s-electron density at the nucleus, which in turn depends upon the nature of the chemical combination. Energy levels are *raised* by this correction term, and the transition energy is also increased if the radius of the excited state exceeds that of the ground state*. The effect of increasing $\psi^2(0)$ is therefore to increase the transition energy, and a decrease of $\psi^2(0)$ (as in compounds of increased ionicity) is associated with a *negative shift* in E_γ. This effect is observable under recoilless conditions, and has given valuable information on the ionicity of tin atoms in compounds of tin. We refer to this again in chapter 8 in connection with the closely related chemical X-ray shift.

One of the most interesting consequences of the large magnitude of the Mössbauer effect and its resonant nature has been the possibility of studying interference effects *vis à vis* other elastic scattering processes. There is a change of phase in resonantly scattered radiation as the energy of the incident radiation is swept through resonance (by mechanical motion of the source), and a phase difference between incident and scattered radiation of $\pi/2$ radians on exact resonance. Any interference with, for example, Rayleigh scattering is expected to change from constructive to destructive interference as the resonance is traversed. A discussion of this situation is given by Moon (1961) and its experimental realization is described by Black *et al.* (1962).

5.9 Attenuation coefficients in practical conditions

It is clear that the principal 'removal' mechanisms are Compton scattering, pair production, and photoelectric absorption, in the energy region normally associated with X-rays. Rayleigh scattering is important in certain circumstances. Having examined these processes in detail we might now be tempted to group them differently. It would be reasonable to assert that Compton scattering is the basic interaction process at energies well above the electronic binding energies, that Rayleigh scattering plays a similar basic rôle at energies well below the binding energies, and that the photoelectric effect represents a somewhat complex situation in the

* This is the case in ^{119m}Sn, but not, for example, in ^{57}Fe.

intermediate region. In this connection it is interesting to note that expression (5.11b) for Compton scattering reduces to the differential Thomson scattering cross-section if ν' is allowed to approach ν. The second bracket of (5.11b) becomes $(2-\sin^2\theta)$ or $(1+\cos^2\theta)$, and the equation becomes

$$d\sigma = r_e^2 \frac{1+\cos^2\theta}{2} d\Omega \tag{5.46}$$

which is identical with (5.23). It should be noted, however, that this simple procedure obscures one important point – in Compton scattering the electron may remain unchanged in spin direction after the interaction, or it may alternatively have experienced a 'spin–flip' process*. In order to infer the magnitude of coherent scattering at low energies it would be logical to delete the spin–flip term as this is clearly not compatible with coherent scattering. However, the spin–flip term itself can be seen to approach zero at low photon energies and indeed is small for all photon energies $\ll mc^2$, so the simple procedure we adopt yields the correct result and for the correct reason also.

The absorption of energy by these processes has been seen to be important in the study of all radiation detectors including the proportional counter (photoelectric effect) solid-state detector (photoelectric effect) and scintillation counter (photoelectric effect, Compton effect, pair production). Their main properties regarding response to radiation can be understood by detailed reference to these interactions.

The β-filter in crystallography, and also the Ross filters referred to in chapter 4, are seen to be applications of the photoelectric effect. Practical radiation shielding of course involves all these processes. Attenuation of unwanted radiation is achieved essentially by photoelectric absorption. High energy radiation is reduced in the first place by pair production followed by the production of annihilation radiation. This, along with other radiations in this intermediate ($\sim mc^2$) region, is then degraded by successive Compton scattering events until the photoelectric absorption finally absorbs the remaining energy. We should note, however, that the fluorescent radiation may be troublesome, and may be removed by the use of a 'graded filter' in which X-rays are absorbed by layers of material of progressively decreasing atomic number. A convenient series consists of

* The α^2 term in (5.11) represents the 'spin–flip' contribution to Compton scattering. An illuminating discussion of this and earlier expressions for Compton scattering is given by Compton and Allison (1935) in an extended footnote on pages 237–8 of their book.

lead, tin (or cadmium), copper, and aluminium, all of which are readily available in thin metallic sheets.

The choice of shielding in special circumstances, for example when small amounts of radioactivity are being observed, may be dictated by rather different considerations from those just applied. Lead may be unacceptable because of its appreciable radioactive content. Steel or 'heavy alloy' would be more appropriate in such a case.

The presence of scattered radiation from the Compton process is an inevitable consequence of the absorption of X- or γ-rays, and greatly affects the design of shielding suitable for wide beams of γ-radiation, or radiation emanating from extended sources, such as nuclear reactors. The attenuation found in practice is much less than a simple application of removal cross-sections would suggest, because of the 'scattering-in' effect. This causes the radiation to exceed the calculated value, by a factor which may be very considerable in the case of transmission through, e.g. thick shielding walls. This so-called 'build-up' factor has been studied in detail theoretically and experimentally. The attenuation of X-rays in 'broad-beam' conditions requires careful study in radiobiology and radiotherapy, when accurately known radiation doses have to be delivered to tissues which may be deep-seated. Such quantitative information is invariably obtained by experimental determination in a tissue-equivalent medium, and the 'depth-dose distributions' obtained in this way for fields of different sizes are an indispensible aid for calculations of this type. Examples of such distributions, for identical radiation, but three different field sizes are shown in fig. 5.17, where the effect of 'scattering-in' is clearly visible. The radiation is

Fig. 5.17. Depth-dose distributions for 0.662 MeV γ-radiation from ^{137}Cs. Square fields 5×5 cm, 8×8 cm, 10×10 cm. Distance from source to surface of medium 35 cm. (after Johns, 1964).

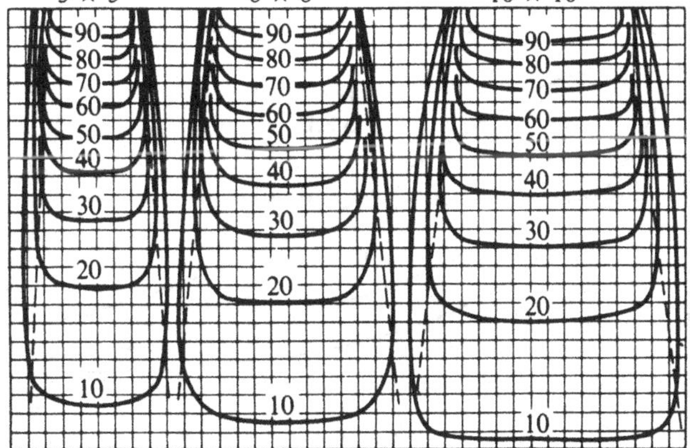

attenuated more rapidly in the smaller field sizes. In fig. 5.18 the contribution due to scattered radiation is shown separately and is seen to be dominant over a considerable range of depths.

In connection with radiation shielding, the radiation dose from broad beams, or extended sources, assumes considerable importance, and the behaviour of γ-radiation in these circumstances has been subjected to considerable theoretical and practical investigation. By Monte Carlo and other methods, it is possible to compute the progressive change in spectrum as γ-radiation penetrates an extended solid, and analytical formulations are also possible. The subject of γ-ray transport has been treated by, for example, Goldstein (1959). In order to make calculations for practical use, the concept of the *build-up factor* is useful and is widely used. This is defined as the factor by which the quantity being studied exceeds the quantity measured under narrow beam conditions, the latter being due to the 'uncollided' flux. E.g.

$$F = F_0 B e^{-\mu x}.$$

where B is a function of μx.

F may be the radiation dose, the energy flux, or the photon flux, and the build-up factors, B, may, in principle, be different for these quantities. In practice, the radiation dose (expressed as exposure, for example) is the most commonly desired quantity. The nature of the source also determines the

Fig. 5.18. Primary and scattered radiation for two different circular field sizes. The larger field size introduces the greater contribution from scattered radiation. Distance from source to surface of medium 50 cm (Radiation specified in terms of 'half-value layer', 1 mm Cu, i.e. the distance which if travelled in copper, would reduce the intensity by 50%) (after Johns, 1964).

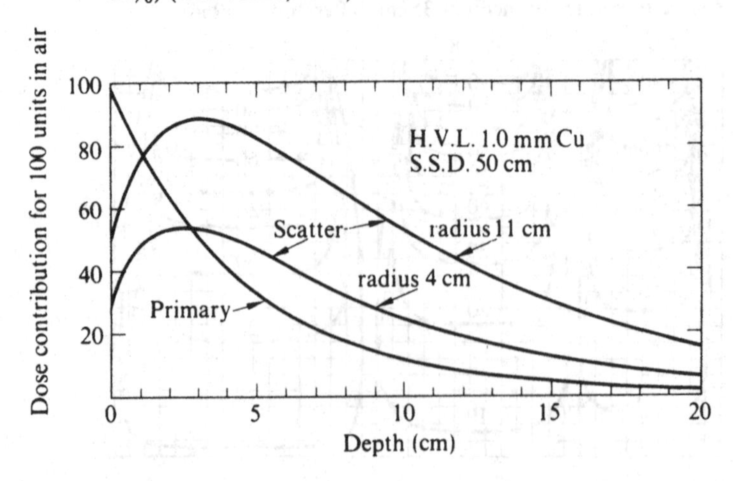

build-up factor, and we distinguish between a '*point-isotropic source*' (a point source radiating isotropically), a '*plane isotropic source*' (a plane source, of which the individual elements radiate isotropically) and a '*plane monodirectional source*' (a plane source which by collimation, or other features of design, radiates a parallel beam). We see that the situation is complex.

Analytical formulations of the build-up factor are possible, which are often expressed in terms of mean free paths (i.e., multiples of μx, where μ is an attenuation coefficient. One such formulation takes the form

$$B = A\mathrm{e}^{-\alpha_1 \mu x} + (1-A)\mathrm{e}^{-\alpha_2 \mu x}$$

with adjustable parameters A, α_1 and α_2.*

For a small number of mean free paths, a simpler formulation

$$B = 1 + C(\mu x)$$

can be made. B rises progressively with μx, as expected from the physical 'scattering-in' of radiation from peripheral parts of the incident radiation field. B tends to unity as $\mu x \to 0$.

Further detail would take us away from the subject of X-ray physics, so, in order to pursue this topic, specialised texts must be consulted.

Finally, in the light of the principal interactions examined in this chapter, we consider the degree to which 'narrow-angle' geometry is required, if scattered and other secondary radiation is not to impair the accuracy of measurements of removal cross-section. We use the symbols of fig. 5.1(*b*) in our discussion.

Photoelectric absorption will produce fluorescent radiation which will be emitted isotropically. The fraction of photoelectric absorptions which will result in a fluorescent X-ray entering the detector is given by $\frac{1}{4}\omega\theta_0{}^2$ where θ_0 is the semiangle subtended by aperture B at the absorber and ω the fluorescence yield. For example, if $\theta_0 = 10^{-2}$ rad and $\omega = 0.4$ (i.e. for $Z \sim 30$), this fraction would be 10^{-5}. This has to be compared with the fraction of incident radiation transmitted through the absorber. In an absorption experiment this would hardly be less than 0.1 of the incident radiation, and so the correction necessitated by fluorescence would be of order 10^{-4}, which would be negligible. If, however, some primary radiation strikes the absorber which is outside the solid angle defined by aperture B in fig. 5.1(*b*), fluorescent radiation from the absorber could be very much stronger. If the detector is able to discriminate against fluorescent radiation, errors caused by this can, however, be reduced to a low level. If no such discrimination is

* α_1 is normally *negative* in this formulation.

possible, the effect of fluorescent radiation may be appreciable, and if the detection efficiency for the fluorescent radiation is greater than for the primary radiation (as would probably be the case if the detector were, for example, a Geiger counter), the effect of the fluorescent radiation would be even more pronounced. In situations such as this the detection efficiency as a function of energy would need to be known at least approximately in order to apply a correction

In the case of Compton interactions, effects due to scattered radiation will often be greater. The *electronic* cross-section for Compton scattering into a narrow cone of semiangle θ_0 is given by

$$\sigma_{\theta_0} = \pi r_0^2 \theta_0^2 \tag{5.47}$$

to be compared with the removal cross-section given by (5.12). The fraction of Compton interactions which will result in scattered radiation entering the counter thus approaches $\frac{3}{8}\theta_0^2$ at low energies, and rises at higher energies. Note that Compton scattering at low angles produces a scattered radiation only slightly reduced in photon energy, which may not be discriminated against by the detector.

The effect of pair production in the absorber followed by annihilation is somewhat similar to photoelectric absorption, in that both processes produce a secondary radiation which is isotropic and which can readily be allowed for. In the case of pair production, annihilation will normally take place within the absorber, and will be by the two-quantum process, producing photons of 511 keV. Each pair production will thus produce, after annihilation, 2 secondary photons.

When we turn to the elastic scattering processes, we realise that the Rayleigh scattering could in principle give rise to considerable scattering through small angles in some circumstances, and that this could involve appreciable correction for radiation entering the detector aperture.

If the removal cross-section of an absorbing and scattering material is being investigated, and if the order of magnitude of the Rayleigh scattering is known, we can use the data of Franz (1935) to estimate the fraction falling within the aperture. Representative plots of differential cross-section as a function of angle are shown in fig. 5.12.

We may also refer back to (5.33) giving the angle within which 75% of the scattered radiation falls. This expression shows that this angle will be much larger than the angular aperture of an experimental arrangement to determine removal cross-section for γ-radiation below ~1 MeV. Above 1 MeV we may expect correction for small angle scattering to be appreciable. In fig. 5.19 we have displayed in the form of a contour diagram, the calculated Rayleigh scattering in relation to other scattering processes, showing the regions of energy and atomic number in which the Rayleigh scattering exceeds certain fractions of the whole. For substantial ranges of energy and atomic number the Rayleigh scattering exceeds 5%, and at the higher energies (> 500 keV) within these domains most of the scattering will take place within about 10 degrees.

It is interesting to note that the Rayleigh cross-section becomes relatively important just below the K and L critical absorption edges of the heavier elements, and that there is one small domain in which Rayleigh scattering exceeds 20% of the total. However, the scattering will then take place over a large angular range and so will not necessitate unduly large corrections to absorption experiments. 'Scattering-in' of radiation could be appreciable in these circumstances and this points to the importance of a small aperture A in fig. 5.1(*b*).

A satisfactory way of avoiding corrections for small-angle scattering is to use apertures of various sizes, and to extrapolate to zero aperture, as referred to in section 4.4. At energies above 1 MeV this will require the use of very small apertures for reliable extrapolation. Equation (5.33) provides a valuable guide for this procedure.

The availability of Mössbauer radiation opens up the possibility of discriminating against Rayleigh scattering in an absorption experiment. If Mössbauer radiation is used for a removal cross-section determination (in a non-resonant sample), it can be analysed by means of an additional, resonant, absorber (or possibly a resonant *scatterer*, in a suitable geometry) before entering the detector. All modified radiation in the beam which

Fig. 5.19. Percentage of Rayleigh scattering in the total attenuation cross-section, as a function of Z and E.

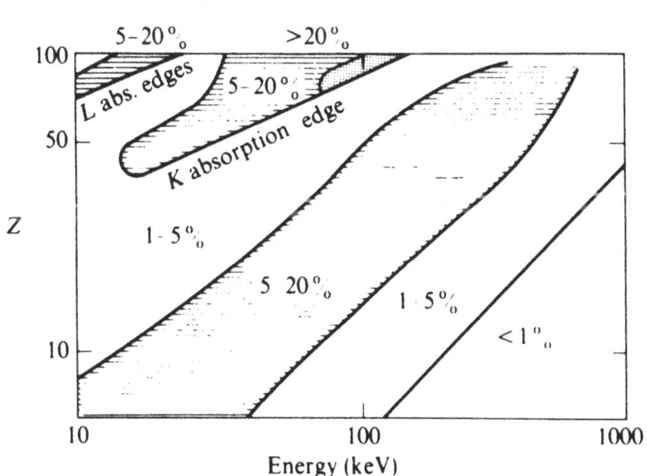

would inadvertently enter the detector, including the incoherent part of the elastic scattering (Rayleigh scattering with recoil) can, in principle, be measured and allowed for in such an experiment. We note also the possibility of observing separately the effect of recoilless Rayleigh scattering by giving a transverse velocity to the absorber during the measurement, as demonstrated by Champeney and Woodhams (1966), who observed that high speed rotation of a (non-resonant) absorber in its own plane will cause a detectable loss of resonance in Mössbauer radiation passing through the moving sample to a suitable resonant foil absorber. The effect is attributed to the radiation acquiring a Doppler shift equal to $E_\gamma(v/c)\sin\theta$, where v is the transverse velocity of the moving absorber and θ the scattering angle.

A further difficulty arises if there is any small-angle scattering at very low Bragg angles, due to macrostructure in the absorber. Such scattering can occur at angles of only a few minutes of arc, and would, of course, be structure-dependent. Chipman (1955) has demonstrated the importance of this in the present context; in the course of measuring attenuation coefficients in carbon and some compounds containing carbon, small-angle scattering at angles of less than 20 min of arc was demonstrated in graphite, and it was shown that the mass attenuation of graphite for copper K_α radiation depends markedly on whether this scattered radiation is rejected from the detector or not. The small-angle scattering was not present in the case of a paraffin absorber, which accordingly gave a more useful value of (μ/ρ) for carbon in compounds.

6

X-ray production by protons, α-particles and heavy ions

6.1 Ionization cross-sections (experimental) for protons and α-particles

The observation that characteristic X-radiation can be produced by α-particle bombardment seems to have been first made and reported by Rutherford, Chadwick and others during the years immediately following the discovery of characteristic radiation. Since that time, the subject has been re-examined at intervals. In 1928, Bothe and Franz reported a series of measurements in which various elements were irradiated with α-particles from radioactive decay. Cork (1941) carried out a series of measurements of characteristic radiation produced by deuteron bombardment, using a maximum deuteron energy of 10 MeV. He observed that K X-rays were detectable up to an atomic number of 38. Above this value, no K X-rays were observable, but between atomic numbers of 52 and 78 the L X-rays could be detected. Evidently the L electrons can be ejected from the L shell in circumstances where the K electrons are too strongly bound to be ejected by deuterons of this energy.

For detailed measurements we must wait until the 1950s, when the increasing availability of small accelerators opened up the possibility of experimental work on an absolute basis over a wide range of particle energies. From investigations for example by Messelt (1958) and by Khan and Potter (1964) it became clear that the X-ray yield increased rapidly with proton energy. The data of fig. 6.1 illustrate this point, and also show that the yield at these energies is several orders of magnitude below that obtained from electrons used under the conditions of fig. 3.20. It was also found that there is a strong inverse variation with atomic number (confirming the earlier work of Cork) and this was investigated systematically at about this time. Examples of such work are given by Merzbacher and Lewis* (1958; fig. 12 of their paper), illustrating these two main features of X-ray production by proton bombardment. We note also that at these somewhat higher proton energies the yield of X-ray photons per proton can be quite comparable with the yield under electron bombardment. This is of importance in connection with the use of proton-induced X-rays for elemental analysis, to be described below.

Inner-shell ionization is defined by a cross-section, as for electron bombardment, and the experimental study of cross-sections has been carried out intensively in recent years. This may be approached by the use of either thin or thick targets; for if the intensity of K-shell characteristic X-ray emission (expressed here as photons per steradian per unit time) be given by $I(E)$, we note that, for a thin target

$$I(E)\mathrm{d}\Omega = \omega_K n \sigma_K(E) x \mathrm{d}\Omega \tag{6.1}$$

where n is the number density of target atoms, x the target thickness, σ_K the ionization cross-section for the K shell and ω_K the fluorescence yield. The cross-section may therefore be determined by preparing a thin sample of

* These authors give a detailed account of experimental work carried out up to the time of their paper, together with a description of associated theoretical analysis; in the present account it has therefore been felt unnecessary to cite many publications prior to the date of their paper.

Fig. 6.1. Yield of X-rays from proton bombardment of thick targets (Khan and Potter, 1964). [● Jopson *et al.*, unpublished; ▲ Messelt, 1958; ■ Khan and Potter.]

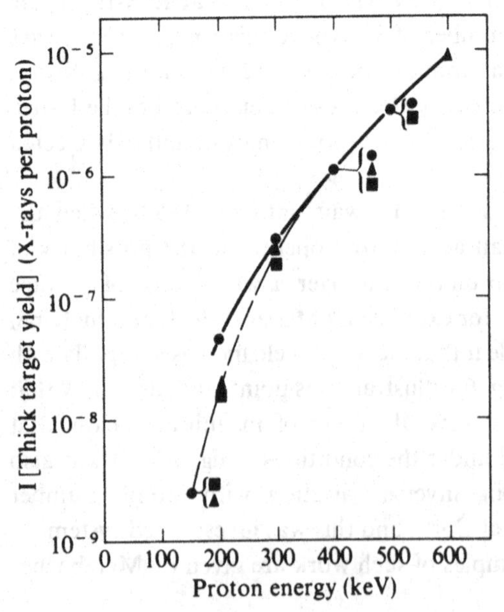

target material of known mass–thickness, and determining the photon emission rate into a defined solid angle. (The radiation may be assumed to be emitted isotropically.)

In the case of a thick target, let the radiation in the measured direction be emitted at an angle θ to the target normal. If the proton energy be increased by $\mathrm{d}E$, the proton range increases by $\mathrm{d}x$, and we may write

$$[I(E+\mathrm{d}E)-I(E)]\mathrm{d}\Omega=\frac{\omega_K\sigma(E)n\mathrm{d}x\,\mathrm{d}\Omega}{4\pi} -I(E)\left(\frac{\mu}{\rho}\right)\rho\sec\theta\,\mathrm{d}x\,\mathrm{d}\Omega \tag{6.2}$$

where μ is the linear attenuation coefficient of the target material and ρ its density. The first term on the right-hand side represents the increase in intensity due to the increment of thickness traversed by the proton beam on entering the target, and the second term represents the extra attenuation of the remainder of the radiation which is now produced at a greater depth than before.

Hence

$$\frac{\mathrm{d}I}{\mathrm{d}x}=\frac{\omega_K\sigma(E)n}{4\pi}-I\,\mu\sec\theta \tag{6.3}$$

and

$$\sigma(E)=\frac{4\pi}{\omega_K n}\left\{\frac{\mathrm{d}I}{\mathrm{d}E}\cdot\frac{\mathrm{d}E}{\mathrm{d}x}+I\,\mu\sec\theta\right\} \tag{6.4}$$

Cross-sections may thus be determined from thick targets by carrying out determinations of X-ray yield over a range of closely-spaced energies, incorporating tabulated values of $\mathrm{d}E/\mathrm{d}x$ and of X-ray attenuation coefficients in the calculation.

In our review of experimental data, we distinguish between thick targets, in which the incident particles are brought completely to rest, and thin targets, through which the incident particles pass with small energy-loss. The work so far referred to was carried out with thick targets.

Direct measurements, using thin targets, were reported by Bissinger *et al.* (1970). They had available a proton beam of 2–30 MeV energy, and they bombarded targets of calcium, titanium and nickel. A proportional counter was used as the detector. It was found that the cross-section rose steadily as the proton energy was increased, reaching a maximum, and then beginning to decline slowly at higher proton energies. This confirmed the general trend of theoretical prediction. Absolute fitting with theory was rendered difficult by their use of an empirical expression for the fluorescence yield which is now thought to give predictions on the low side. Their method of data plotting follows that of Merzbacher and Lewis (1958), but simple methods have been subsequently devised, and these and other data have been presented in a simple form by Garcia (1970b) using the same parameters ($\sigma_K E_K^2$ and $U = E/E_K$) as are widely employed for electron data. Figure 6.2 illustrates these results, from which we see that the cross-section for ionization by proton bombardment attains a value about three times that obtainable by electron bombardment – peak cross-sections of 1000

Fig. 6.2. Cross-sections for inner-shell ionization as a function of the parameter E/E_k (Garcia, 1970b). The theoretical curve follows the theory of Garcia (1970a). (The units are the same as for fig. 3.13.)

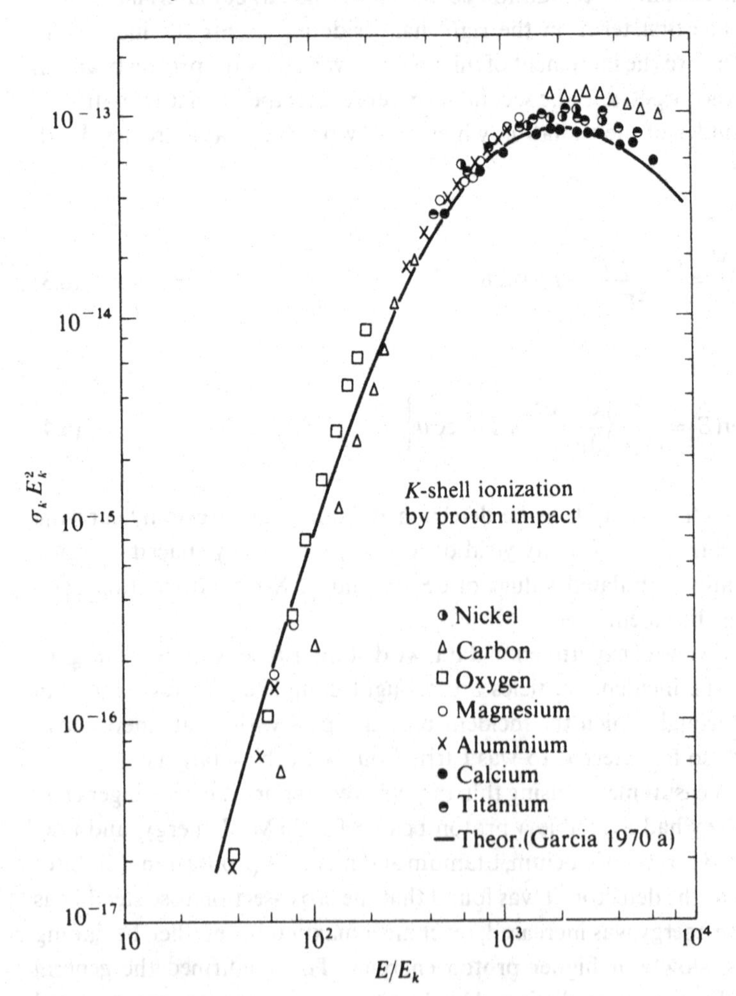

barns (Ni) and 3×10^4 barns (Al) may be compared with 300 and 10^4 barns respectively estimated from the curves in fig. 3.13. We note from Garcia's data that $\sigma_K E_K{}^2$ is the same function of U for all elements studied, although in the case of carbon the agreement is not good. However, the data for carbon were obtained from thick targets and the data reduction therefore involved the use of stopping-power values, which may have introduced uncertainties into the calculations.

A more recent presentation of data in the same form, but with more data, has been given by Garcia, Fortner and Kavanagh (1973) (fig. 6.3). Although the data do not lie entirely on a smooth curve, the discrepancy noted in the

Fig. 6.3. Cross-sections for inner-shell ionization as a function of $\frac{E}{E_k} \cdot \frac{m}{M}$ (Garcia *et al.*, 1973).

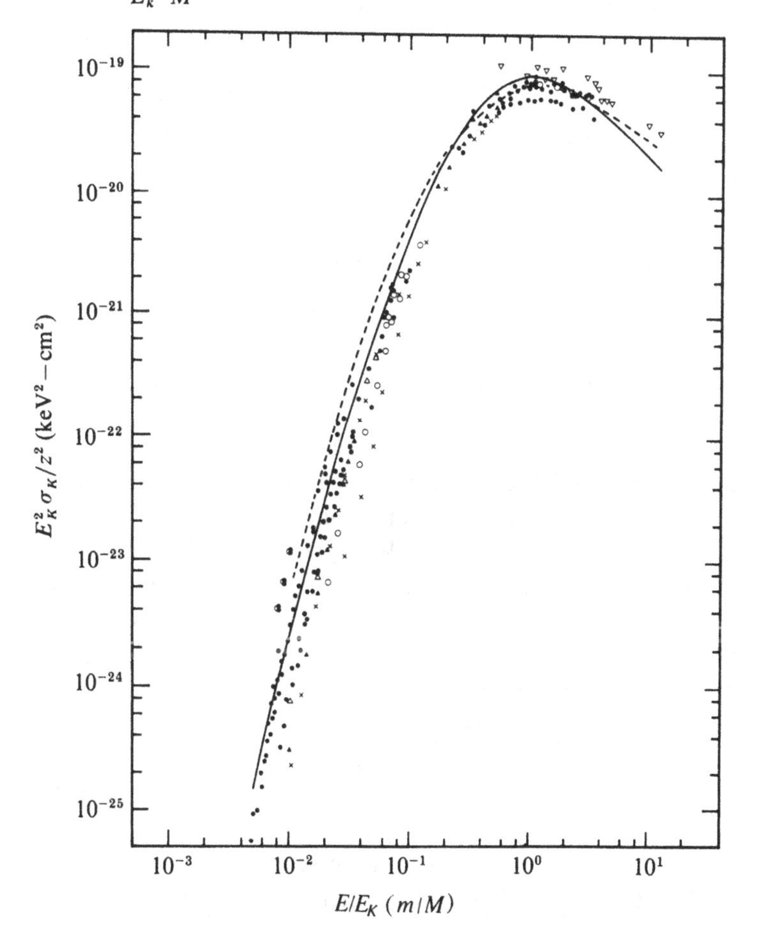

case of carbon in fig. 6.2 has been reduced, and a better overall fit to a smooth curve, and to Garcia's theory, is obtained.

Still more recent data have been reported by Khan *et al.* (1976, 1977) and are discussed by them. In fig. 4 of their 1977 paper they plot their data in the same manner as Garcia. This procedure reveals a close adherence to a universal curve over the range of proton energies and target atomic numbers investigated by them, and good agreement with Garcia's theory.

We shall return to a discussion of universal curves later in this chapter.

6.2 Theoretical considerations

Throughout this chapter the discussion is confined to protons or α-particles which are slow, in the sense that the energies T_m, transferred in a head-on collision with a free electron are either less than, or do not greatly exceed, the K-shell binding energy, i.e.

$$T_m \ngtr E_K \tag{6.5}$$

If the masses of the heavy incident particle and the electron are M and m respectively, the quantity T_m may be evaluated to be

$$T_m = 2mv^2\left(\frac{M}{m+M}\right)^2 \tag{6.6}$$

where v is the velocity of the incident particle.

The term in brackets may be closely approximated to 1, so we may write

$$E_p \ngtr \frac{M}{4m} E_K \tag{6.7}$$

We follow Merzbacher and Lewis in setting an upper limit of 5 MeV, and in excluding elements with atomic numbers less than 10 (i.e. $E_K \sim 1$ keV). Under these circumstances the nature of the interaction is very different from that presupposed in, for example, (3.22) where the projectiles (electrons) are swift particles. The interactions are therefore *quasi-adiabatic*, in which little perturbation of the electron orbits occurs, and the probability of K-electron ejection is small.

Merzbacher and Lewis also set a *lower limit of 100 keV* for the incident particle energy. This allows the inequality

$$\frac{1}{4\pi\varepsilon_0} \cdot \frac{ze^2}{hv} \ll 1 \tag{6.8}$$

to apply, where z is the atomic number of the projectile.

This is the condition of the Born approximation, and is the basis of the calculation of Merzbacher and Lewis for inner-shell ionization by protons and α-particles.

We have already noted from (6.6) that the maximum energy which may be transferred to an electron in collision with a proton is given very closely by

$$T_m = \frac{4m}{M} E_p \tag{6.9}$$

If the view is taken that inner-shell ionization occurs because of collisions of this kind, we would expect a threshold to occur, when $T_m = E_K$, i.e. when

$$E_p = \frac{M}{4m} E_K \tag{6.10a}$$

i.e. when

$$U = \frac{M}{4m} \tag{6.10b}$$

However, the data of figs. 6.1–6.3 show that X-rays are produced at values of U considerably less than this. It is clear the ionization occurs not by collision with a free electron but by collision with the atom taken as a whole. Under these circumstances, considerably greater amounts of energy can be transferred. Such collisions therefore play a significant part in the process. Furthermore, it is clear from the data that no threshold for X-ray production exists, and in this respect the process differs sharply from X-ray production by electron bombardment.

We have defined E/E_K as the *excitation ratio*, U. T_m/E_K may be defined as the *reduced excitation ratio*, U_R. We note that the data of figs. 6.2 and 6.3 pass through a maximum when $U = M/m$. It may also be noted that for $U_R \ll 1$, the data may be expressed approximately in the form

$$\sigma_K {E_K}^2 \propto \left(\frac{E}{E_K}\right)^4 \tag{6.11a}$$

or

$$\sigma_K \propto \frac{E^4}{{E_K}^6} \tag{6.11b}$$

If we take E_K as being approximately proportional to Z^2, this implies a Z^{-12} dependence. This conforms with the theoretical treatment of Merzbacher and Lewis at low proton energies, for which $T_m \ll E_K$.

At low energies, their expression is

$$\sigma_K = \frac{2^{10}\pi}{45} \frac{T_m{}^4}{E_K{}^6} \left(\frac{ze^2}{4\pi\varepsilon_0}\right)^2 \tag{6.12}$$

This may be written

$$\sigma_K = \frac{2^{10}\pi}{45} \left(\frac{ze^2}{4\pi\varepsilon_0}\right)^2 \left(\frac{4m}{M}\right)^4 \frac{E^4}{E_K{}^6} \tag{6.13}$$

showing that the cross-section varies directly as the *fourth* power of the incident particle energy, and inversely as the *sixth* power of the ionization potential. (This leads to their prediction of a Z^{-12} dependence on atomic number.)

For higher energies the rise of cross-section with particle energy is less strong, and a maximum occurs when velocity matching of the incident particle and the K electron takes place. A quantity η_K was defined, given by

$$\eta_K = \frac{v_p{}^2}{v^2{}_{\text{eff}}} \tag{6.14}$$

where v_{eff} is the mean Bohr characteristic velocity of the K electrons. Velocity matching thus occurs when $\eta_K \approx 1$.

We have seen that the method of data plotting used by Garcia leads to a universal curve, in which a plot of $\sigma_K E_K{}^2$ v. U leads to approximately the same curve for all elements. The plotting of data in such a way as to obtain a universal curve is an important aspect of data reduction, as it leads to numerical methods of computing cross-sections for any element at any energy. Methods of achieving this, if possible improving on the method already described, have been examined by Basbas *et al.* (1973) using data newly obtained by them, together with some additional determinations available at that time. Instead of the parameter U, they introduced the quantity $\eta_K/\theta_K{}^2$ as abscissa. The theoretical significance of this parameter is discussed by Merzbacher and Lewis (1958). η_K is as already defined. We may note at this point that

$$v_{\text{eff}} = Z_{\text{eff}}\alpha c \tag{6.15}$$

where α is the fine structure constant, c the velocity of light, and Z_{eff} is the effective atomic number taking into account *inner* but not *outer* screening. θ_K is the *screening number* defined by

$$E_K = \theta_K \cdot \tfrac{1}{2}mc^2(Z_{\text{eff}}\alpha)^2 \tag{6.16}$$

and, in effect, allows for the non-hydrogenicity of the atom, that is, it expresses the effect of outer screening. Values of θ_K have been compiled by

Walske (1956) as a function of Z. They range from approximately 0.7 for $Z = 10$ to 0.9 for $Z = 100$.

When applied to the data of Basbas *et al.*, their modified method of plotting (fig. 5 of their paper) does not lead to an improved universal curve, but when applied by Khan *et al.* (1977) to their own data, the data lie well on a smooth curve, and also provide confirmation of a curve predicted by theoretical considerations (Plane-wave Born approximation). Evidently the experimental data obtained by the latter authors and the values of the fluorescence yield used by them to calculate cross-sections from their experimentally determined X-ray emission rates are internally more consistent than the data of Basbas *et al.*, although the data of Basbas *et al.* do include data obtained by deuteron, ^{3}He and α-particle bombardment as well as by proton bombardment, thereby providing a more stringent test of the theory.

A further factor influences the behaviour of slow protons which has not yet been referred to. This is the Coulomb deflection of the incident particle by the atomic nucleus. This becomes significant for protons at or near the lower limit of energy under consideration here, and has been taken into account by the treatment known as the *semi-classical approximation.* This allows the projectile to be treated as a point particle. A full treatment of this approach is given by Bang and Hansteen (1959), and by Hansteen and Mosebekk (1973). Their treatment of inner-shell ionization in terms of the impact parameter of the incident particle is important, particularly at relatively low bombardment energies, and for details of this treatment, their original papers must be consulted.

A classical treatment (see fig. 6.4) sheds light on the ionization process by drawing attention to the time spent by the projectile in exerting its Coulomb force on the electron being ejected. This may be compared with

Fig. 6.4. Illustrating the semi-classical approximation.

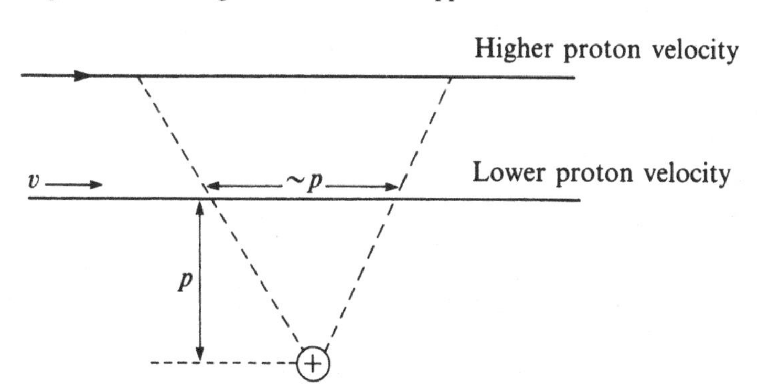

the period of the orbit. In order for the impulse to be effective, the duration of the impulse must not exceed the period of the electronic orbit. The duration is of order p/v, and the period is $\sim h/E_K$. It is usual to divide this quantity by 2π, i.e. to take $1/\omega$ as the limiting time. Hence the most important impact parameters are those for which

$$\frac{p}{v} \not> \frac{\hbar}{E_K}$$

or

$$p \not> \frac{\hbar v}{E_K} \tag{6.17}$$

where E_K is the transition energy.

The quantity $\hbar v/E_K$ therefore plays a significant part in determining the most important impact parameters for this process, and is known as the *adiabatic radius*, r_{ad}. The greatest contribution to the ionization probability

Fig. 6.5. Coincidence measurements as a function of impact parameter (Laegsgaard *et al.*, 1972).

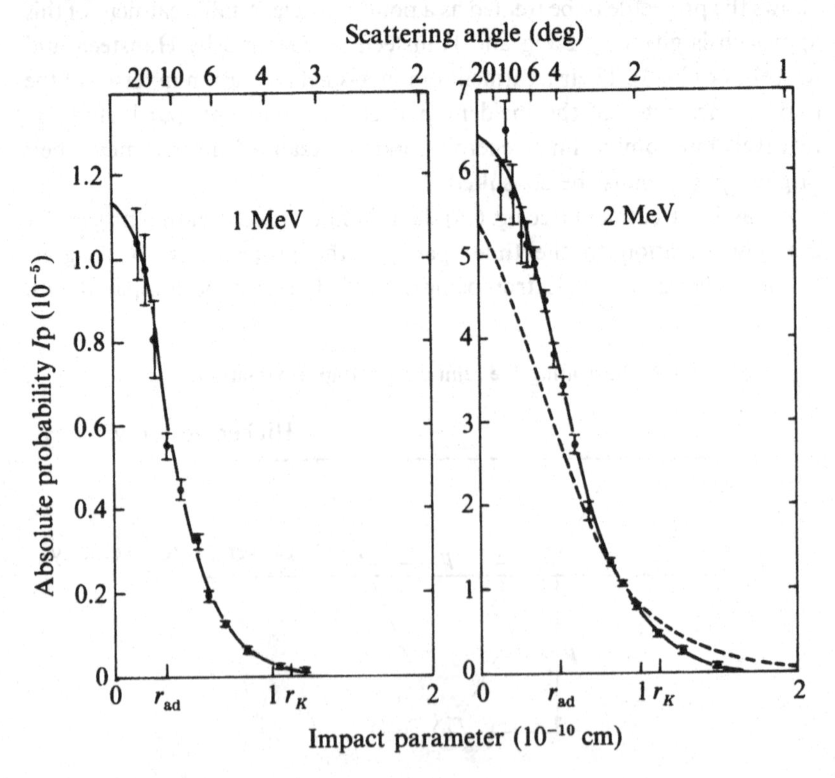

comes from impact parameters equal to the adiabatic radius or less. This has been demonstrated in an elegant fashion by Laegsgaard *et al.* (1972). These authors describe an experiment in which targets of selenium and silver were bombarded with protons of energy 1 and 2 MeV, the special feature being that the scattered protons were observed in coincidence with the characteristic X-rays. By fixing the proton detector at a series of selected scattering angles, it was possible to measure the X-ray yield as a function of impact parameter, assuming a classical trajectory for the protons. Their data (fig. 6.5) illustrate clearly the importance of the adiabatic radius, and also show that it rises with increasing proton velocity, in accordance with the expectation of (6.17).

It is instructive to compare this with the Bohr radius for the K-shell, and it is not difficult to establish that

$$\frac{r_{\mathrm{ad}}}{r_{\mathrm{Bohr}}} = \frac{2v_{\mathrm{p}}}{v_{\mathrm{B}}} \tag{6.18}$$

i.e. for particle velocities which are slow compared with the Bohr characteristic velocity (i.e. for $\eta \ll 1$), the important impact parameters lie well inside the K-shell; the adiabatic radius gives a truer indication of the effective impact parameter than the K-shell Bohr radius. As the particle energy increases, the effective impact parameter increases also, and it can be established that the adiabatic radius approaches the Bohr radius as $T_m \rightarrow E_K$.

The adiabatic radius therefore becomes equal to the shell radius as the region of maximum cross-section in fig. 6.2 is approached.

6.3 Relative intensities of lines in the K and L spectra

When an atom is singly ionized in the K shell, the relative intensities of K lines are expected to be independent of the mode of ionization of the shell. Comparisons have been made between spectra obtained by electron bombardment, proton bombardment and fluorescence excitation, and the broad structure has been found to be the same in all cases. Most of these studies have been carried out using detectors of moderate resolution only, and we defer, for the moment, any consideration of the satellite lines which can be seen at higher resolution.

The relative intensities $K_{\alpha_2}:K_{\alpha_1}$, and $K_\beta:K_{\alpha_{1+2}}$ have already been discussed in section 3.3. These data were in fact obtained by proton bombardment, because the low level of Bremsstrahlung (see section 6.4 below) obtained under these conditions is a distinct advantage in the data-processing. These data are essentially independent of bombarding energy.

As an example of L spectra, we illustrate in fig. 6.6 the L spectrum of tantalum obtained by proton bombardment at 480 keV using a Si(Li) detector. The main lines of the spectrum are clearly resolved, and the transitions to which they correspond may be seen from the energy-level diagram of fig. 3.3. The lines of the L_2 and L_3 sub-spectra predominate, and the lines of the L_1 spectrum are weak, the strongest of these being the $L_{\gamma 2+3}$ lines. This is qualitatively similar to the situation found in electron bombardment (see table 3.5) and occurs because L_1 vacancies migrate with high probability to the L_2 or L_3 shells before being filled radiatively. That is to say, the Coster–Kronig coefficients (especially f_{13}) tend to be high. This, in effect, reduces the fluorescence yield of the L_1 sub-shell to a low value.

Fig. 6.6. L spectrum of tantalum.

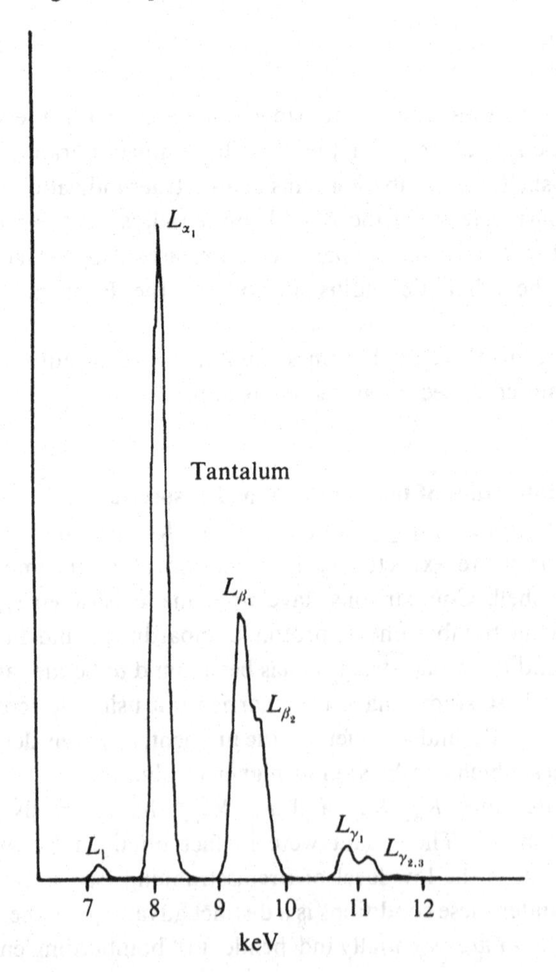

Table 6.1. *L-shell fluorescence yields and Coster–Kronig coefficients* (*Avaldi et al., 1983, after Krause* (*1979*).)

	ω_1	ω_2	ω_3	f_{12}	f_{13}	f_{23}
$_{48}$Cd	0.018	0.056	0.056	0.10	0.59	0.155
$_{50}$Sn	0.037	0.065	0.064	0.17	0.27	0.157
$_{52}$Te	0.041	0.074	0.074	0.18	0.28	0.155
$_{53}$I	0.044	0.079	0.079	0.18	0.28	0.154
$_{56}$Ba	0.052	0.096	0.097	0.19	0.28	0.153

These facts may be seen from the representative data of table 6.1 (Avaldi *et al.*, 1983).

Turning to the ionization cross-sections for the L shell, we may look at the data for the L_3 sub-shell illustrated by Garcia *et al.* (1973) (fig. 6.7(a)).

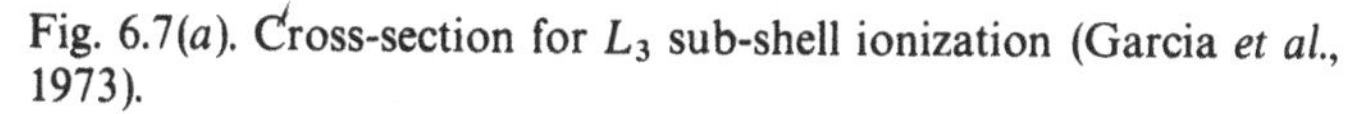

Fig. 6.7(a). Cross-section for L_3 sub-shell ionization (Garcia *et al.*, 1973).

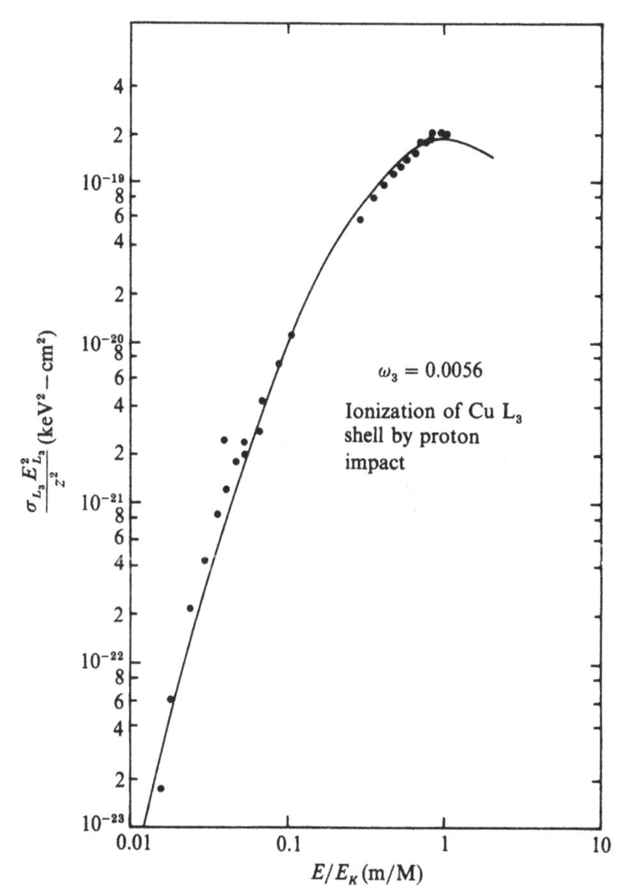

We note that the curve has the same form as fig. 6.2, but with cross-sections higher by a factor of 2, because there are 4 electrons in this sub-shell compared with 2 in the K shell. The points lie on a smooth curve which follows closely the theoretical prediction. Fig. 6.7(b) illustrates the total L shell ionization cross-section for various elements plotted in a similar way. These data were obtained essentially by summing the total L shell production, and dividing it by a mean fluorescence yield for the L shell in order to obtain an ionization cross-section. Again the shape of the curve is very similar but increased by a further factor of 2, showing that the L shell may be treated as consisting of eight equivalent electrons.

When considering all three sub-shells separately, a more complicated picture emerges. By observation of the relative intensities of the lines

Fig. 6.7(b). Cross-section for L shell (total) ionization (Garcia *et al.*, 1973).

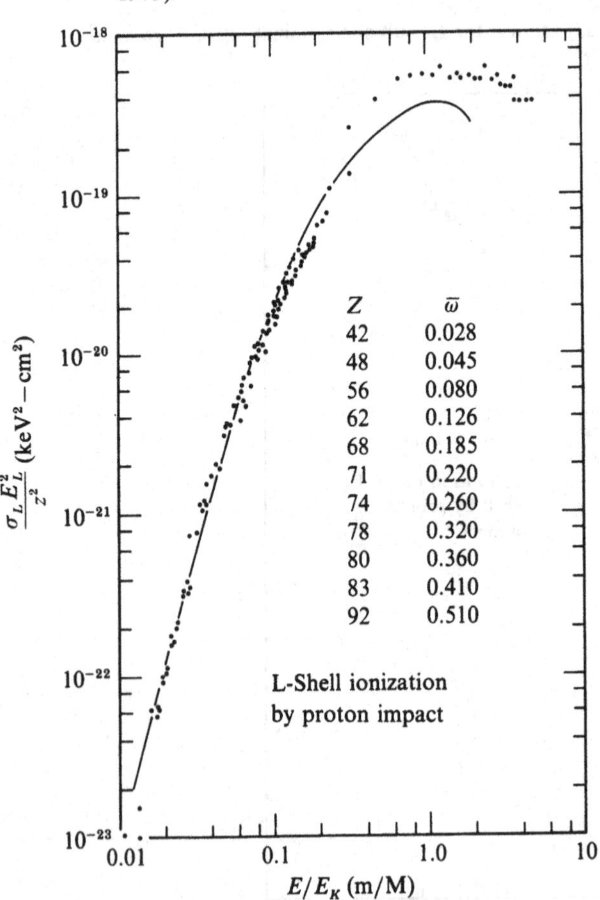

emitted from a thin target, the X-ray *production* cross-section may be determined. From this the *ionization* cross-sections may be calculated, if the fluorescence yields and Coster–Kronig transition probabilities for the three sub-shells are known. The procedure for this is described by, for example, Sokhi and Crumpton (1986).

The existence of the three sub-shells implies that the ionization cross-section may vary in a different way with incident bombarding energy, giving rise to some degree of complexity. However, lines of the same sub-spectrum are expected to show a relative intensity which is independent of bombarding energy. This is illustrated by the data of Chen *et al.* (1976) shown in fig. 6.8, where the ratio of the $L_{\gamma_1} : L_\eta$ lines (L_2 sub-spectrum) is seen to be independent of proton bombarding energy for the three elements investigated.

Data for the three sub-shell ionization cross-sections for gold are given by Chen (1977), and reproduced here in fig. 6.9. The forms of the σ_{L_2} and σ_{L_3} curves resemble each other closely. However, the curve for σ_{L_1} has a 'knee', which causes the ratio of the $L_1 : L_2$ line intensities to vary with proton energy. The ratio of $L_1 : L_2$ ionization cross-sections is plotted by Chen, and reproduced here in fig. 6.10. A noteworthy feature of this ratio is its steady rise with *decreasing* proton energy. A simple experimental illustration of this point may be given. Fig. 6.11 illustrates the L_{γ_1}(L_2 sub-spectrum) and

Fig. 6.8. Ratio of L_{γ_1}:L_η line intensities as a function of bombarding energy. (Chen *et al.*, 1976).

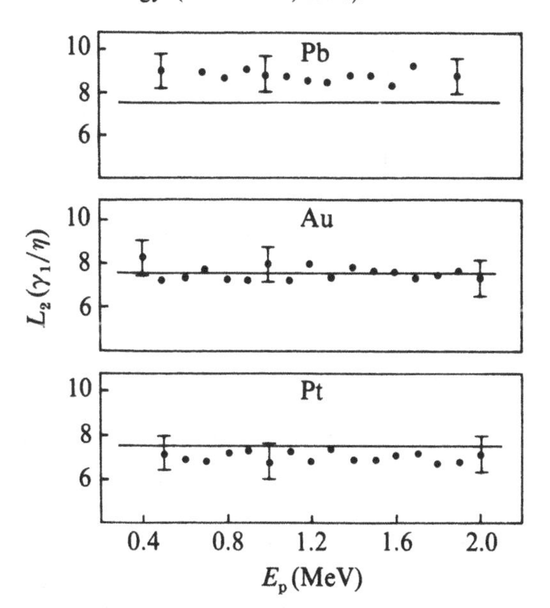

Fig. 6.9. Ionization cross-sections for L_1, L_2 and L_3 sub-shells in gold (Chen, 1977).

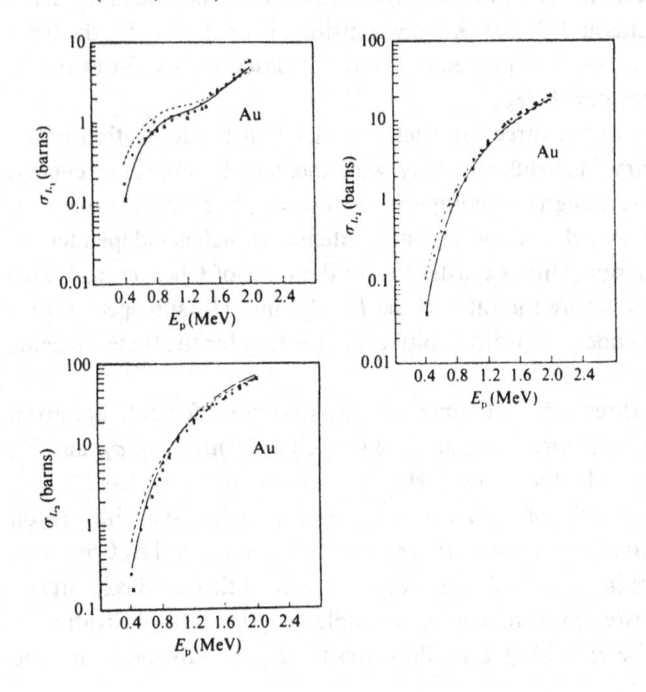

Fig. 6.10. Ratio of L_1:L_2 ionization cross-sections in gold (Chen, 1977).

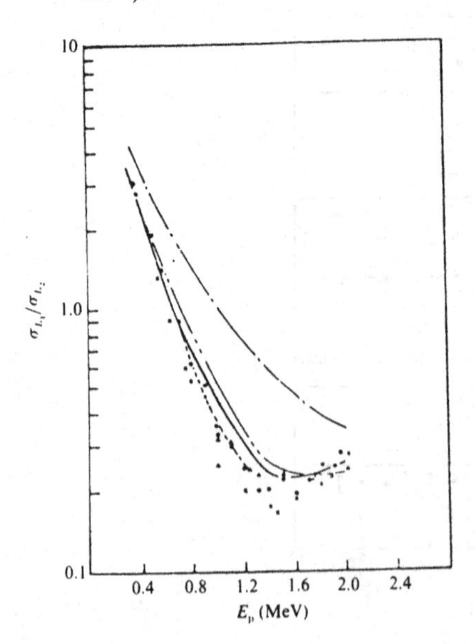

Fig. 6.11. Part of the spectrum of tantalum under proton bombardment at (a) 480 keV and (b) 250 keV to illustrate the change in the relative intensity of $L_{\gamma 1}$ and $L_{\gamma 2.3}$ lines with change of bombardment energy.

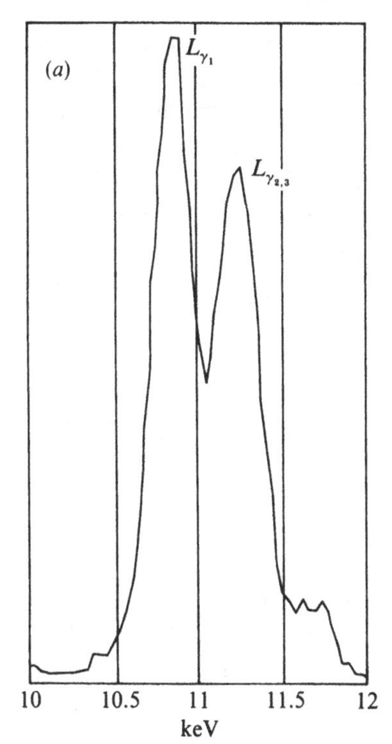

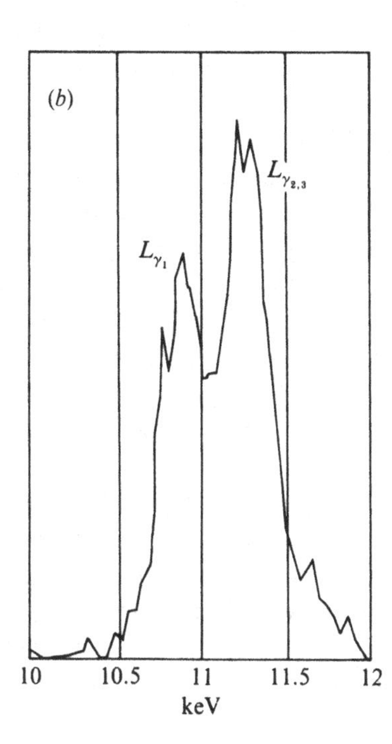

Fig. 6.12. Wave-functions in the L shell for a hydrogen-like Au atom. The impact parameters are shown which for different proton energies give maximum contribution to 'straight-line' (i.e. neglecting Coulomb deflection) ionization cross-sections for L_1, L_2 and L_3 sub-shells (Hansteen and Mosebekk, 1973).

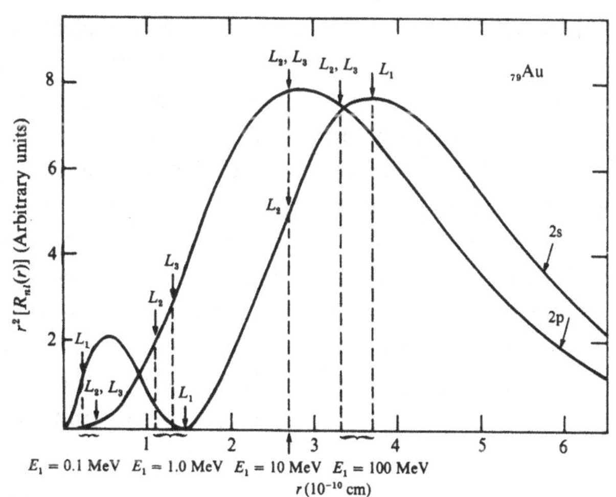

$L_{\gamma_{2,3}}$ (L_1 sub-spectrum) lines obtained from a thick target of tantalum using proton energies of 480 keV and 250 keV respectively. At 480 keV, the L_{γ_1} line predominates over the $L_{\gamma_{2,3}}$ line. At 250 keV the relative intensities are reversed.

This phenomenon may be explained qualitatively, as illustrated in fig. 6.12, in terms of the radial wave-functions of the L_1 and L_2 sub-shells and the changing impact parameter of the protons as the bombarding energy is changed. It will be seen that there is a node in the L_1 radial wave-function, and this is the general region of interest in the present context. On reducing the proton energy from about 1.0 MeV, the impact parameter enters a region where the L_1 wave-function (hence the L_1 ionization probability) predominates over the L_2 wave-function. Hence at lower energies the ionization in the L_1 shell becomes relatively greater, and the lines in the L_1 sub-spectrum relatively more intense.

A substantial body of data now exists for L shell X-ray production by proton bombardment. A review has been given by Sokhi and Crumpton (1984), who also include comprehensive references.

6.4 Bremsstrahlung production

In addition to the inner-shell ionization just described, there is some production of Bremsstrahlung, due to the Coulomb interaction between the incident charged particles and the target nuclei. This is of low intensity, essentially because of the much greater mass of protons as compared with electrons. Detailed calculations have been carried out by Drell and Huang (1955).

As in the case of electrons, the Bremsstrahlung produced is a continuous spectrum, with an intensity varying slowly with photon energy and extending upwards towards a high energy limit. At the lower end of the spectrum it is, however, dominated by a process giving rise to relatively much greater intensities. This results from electrons which are 'knocked-on' by the protons as they slow down in the target material. Some reference to this has been made in chapter 2, section 11. We have seen that the maximum energy of *free* electrons knocked on in this way is T_m. However, electrons which are bound in inner-shells may be ejected with larger energies, and the continuous spectrum produced by these electrons as they slow down is the dominant feature of the continuum at photon energies of a few keV and below. These electrons are of course those which are ejected during inner-shell ionization, and so the Bremsstrahlung they emit is normally inseparable from the characteristic radiation described in the preceding three sections of this chapter. Detailed experimental data have been obtained by Ishii *et al.*, (1976) for protons and ^3He projectiles. From this work it emerges that the intensity of this secondary Bremsstrahlung is proportional to z^2, where z is the *projectile* atomic number, and that it extends up to photon energies well above T_m. Work has also been carried out by Folkmann *et al.* (1974) in connection with elemental analysis by proton-induced X-ray emission (see section 6.6, below).

In order to calculate the secondary Bremsstrahlung effect, some knowledge of the energy distribution of knocked-on electrons is needed. The work of Huus *et al.* (1956) establishes that the number of secondary electrons with energies in excess of T_m falls very rapidly with increasing electron energy. Above a few keV the secondary Bremsstrahlung produced by these is undetectable, and only the primary proton Bremsstrahlung remains. This is illustrated in fig. 6.13. The experimental data illustrate the

Fig. 6.13. Bremsstrahlung from aluminium (thick target; 450 keV protons). ⫿ Experimental observations; - - - calculations based on Drell and Huang for primary radiation; -·-·-·- Drell and Huang calculations plus the effect of secondary electrons from the 1*s* (*K*) shell; ——— Drell and Huang calculations plus the effect of all *s* electrons (Ward and Dyson, 1978).

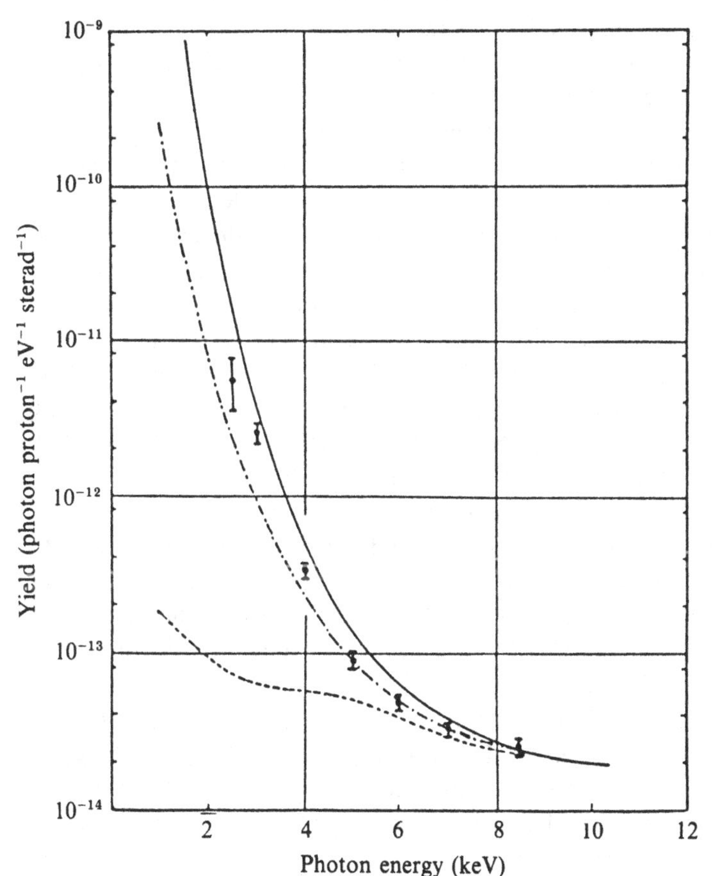

secondary Bremsstrahlung clearly, and the question arises as to which electrons are mainly responsible for this process. The contribution from the 1*s* (*K* shell) electrons is illustrated in the diagram, and this appears to be insufficient to account for the observed intensity. From the theoretical work of Huus *et al.*, it is possible to calculate the additional contribution of the remaining *s* electrons (L_1, M_1, etc., sub-shells). Evidently, from the diagram, some additional electrons are also involved.

6.5 X-ray production by heavy-ion bombardment

A much more complex situation arises here, and only the essential features of the varied phenomena will be outlined. The additional complexity arises for two reasons – first, the projectile has a greater mass and charge, and would therefore be expected to exert a more disruptive effect on the target atom; and secondly, the projectile has an electronic level structure of its own, creating new phenomena when the two nuclei approach each other during the collision process.

When the proton-induced X-ray spectra are examined at higher resolution, some structure is observed in the lines. Experiments have been carried out by McWherter *et al.* (1973) using a LiF crystal spectrometer together with a proportional counter to serve as a detector. One of their published graphs is illustrated in fig. 6.14. In the spectrum produced by proton bombardment, there is clear evidence of multiple ionization, that is, evidence that one or more *L* electrons are ejected simultaneously with a *K* shell electron. In their terminology, $(2p)^6$, $(2p)^5$, $(2p)^4$, etc., means that 6, 5, 4, etc. electrons remain in the 2*p* shell, so that the $K_\alpha(2p)^6$ line is the main line, emitted with the 2*p* shell intact; satellite lines corresponding to double (*KL*) and triple (*KLL*) ionization may be seen on the high energy side of this line. Evidently the $K_\alpha(2p)^5$ line may be identified with the K_{α_4} line of Parratt (1936a) (see fig. 3.33) though with somewhat higher intensity relative to the parent line. A similar structure may be seen in the K_β line.

When we look at the spectrum obtained by α-particle bombardment, the satellite lines are seen to be stronger, and in the case of bombardment by oxygen ions, the satellites are stronger than the parent, and the median position of the whole group of lines has shifted towards higher energies by an amount in the region of 1%. The loss of from 0 to 5 *L* electrons can be seen.

A further feature of this spectrum is the appearance of a new set of lines, the so-called hyper-satellites. These are satellites of the K_α line, shifted by the order of +200 eV, and are attributed to the loss of both *K* electrons in the bombardment. The lifetime of these ions is known to be sufficiently short to show conclusively that these multiple ionizations are due to simultaneous ejection of the two electrons by one incident particle.

Double K ionization has also been observed in radioactive decay by electron capture and simultaneous 'shake-off' of a second electron from the K shell, as observed by Briand *et al.* (1971) (see section 7.2.).

Similar work has been reported by Moore *et al.* (1972) on titanium, and also by Kaufmann *et al.* (1973).

The relative intensities of satellites produced in this way are demonstrated very clearly by Hopkins *et al.* (1973) using chlorine ions bombarding aluminium (fig. 6.15). The intensities follow approximately a binomial distribution. This has given rise to an 'independence hypothesis'

Fig. 6.14. Spectra of calcium at high resolution, illustrating satellite emission due to multiple ionization (McWherter *et al.*, 1973).

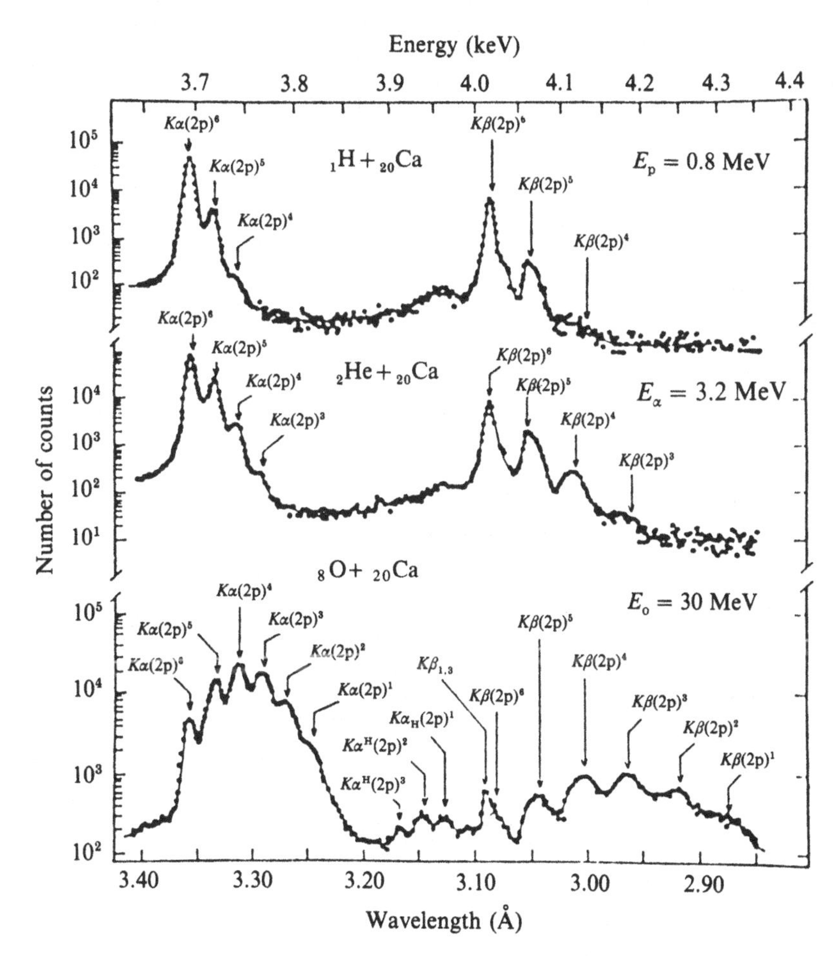

Fig. 6.15. Satellites in aluminium bombarded by chlorine ions showing binomial distribution (Hopkins *et al.*, 1973).

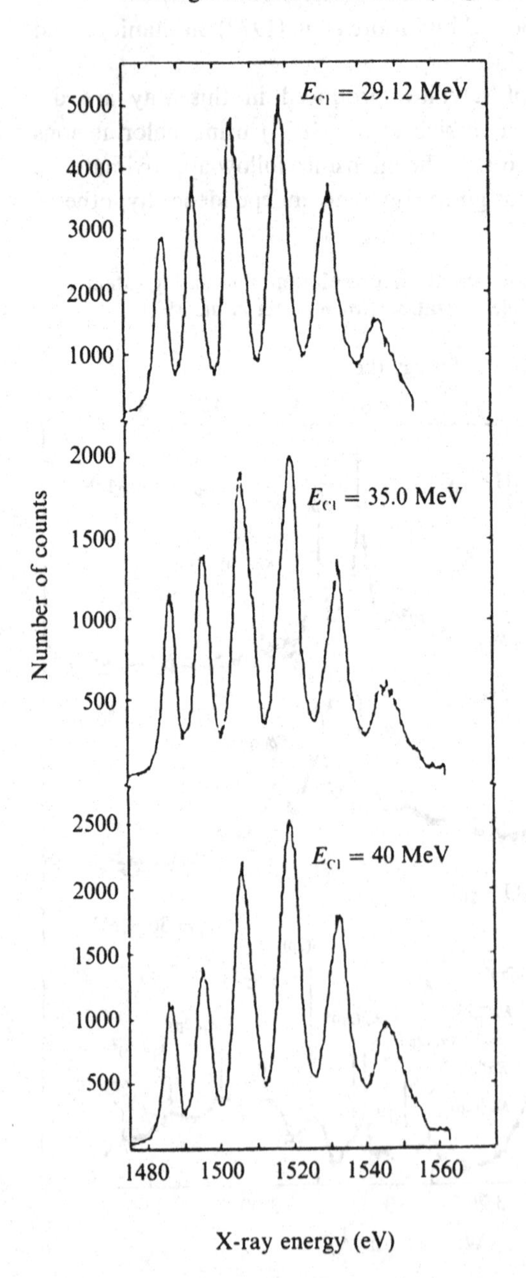

for ionization in the $2p$ shell, i.e. the hypothesis that the probability of ejection of any $2p$ electron is independent of the presence or absence of any other $2p$ electron. On this basis the relative intensities of the lines in any one group can be explained. It should be added, however, that ionization may take place in outer shells also, affecting the relative intensities of X-ray lines through modified transition probabilities. This in turn leads to changes in the fluorescence yields of atoms in which multiple ionization has occurred.

Finally, we look at the effect of the energy levels of the projectile in determining the intensity of characteristic lines emitted by target atoms. The interplay between these two atoms is complex, and as the nuclei approach, all shells of both atoms are involved, and the energy levels move over to levels appropriate to the 'quasi-molecule' formed by the two systems. At sufficiently close distances of approach, the levels become those of an atom of atomic number $Z_1 + Z_2$. At intermediate separations new X-rays seen from neither atom individually may appear. In general, the energy levels change adiabatically as the two atoms approach, but some electrons may be promoted to higher levels, leaving vacancies in the inner shells as the atoms recede. We would therefore expect that the cross-section for these inner-shell ionizations in, say, the target atom will depend upon the atomic number of the projectile, and *vice versa*. An illustration is given in fig. 6.16, reproduced from the paper by Saris *et al.* (1971). The target is argon, and a

Fig. 6.16. Cross-sections in heavy ion bombardment (Ar target; Z is projectile atomic number; the impact energies are shown in c.m. (system) (Saris *et al.*, 1971).

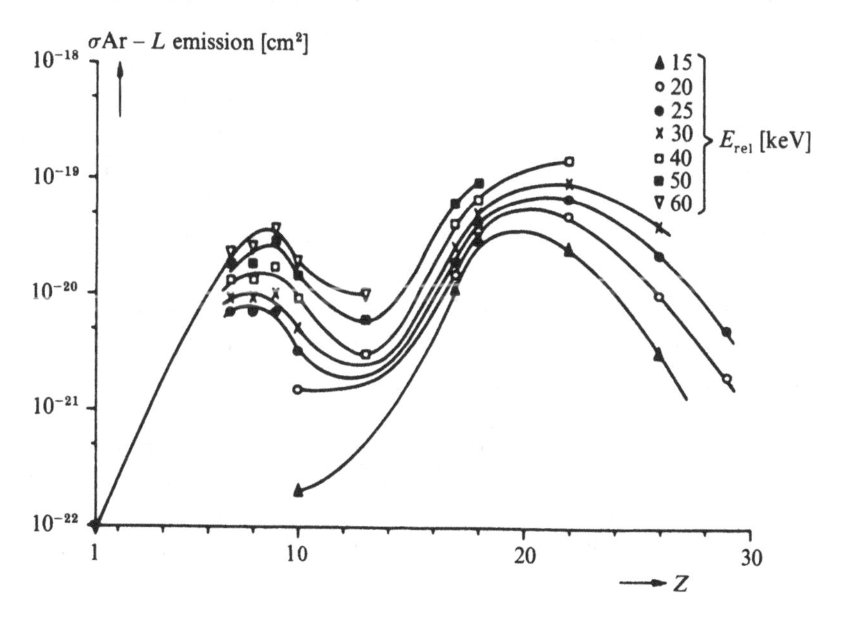

peak in the cross-section for L ionization in this gas is seen when the projectile has an atomic number in this region. A second peak appears at lower Z, when the K-shell energy of the projectile approximately equals the L shell of argon. This peaking of cross-sections when level-matching occurs is also reported by Kavanagh *et al.* (1970). In this work the cross-section for Cu L ionization in a copper target is seen to exhibit resonances as a function of projectile atomic number, and a similar effect is observed when the situation is reversed, using copper ions as projectiles on to targets of varying Z.

A new theory of X-ray production in these circumstances, involving quasi-molecular orbitals, has been given by Fano and Lichten (1965). This is reviewed by Garcia *et al.* (1973) and details, including molecular orbital energy-level diagrams, are given in both these references.

6.6 Proton-induced X-rays for elemental analysis

The study of X-rays produced by proton and positive-ion bombardment has received a great impetus by the development of an analytical technique for elemental analysis using these X-rays. We have seen that the characteristic X-rays are produced in abundance, and that the Bremsstrahlung background is low. This has led to the widespread development of *proton-induced X-ray emission analysis*, or PIXE. Protons are normally used as projectiles, obtained from small accelerators – usually Cockcroft–Walton machines or van de Graaff generators. A beam current of a few tens of nanoamperes is usually sufficient, although for certain kinds of work more current can be usefully employed. The beam is transported via suitable bending magnets to a target chamber in which the material to be analysed is mounted. The X-rays emerge through a thin window, of mylar or other plastic material, and are detected using a Si(Li) detector in association with a multichannel analyser. The X-ray peaks stand strongly against the background, and under favourable conditions it is possible to detect 1–2 parts per million of a wide range of elements. The X-rays from the lightest elements are absorbed too strongly by intervening materials and the X-ray energy resolution is not adequate. But for $Z=12$ upwards the technique is well established. The yield of K X-rays falls rapidly with increasing atomic number, but the L X-rays are available from the heavier elements and are much used. Normally the range of X-rays used for PIXE is ≈ 3–30 keV.

If the Si(Li) detector is mounted directly on to the beam line and is connected to the vacuum system, no intervening windows are necessary, and this is occasionally done to facilitate analysis in the $12<Z<15$ range.

Samples may be drawn from a very wide variety of sources for this type of work. Metallurgical and geological analyses are possible and are very successful for minor or trace elements. Biological work has been carried out extensively. Samples may be 'thin' or 'thick' as defined earlier. Thin targets can be prepared by evaporating a drop of fluid on a mylar film. Thick samples may be prepared by freeze-drying, grinding and then pelletizing, to form samples with adequate mechanical strength. When liquid samples are evaporated (perhaps from an initial volume which is quite large) the sensitivity may of course be greatly improved and detection limits down to a few parts per billion have been reported.

Corrections for self-absorption of the emergent X-rays can be applied where necessary, and suitable samples for standardisation of concentration measurements can be made.

In recent years several developments of the technique have become possible, of which the most notable is the production of proton micro-beams, which, by means of suitable magnetic lenses, can be produced down to a few micrometres in diameter. The technique then becomes somewhat related to the electron microprobe technique described in chapter 8 (section 1) but with much greater sensitivity. The PIXE technique has been widely described, and very many PIXE applications have been reported in the literature. The paper by Folkmann *et al.* (1974) cited earlier provided one of the earliest definitive accounts of the technique including calculations of the lower limits of detection under a variety of conditions. The review by Cahill (1980) is comprehensive.

7

X-rays in radioactive decay

7.1 Introduction

The study of radioactivity has taken place continuously since the beginning of the century, and the great wealth of artificially radioactive nuclides which has become available since the invention of the cyclotron and the nuclear reactor has given great impetus to this work during the latter half of this period. The phenomenon of α-decay was first recorded by Rutherford in 1899 in uranium minerals, and is a process confined, with very few exceptions, to heavy elements, that is, to elements with mass numbers greater than 200. The existence of natural decay series denoted by the mass numbers $4n$, $4n+2$, $4n+3$, was recognised during the early part of this period, but the fourth series ($4n+1$) was not discovered until much later, because its longest lived member (^{237}Np) has a half-life of only 2.2×10^6 years, which prevents the occurrence of members of this series in nature.

The existence of a more penetrating radiation was also recognised at an early stage, and had in fact been responsible for the original discovery of radioactivity by Becquerel in 1896. This more penetrating radiation consisted of the β-particles, soon to be identified with the negative electrons produced in discharge-tube experiments.

It was soon realized that the phenomenon of β-decay is not confined to the heavy elements – Campbell in 1907 detected β-activity in potassium and rubidium, and many hundreds of β-emitters (mostly produced artificially) are now known and have been investigated in some detail.

The emission of a negative electron is the normal mode of decay of nuclei with an excess of neutrons, such as are produced by the processes of nuclear fission, or thermal neutron capture in nuclear reactors, but this form of β-decay is not the only one involving weak interactions between leptons* and

* The *leptons*, or light particles, include electrons, positrons, μ-mesons and neutrinos.

nuclei. The decay of neutron-deficient nuclides (such as are formed in accelerators by charged particle reactions) proceeds either by the emission of a positron, or by the process of orbital electron capture. Since the latter alternative creates an inner-shell vacancy we see that the subsequent production of X-rays will be an inevitable consequence of decay by this process, and we would therefore expect that the emission of characteristic X-rays will be a widespread phenomenon, except perhaps in the lightest elements, where the fluorescence yields are known to be very low.

The existence of γ-radiation associated with radioactivity was reported in 1900. This is produced when the daughter nucleus is formed in an excited state, and the subsequent decay to the ground state will normally be a radiative process producing the γ-radiation. An alternative method of decay is possible, however, in which an orbital electron is ejected from the atom. This internal conversion of energy creates an inner-shell vacancy, which then decays with the emission of characteristic X-rays. We thus recognise two primary processes by which X-rays are produced in radioactive decay.

When β-decay takes place, the acceleration or deceleration of the β-particle in the Coulomb field of the nucleus (or, in the case of electron capture, the change in dipole moment of the electron/nucleus system) results in the production of Bremsstrahlung, the end-point of which is determined by the energy of the transition. This 'inner' Bremsstrahlung is weak, but can be detected when there is little or no γ-radiation present, and when the source material is thin enough to prevent it being masked by external Bremsstrahlung produced by the slowing down of β-particles in the source material.

A further possibility is the 'inner ionization' of an atom by a β-particle emitted from its nucleus.

In addition to these internal processes, which are quite independent of the physical nature of the radioactive source preparation, external processes will also take place. These include the production of the external Bremsstrahlung already referred to in this context, and the external inner-shell ionization of atoms of the source material. These externally ionized atoms need not of course be of the same element as the emitting nuclei, so there exists the possibility of radiation being produced which is not necessarily characteristic of the radionuclide present. Furthermore, we would expect X-rays to be produced as a direct consequence of α-decay, and also following the photoelectric absorption in the source of γ-radiation produced by the decay of nuclear excited states. So the X-rays associated with radioactive decay take on some degree of complexity. We shall discuss each of these processes in turn.

7.2 X-rays produced by electron capture

We can devise some simple expressions to determine whether β^-, β^+ or electron capture decay will take place from a particular parent radionuclide. In β-decay by the emission of an electron or positron, the energy available is divided between an anti-neutrino (or neutrino) and the charged particle (fig. 7.1). The recoil energy of the nucleus is small enough to be neglected in the present context. We shall denote the neutrino energy by E_ν (or $E_{\bar{\nu}}$) and the kinetic energy of the charged particle by E_β. It is found experimentally (and is also one of the assumptions of the Fermi theory of β-decay, to be outlined below) that the emitted charged particle may have a kinetic energy of any value from zero up to a limit denoted here by E_0, in which case the neutrino energy approaches zero. We can write

$$E_\nu = E_0 - E_\beta \tag{7.1}$$

The energy available for the decay process depends on the change of mass during the disintegration. This is expressed as the change of mass-energy of the neutral atoms of parent and daughter, and is denoted by Q. For β^--decay we can write:

Total mass-energy before disintegration
$= (M_z + Zm)c^2$
Total mass-energy after disintegration
$= (M_{z+1} + Zm)c^2 + E_\beta + E_{\bar{\nu}} + m^- c^2$

where m^- is the mass of the created β^--particle (equal in this case to m the electronic mass) and M_z, M_{z+1} are the *nuclear* masses.

$$\therefore\ E_\beta + E_{\bar{\nu}} = (M_z + Zm)c^2 - (M_{z+1} + (Z+1)m)c^2 = Q = E_0$$

In the case of β^+-decay we obtain, by a similar argument

$$E_\beta + E_\nu = (M_z + Zm)c^2 - (M_{z-1} + Z(m-1))c^2 - 2mc^2 = Q - 2mc^2$$
$$= E_0$$

Clearly the conditions that β^-- or β^+-decay may take place are

$$Q > 0 \tag{7.2a}$$

and

$$Q > 2mc^2 \tag{7.2b}$$

for β^-- and β^+-decay respectively.

In the case of electron capture, we may write:

Total mass-energy before decay
$= (M_z + Zm)c^2$

Total mass-energy after decay

$$= (M_{z-1} + Z(m-1))c^2 + E_\nu + E_{K,L}$$

where $E_{K,L}$ represent the X-rays (or the kinetic energy of Auger electrons) emitted subsequent to the capture. The total energy emitted in this way is equal to the binding energy (in the $Z-1$ atom) of the electron which is captured (E_K,E_L, etc.)

$$\therefore\ E_\nu = Q - E_{K,L} \ldots$$

i.e. the condition for electron capture decay to take place from the K,L . . . etc. shell is

$$Q > E_{K,L} \ldots \tag{7.2c}$$

The difference between β^-- and β^+-emission may be understood by noting that in β^+-emission (in which the nucleus reduces its atomic number by one), there is one superfluous electron as well as the created positron. These two particles represent a mass-energy of $2mc^2$ which has to be subtracted from Q in order to obtain the available kinetic energy and neutrino energy.

Fig. 7.1. Illustrating β-decay.

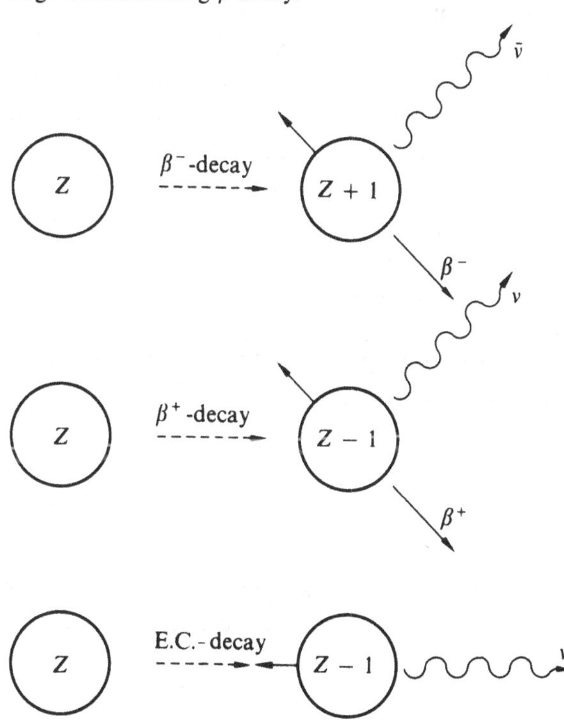

In setting out the theory of β-decay these results are often written in terms of the *total* mass-energy W_0 of the emitted particle and the neutrino, in units of mc^2, in which case we have

$$W_0 = \frac{E_0}{mc^2} + 1$$

In terms of Q we have

$$\text{(a) for } \beta^-\text{-emission } W_0 = \frac{Q}{mc^2} + 1 \qquad (7.3a)$$

$$\text{(b) for } \beta^+\text{-emission } W_0 = \frac{Q}{mc^2} - 1 \qquad (7.3b)$$

By making allowance for the difference in the number of electrons in the neutral atoms, we may see that W_0 is simply equal to the difference between *nuclear* mass-energies in each case.

The theory of β-decay is based on that of Fermi. This enables the probability of the emission of a β-particle in a defined energy range to be calculated, and, in the case of neutron-deficient nuclides, the probability of electron capture taking place. The first basic assumption is that there is a continuous spectrum of β^- or β^+ kinetic energies ranging from zero up to an end-point E_0, determined by (7.1).

The second basic assumption of the Fermi theory relates to the momentum of the two emergent particles. It is supposed that all regions of phase space are equally accessible to the β-particle and the neutrino. It should be noted that because of the existence of a third particle (the recoiling nucleus), there are no restrictions on the total amount of momentum carried off by the β-particle and the neutrino, except those implied by the conservation of energy.

Let the β-particle and the neutrino have momentum components $p_{x,y,z}$ and $q_{x,y,z}$ respectively.

The volume of momentum space corresponding to momentum values between p and $p + dp$ is given by

$$dV_p = 4\pi p^2 dp$$

and, for the neutrino,

$$dV_q = 4\pi q^2 dq$$

The probability of the β-particle and the neutrino occupying values in these two regions simultaneously is therefore given by

$$dn = \frac{4\pi p^2 dp}{h^3} \cdot \frac{4\pi q^2 dq}{h^3}$$

The h^3 terms are a consequence of restricting the density of states to one per unit cell (volume h^3) of momentum space.

Now $q = E_\nu/c^* = (E_0 - E_\beta)/c$ and we can write

$$dq = \left(\frac{\partial q}{\partial E_0}\right)_{E_\beta} dE_0 = \frac{dE_0}{c}$$

We must also introduce a statistical weighting factor S_1, equal to 4 for positron or negatron emission, to allow for the fact that there are two particles emitted, each with two possible directions of spin. So we may write, for the density of states leading to a particular final state in the range of electron momentum dp, introducing factor S_1 as the first term on the RHS,

$$\frac{dn}{dE_0} = S_1 \frac{4\pi p^2}{h^3} \frac{4\pi q^2}{h^3} \frac{dq}{dE_0} dp$$

or

$$P(p)dp \propto S_1 \frac{4\pi p^2}{h^3} \frac{4\pi(E_0 - E_\beta)^2}{c^2 h^3} \frac{1}{c} dp \tag{7.4}$$

for the probability of emission of a β-particle in the momentum range $p \rightarrow p + dp$.

This represents the so-called 'statistical factor' in β-decay, and gives rise to distributions of the general shape shown in the curve for $Z = 0$ in fig. 7.2. The constant of proportionality may be written as

$$\frac{2\pi}{\hbar} g^2 |M|^2$$

where g is the strength of the 'weak interaction' between the β-particle and the nucleus, and $|M|^2$ is the square of the matrix element representing the degree of overlap of the initial and final nuclear wave-functions. The magnitude of $|M|^2$ depends strongly on the nature of these wave-functions. In circumstances where $|M|^2$ is large, we speak of 'allowed' transitions; $|M|^2$ can be of the order of unity in such cases. When the transition probability is small, however, $|M|^2$ can be less than unity by several orders of magnitude. Such transitions are known as 'forbidden' transitions, and are very common in nuclear decay. For detailed discussion of the purely nuclear aspects of β-decay, the reader is referred to standard texts on

* The statement that $q = E_\nu/c$ assumes a massless neutrino. This assumption is not a necessary part of the Fermi theory, but its experimental verification through a study of the shape of β-spectra near their end-points is well-known, and we shall adopt the assumption of a massless neutrino here.

nuclear physics. In the present context, we shall see that positron emission and its alternative, electron capture, both depend on the same quantity $|M|^2$, and so their ratio, which will be of interest to us shortly, is independent of the nuclear wave-functions for the transition in question.

In the case of positron emission, the argument leading to the statistical factor is exactly the same, except that a neutrino rather than an anti-neutrino is emitted. In the case of electron capture, there is only one emergent particle, and the statistical weighting factor, S_2, is equal to 2, as only one particle of spin $\frac{1}{2}$ is emitted, so we may write

$$P_{\mathrm{EC}}=\frac{2\pi}{\hbar}g^2|M|^2S_2\frac{4\pi q^2}{h^3}\frac{dq}{dE_0} \tag{7.5a}$$

Bearing in mind that the available energy is $Q-E_K$ (for K capture), the momentum of the neutrino is given by $(Q-E_K)/c$, or $[mc^2(W_0+1)-E_K]/c$, using (7.3b).

Hence

$$P_{\mathrm{EC}}=\frac{2\pi}{\hbar}g^2|M|^2S_2\frac{4\pi}{h^3}\frac{(mc^2(W_0+1)-E_K)^2}{c^3} \tag{7.5b}$$

Up to this point in the argument we have taken no account of the influence of the Coulomb potential barrier upon the probability of β^+- or β^--emission. To do this, (7.4) needs to be multiplied by a 'barrier'

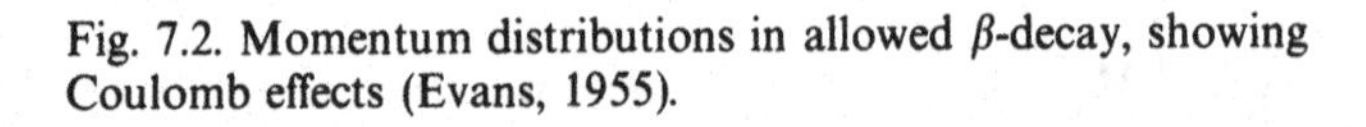

Fig. 7.2. Momentum distributions in allowed β-decay, showing Coulomb effects (Evans, 1955).

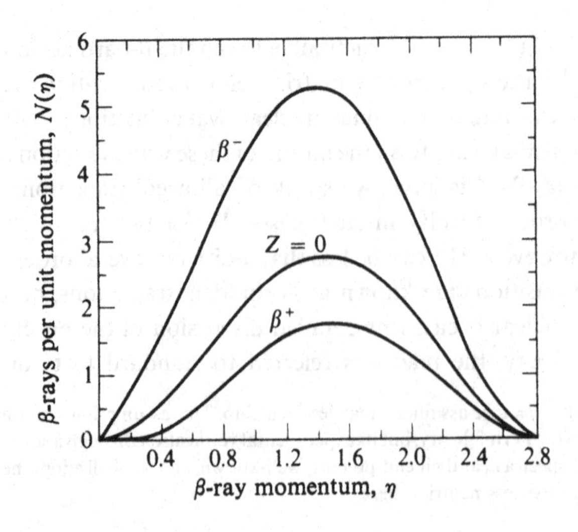

penetration factor, which for positrons is given by

$$F(Z,p)=\frac{2\pi\alpha Z/\beta}{1-\exp(-2\pi\alpha Z/\beta)} \tag{7.6}$$

where β is the velocity of the positron relative to the velocity of light, and α is the fine structure constant ($=1/137$). Z is taken as positive for β^--decay and negative for β^+-decay, and refers to the daughter nucleus. For large Z and small β, (7.6) reduces to

$$F(Z,p)=2\pi\alpha Z/\beta\exp(-2\pi\alpha|Z|/\beta) \tag{7.7}$$

The full expression for the shape of the β-spectrum modified by Coulomb effects is given by

$$P(p)\mathrm{d}p=\frac{2\pi}{\hbar}g^2|M|^2S_1\frac{16\pi^2}{h^6c^3}F(Z,p)p^2(E_0-E_\beta)^2\mathrm{d}p \tag{7.8}$$

and the total probability per unit time of β^+-particle emission by

$$P_+=\frac{2\pi}{\hbar}g^2|M|^2S_1\frac{16\pi^2}{h^6c^3}\int_0^{p_0}F(Z,p)p^2(E_0-E_\beta)^2\mathrm{d}p$$

where

$$p_0=[E_0(E_0+2mc^2)]^{\frac{1}{2}}/c$$

This may be written

$$\frac{2\pi}{\hbar}g^2|M|^2S_1\frac{m^5c^4}{h^6}16\pi^2\int_0^{p_0/mc}F(Z,p)\left(\frac{p}{mc}\right)^2\left(\frac{E_0-E_\beta}{mc^2}\right)^2d\left(\frac{p}{mc}\right)$$

$$=\frac{2\pi}{\hbar}g^2|M|^2S_1\frac{m^5c^4 16\pi^2}{h^6}f(Z,p) \tag{7.9}$$

The Fermi integral $f(Z,p)$, (or $f(Z,E_0)$), is illustrated in fig. 7.3. To obtain the corresponding expression for electron capture, no allowances for Coulomb effects need to be made, but (6.5) has to be multiplied by the probability of finding a K (or L, etc.) electron at the origin. The appropriate wavefunction is

$$u^2=\frac{1}{\pi}\left(\frac{Z^*}{a_0n}\right)^3e^{-2Z^*r/a_0n} \tag{7.10}$$

where n is the principal quantum number (1, 2 etc. for the K, L etc. shells), a_0 is the Bohr hydrogen radius $4\pi\varepsilon_0\hbar^2/e^2m$, and Z^* relates to the *parent* nucleus.

For $r=0$ (i.e. at the origin), and for $n=1$ (i.e. for K electrons), we have simply

$$u^2=\frac{1}{\pi}Z^{*3}\left(\frac{mc\alpha}{\hbar}\right)^3 \tag{7.11}$$

So the probability of electron capture taking place from the two electrons in the K shell becomes

$$\begin{aligned} P_{\mathrm{EC}}&=\frac{2\pi}{\hbar}g^2|M|^2S_2\frac{4\pi}{h^3c^3}\left[W_0+1-\frac{E_K}{mc^2}\right]^2 2(mc^2)^2\frac{1}{\pi}Z^{*3}\left(\frac{mc\alpha}{\hbar}\right)^3 \\ &=\frac{2\pi}{\hbar}g^2|M|^2S_2\frac{m^5c^4}{h^6}64\pi^3\left[W_0+1-\frac{E_K}{mc^2}\right]^2(Z^*\alpha)^3 \end{aligned} \tag{7.12}$$

From this and (7.9) we obtain

$$\frac{P_{\mathrm{EC}}}{P_+}=\frac{2\pi(Z^*\alpha)^3\left[W_0+1-\frac{E_K}{mc^2}\right]^2}{f(Z,p)} \tag{7.13}$$

Electron capture is strongly favoured at high atomic numbers, and there are two reasons for this. First, the K-orbit radius shrinks in proportion to Z^{-1}, so that the 'effective volume' occupied by the K electrons would be

Fig. 7.3. The Fermi integral $f(Z,E)$ as a function of E_0 and Z, (after Feenberg and Trigg, 1950).

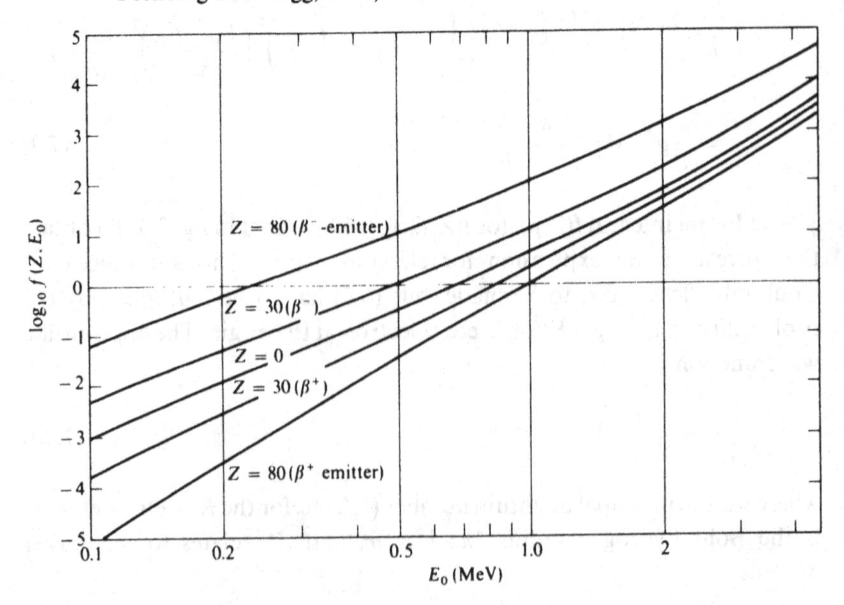

expected to vary as Z^{-3}. The Z^3 term in (7.11) thus expresses the increasing probability of finding an electron in the vicinity of the nucleus as Z increases. Secondly, the Coulomb barrier becomes increasingly difficult to penetrate, as shown by the Z dependence in the curves of fig. 7.3.

Electron capture is also favoured by low disintegration energies. The probability of positron emission varies approximately as $W_0{}^5$*, and this is only partially offset by the approximate $W_0{}^2$ dependence of the electron capture probability.

The ratio of electron capture to positron decay as a function of Z and E_0 is illustrated in fig. 7.4.

For values of W_0 less than 1, positron emission is energetically impossible, as seen from (7.2b) and (7.3b), but electron capture continues to be possible for values of W_0 down almost to -1. An example of such a radionuclide is ^{55}Fe, for which $W_0 = -0.57$. This decays exclusively by electron capture, yielding the characteristic radiation of manganese.

The probability of electron capture taking place from some shell other than the K shell is proportional to the value of the appropriate electronic wave-function in the vicinity of the origin, and to the number of electrons in the shell. Capture from the L_1 sub-shell is governed by (7.10) with $n = 2$, and

* This approximation becomes less useful at low values of W_0. For a detailed discussion of the behaviour of the Fermi integral, nuclear physics texts by R.D. Evans (1955) and Blatt and Weisskopf (1952) may be consulted.

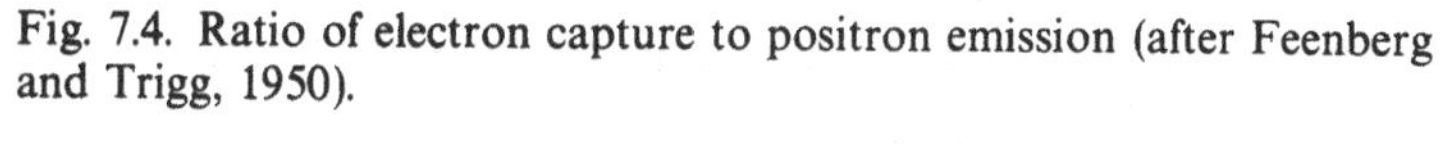

Fig. 7.4. Ratio of electron capture to positron emission (after Feenberg and Trigg, 1950).

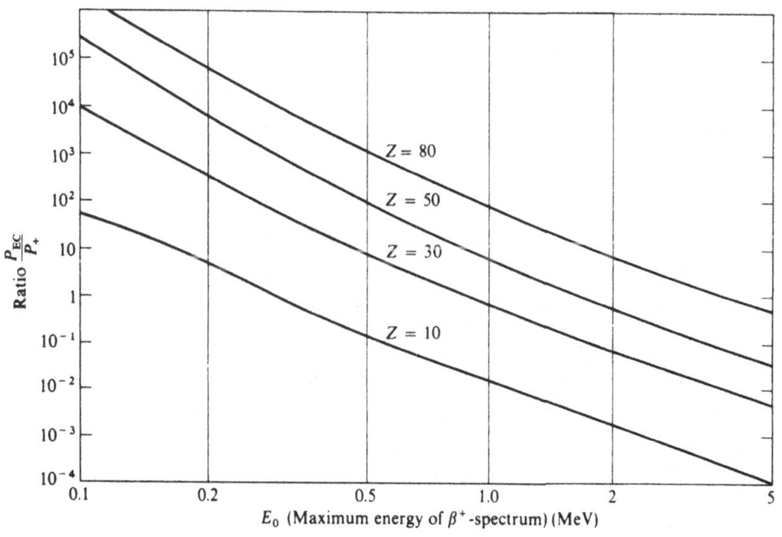

Table 7.1 *Ratios of L : K capture probability*

	E_K (keV) (in $Z-1$ atom)	E_{L_1} (keV)	Q_{EC} (keV)	W_0	L/K_{obs}	$L/K_{calc.}$
^{126}I	31.8	4.95	2150	3.2	$0.142^{+0.005}_{-0.018}$	0.131
^{74}As	11.1	1.42	2564	4.03	0.085 ± 0.020	0.095
^{79}Kr	13.48	1.79	1610	2.16	0.108 ± 0.005	0.109
^{125}I	31.8	4.95	149	-0.7	0.23 ± 0.03	0.23
^{179}Ta	65.37	11.26	119	-0.77	1.4 ± 0.4	
^{235}Np	115.6	21.8	123		$\{30 \pm 2$; 36.7 ± 4	
^{205}Pb	85.5	15.3	30	-0.94	∞	∞

is thus expected to be less than the probability of K capture by a factor of 8. Using (7.2b) and (7.5), the ratio of $L:K$ capture can be seen to be given by the expression

$$\frac{P_L}{P_K} = \frac{1}{8}\frac{(W_0+1-E_L/mc^2)^2}{(W_0+1-E_K/mc^2)^2} \tag{7.14a}$$

from (7.5b).

Tables of decay schemes normally use the difference in mass-energy of neutral atoms, Q, so it is convenient to write (7.14a) in the form

$$\frac{P_L}{P_K} = \frac{1}{8}\frac{(Q-E_L)^2}{(Q-E_K)^2} \tag{7.14b}$$

In cases where Q is rather near to the lower limit, we expect P_L/P_K to rise. If Q should happen to lie within the rather narrow limits given by

$$E_K > Q > E_L \tag{7.14c}$$

only L capture will be possible. This has been observed for the radionuclide ^{205}Pb, for which $E_K = 88.0$ keV, $Q = 30$ keV, and $E_{L_1} = 15.86$ keV.

Capture from the L_2 or L_3 sub-shells is negligibly small for all practical purposes, on account of the very small probability of these electrons being found within the nuclear volume. Calculation of the capture probability of these electrons is referred to by Siegbahn (1965).

A few examples of the probability ratio of $L:K$ capture are listed in table 7.1. For ^{79}Kr, ^{74}As and ^{126}I, where the effect of the electronic binding energies is negligible, the ratios do not deviate much from the value of 1:8

expected on simple theoretical grounds. For ^{125}I, the binding energy effect is sufficient to increase the ratio appreciably above this value.

In the case of nuclides such as ^{179}Ta and ^{235}Np, in which K capture is energetically only just possible, the experimental $L:K$ ratios provide one of the most sensitive methods of determining the disintegration energy. The nuclide ^{205}Pb decays by L-capture only, as explained above.

Recently the radionuclide ^{163}Ho has been studied in some detail. In this case, capture from the K and L shells is energetically forbidden, but capture from the M and N shells has been observed by Hartmann and Naumann (1985). The Q value of this radionuclide is 2.3 ± 1.0 keV, making this the lowest Q value of any known β-decaying nuclide.

In nuclides for which the Q values are large enough ($> 2mc^2$) to enable β^+ emission to be energetically possible, the effects of electronic binding will usually be negligible by comparison, and the ratio of $L:K$ capture will be near to the value of 1:8.

Determinations of the $(L+M)/K$, $(M+N)/L$ and L/K ratios have been reported by Bosch *et al.* (1969), and a considerable amount of existing data are tabulated by them. Orbital electron capture has been reviewed by Robinson and Fink (1960), and experimental data are also presented. A detailed review of all aspects of orbital electron capture has been given by Bambynek *et al.* (1977) with an extensive bibliography. A recent study of ^{207}Bi has been reported by Mandal and Patro (1985).

Few electron capture nuclides have been subjected to a detailed examination of their X-ray spectra, apart from the measurement of $K:L$ capture ratios. Abelson (1939) has used a bent-crystal spectrometer to study the characteristic X-rays of zinc emitted during the decay of ^{67}Ga. His observations were directed solely towards element identification, and the detailed shape of the spectrum is not discussed. In the case of pure K capture the spectrum is expected to be identical with that produced by electron bombardment. More detailed considerations suggest that slight differences could in fact exist. We have seen (section 3.7) that the line shape in spectra produced by electron bombardment is affected by the need for the electron orbitals to readjust to the reduced screening (i.e. higher effective Z) caused by the ejection of a K electron from the atom. This slightly increases the transition energies, and the partial failure to readjust in the time available ($\sim 10^{-16}$ s) is thought to be responsible for the low energy tail observed on some lines in the K series (Parratt, 1959). In the case of K electron capture the new orbitals are those appropriate to a reduced nuclear charge, and partial failure to readjust during the lifetime of the K-ionized atom would be expected to produce a high energy excess in the line profile. Such an effect would require a very strong radioactive source for its observation, in view of the high resolution demanded of the spectrometer and its consequent low efficiency, and has yet to be reported.

In the case of L capture we would expect to observe large differences in the relative intensities of the three sub-series; the relative probability of capture from the three sub-shells is quite different from the ratios of the number of ionizations produced in electron bombardment. In general, the direct production of L_2 and L_3 vacancies in orbital electron capture is rather unlikely and the intensities of lines in these sub-series is accordingly smaller than those observed in electron bombardment. This effect would be easiest to observe in cases where there is no K capture to create L_2 and L_3 vacancies by the normal processes of K_{α_2} and K_{α_1} emission. But these sub-series will always be stronger than implied by the low L_2 and L_3 ionization probabilities because of the occurrence of Coster–Kronig processes in which electrons undergo $L_3 \rightarrow L_1$ and $L_2 \rightarrow L_1$ transitions, thereby creating L_2 and L_3 vacancies with rather high probability.

Although these effects do not seem to have been studied as yet, somewhat similar large effects occur in association with internal conversion, and are referred to below. A few radionuclides decaying purely by orbital electron capture are listed, together with their decay schemes, in appendix 3.

An interesting development has been opened up by the possibility of observing X-ray transitions in atoms doubly ionized in the K shell. When electron capture takes place, the impulse caused by this can cause the loss of a further electron by a 'shake-off' process. If this process ejects a second K electron, a double K ionization results. Briand *et al.* (1971) have examined the K X-ray spectra from doubly ionized gallium as ^{71}Ga following electron capture in ^{71}Ge. The probability of double ionization is given as 1.2×10^{-4} per disintegration. The expectation is that two K X-rays will be emitted, and this is found to be the case, but the first photon belongs to a new group of satellite lines associated with $KK \rightarrow KL$ atomic transitions and these lines are found to be displaced from the parent line by as much as 15–20 times the displacement of the $KL \rightarrow LL$ or $KL \rightarrow ML$ satellites already familiar from ordinary X-ray spectroscopic work. The new satellites have been termed 'hypersatellites', and have subsequently been observed in ionization by proton-, α-, and heavy ion bombardment (section 6.5).

The study of hypersatellites in electron capture decay has also been observed in ^{54}Mn, by the observation of coincidences between chromium K_α X-rays (Nagy and Schupp, 1984). The probability of shake-off is given as 3.6×10^{-4} per K capture. The energy shift of the chromium hypersatellites is given as 254 ± 18 and 319 ± 22 eV for the K_α and K_β lines respectively. This process has also been observed in the electron capture decay of ^{85}Sr

(Schupp and Nagy, 1984) and in several other elements (Briand *et al.*, 1974). Clearly the effect is expected to be widespread.

Shake-off processes have been reviewed by Freedmann (1974) who gives an extensive list of references.

7.3 X-rays following internal conversion

When a nucleus is formed in an excited state, for example, following α- or β-decay of a radioactive parent, it may proceed to a lower (ultimately the ground) state either by the emission of γ-radiation or by the ejection of a bound electron from the same atom. The latter process is known as internal conversion and in a β-spectrometer the conversion electrons appear as sharp lines. If the parent has decayed by β^--emission the conversion lines will be superimposed on the continuous spectrum of the β^--particles. Each converted γ-transition may eject K, L, M etc. electrons, the kinetic energies of which will be equal to $E_\gamma - E_{K,L,M}$, etc., where E_γ is the energy of the transition and $E_{K,L,M}$, etc. are the electronic binding energies. We define the K-shell internal conversion coefficient α_K of a transition as the ratio of the number of electrons emitted from the K shell to the number of emitted γ-ray photons

$$\alpha_K = \frac{n_e}{n_\gamma}$$

Definitions for the other shells follow similarly. Conversion in the K shell is more likely than in other shells, because of the greater probability of K electrons being found within the nuclear volume. The other main factor determining whether internal conversion is likely to occur is the multipolarity of the transition. High multipolarities (corresponding to large changes in the angular momentum of the nucleus) correspond to low probabilities of radiating electromagnetic energy, and hence in such cases the relative probability of internal conversion is high. Low transition energies also favour internal conversion. Graphs of internal conversion coefficients in the K shell are shown, for several multipolarities, in fig. 7.5.

The K-ionized atom formed in this way will decay with the production of X-radiation characteristic of the daughter nucleus, and the relative intensities and positions of the lines will be the same as for an atom ionized by electron bombardment or electron capture. It was early recognised (e.g. Abelson, 1939) that an identification of these X-rays would give some information regarding the primary decay process. X-rays corresponding to element $Z+1$ (where Z represents the atomic number of the radioactive parent) indicate β^--decay followed by internal conversion; X-rays of element $Z-1$ indicate β^+-decay followed by internal conversion* (or simply electron capture, with or without internal conversion); and X-rays of element Z are associated with the decay of a metastable state, with internal conversion. Examples of all these processes are given in fig. 7.6. Abelson examined the K X-rays of zinc emitted from ^{67}Ga in order to identify the decay mechanism (electron capture to ^{67}Zn) of this radionuclide. In the same work the X-rays of molybdenum were observed and led to

* Nuclides decaying by positron emission without electron capture are to be found only among the relatively light elements ($Z<10$) and internal conversion does not occur with sufficient probability to give measurable yields of X-rays. This situation is aggravated by the very low fluorescent yields in light elements. The appearance of $(Z-1)$ X-rays is therefore invariably associated with electron capture decay.

Fig. 7.5. K-shell internal conversion coefficients (after data of Rose *et al.*, 1951) (*a*) $Z=30$ (*b*) $Z=78$.

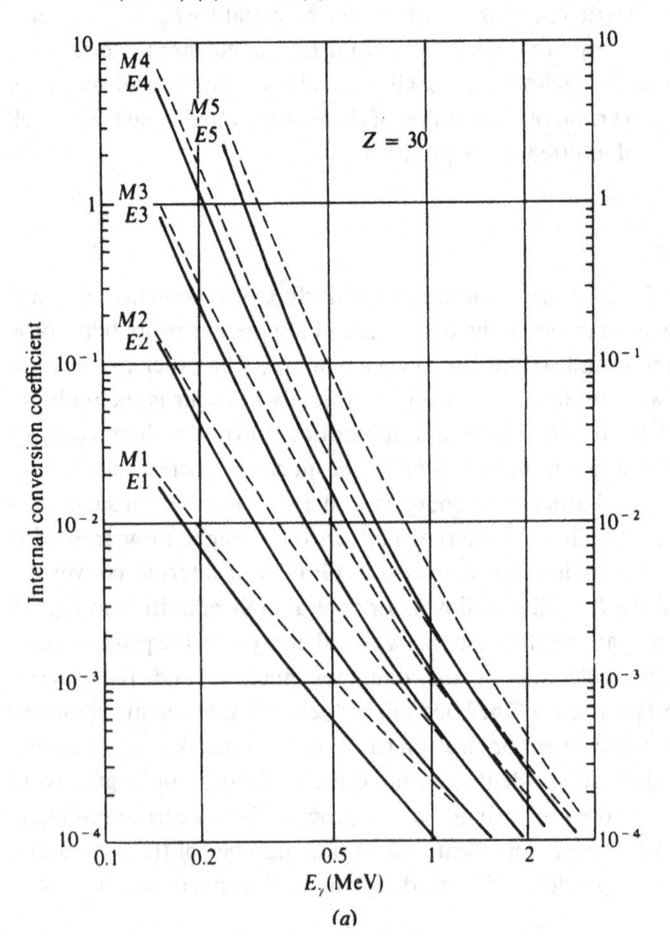

the confirmation of the decay of a radionuclide of technecium ($Z = 43$). As noted above, a bent-crystal spectrometer was used for these identifications, together with photographic recording, but identification of elements can now be carried out more easily with the gas proportional counter or the solid-state detector.

Conversion in the *L* sub-shells is of considerable interest in that the relative probabilities of L_1, L_2 and L_3 conversion depend strongly on the character (magnetic or electric) and multipolarity of the transition. This strongly affects the subsequent X-ray spectrum in that the relative strengths of the three sub-series will depend markedly on the nature of the transition.

Calculations have been carried out by Gellman *et al.* (1950) and by Rose

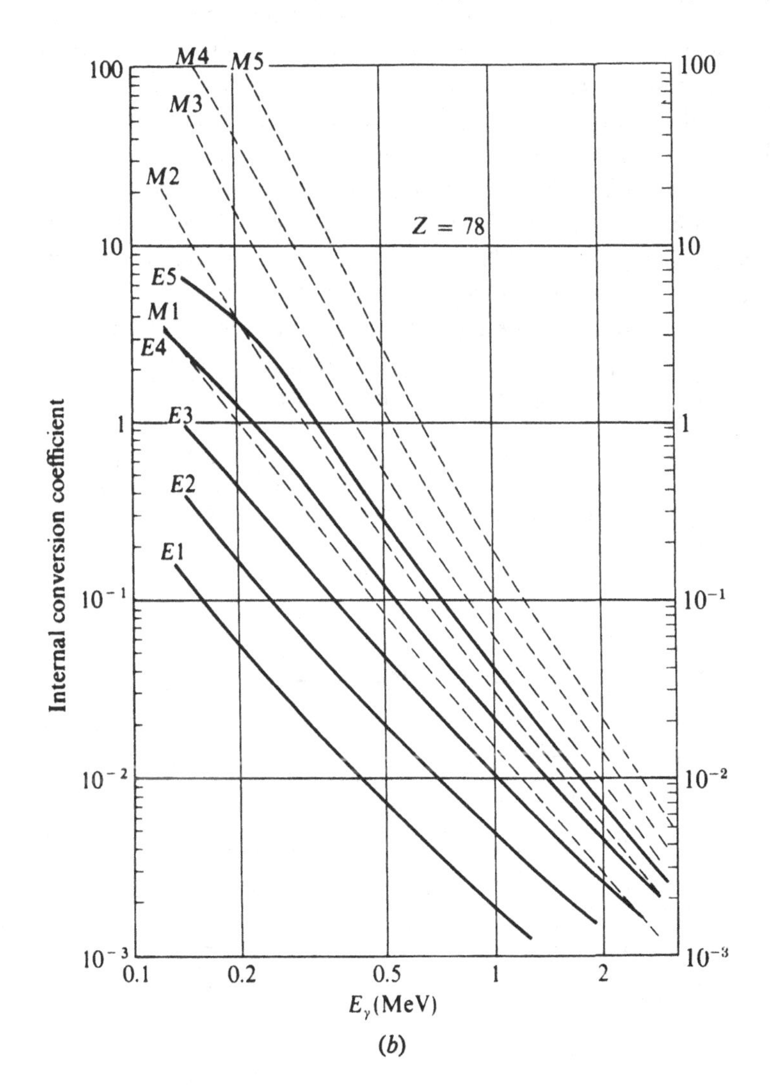

(*b*)

(1955). Victor (1961) had discussed these data in relation to L X-ray spectra produced by various processes.

In electron bombardment, the ratio of ionization in the L_1, L_2 and L_3 sub-shells is approximately 1 : 1 : 2, for bombardment energies well above the excitation threshold. But in internal conversion, calculations lead, for example, to ratios of 90 : 8 : 2 for magnetic dipole radiation and 3 : 55 : 42 for electric quadrupole radiation, for transition energies in the region of 100 keV in nuclei of $Z \sim 90$. Barton *et al.* (1951) have examined the L X-ray spectra from a number of heavy radio-elements, in particular the X-rays arising in the conversion of the 50 keV transition to the ground state of ^{238}Pu following the α-decay of ^{242}Cm. They found that all lines associated with a given sub-shell had the same relative intensities as in the electron bombardment spectrum of uranium (the data for uranium having been recorded many years previously by Allison (1927, 1928), and described by Compton and Allison (1935)), but that the conversion X-rays associated with the L_2 sub-shell were relatively twice as abundant as in the electron bombardment spectra. Further, in the conversion spectrum, no lines associated with the L_1 level were observed and the authors concluded that L_1 vacancies were less than 20% of what would be produced by electron bombardment, that is, would constitute less than 5% of the total L ionization instead of 25%. These results are qualitatively consistent with the known electric quadrupole nature of this transition.

Fig. 7.7 illustrates the L spectrum obtained by the proton bombardment

Fig. 7.6. Production of X-rays in radioactive decay (*a*) β^--decay followed by internal conversion; (*b*) electron capture; (*c*) electron capture with internal conversion; (*d*) metastable state with internal conversion.

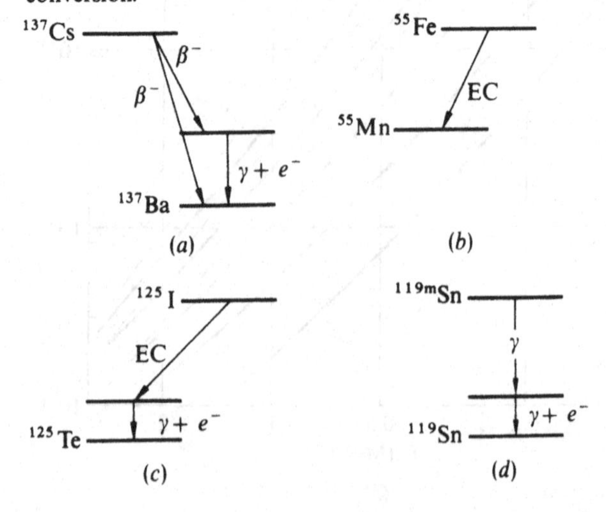

Fig. 7.7. L spectrum of uranium (a) by proton bombardment; (b) by α-decay of plutonium-239 (after Dyson, 1975).

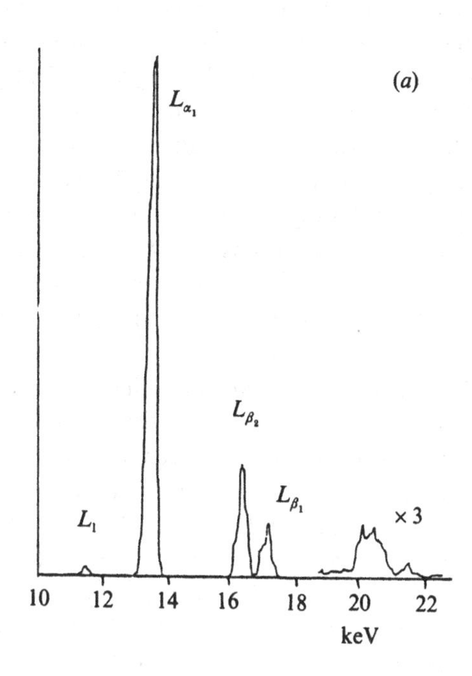

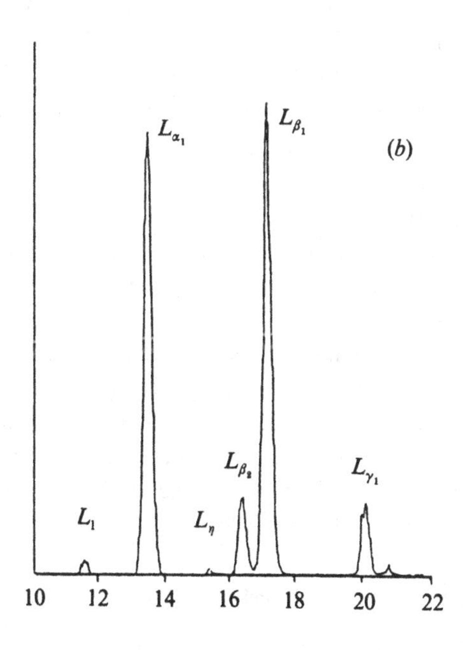

of uranium (as uranyl acetate) and also the L spectrum in ^{235}U following internal conversion associated with the α-decay of ^{239}Pu. The differences in the relative intensities of the lines are marked. Ionization in the case of proton bombardment at this energy (450 keV) is predominantly in the L_1 and L_3 sub-shells, and although there is some migration of vacancies from the L_1 to the L_2 shells, the relevant Coster–Kronig coefficient f_{12} is relatively low, and the lines of the L_2 spectrum (for example L_{β_1}) are therefore rather weak. In the case of decay from ^{239}Pu, the converted γ-rays are mainly electric quadrupole transitions, the L_2 vacancies are considerable, and this is shown by the strong L_{β_1} and L_{γ_1} lines in this case. In each case the L_3 spectrum (e.g. L_{α_1}, L_{α_2}) is strong. The L_1 spectrum is weak because although L_1 ionization takes place to a significant extent in case (a) these vacancies largely migrate to the L_3 sub-shell (f_{13} is large in uranium) reducing the X-ray emission from the L_1 shell.

It is of interest that the X-rays of the L_3 series can, in principle, show some degree of angular correlation with the conversion electrons which precede them, and can also be correlated with any γ-radiation emitted subsequently in cascade. An experimental investigation of such a situation in ^{191}Ir has been reported by Gelberg and Piticu (1971) and the X$-\gamma$ correlation observed. When processes occur which produce changes in both nuclear and atomic states, angular directional correlations are to be expected. In the internal conversion process, nucleus and electron are both coupled to the electromagnetic field, and nuclei with large static deformations (and consequently large electric and magnetic moments) may be expected to show these effects most clearly. Angular correlation between X-rays and the 133 keV γ-radiation in ^{181}Ta has been reported by Avignone *et al.* (1981). A similar effect in ^{169}Tm has been described by Sen *et al.* (1972) and the process has been discussed by Kahlil (1983).

7.4 Inner Bremsstrahlung

A radioactive source decaying by the emission of β^- particles would be expected to show a weak continuous electromagnetic spectrum on account of the external Bremsstrahlung produced by the slowing down of the β-particles during their passage through the source material. But even when this is allowed for, a weak residual radiation remains, which is known as inner, or internal, Bremsstrahlung*. This was first recognised in the radiation from radium E but has subsequently been observed in many radioisotopes, for example ^{32}P, ^{35}S. One of the earliest systematic investigations of this effect was by Wu (1941), who measured the γ-

* We shall normally abbreviate this to IB, and external Bremsstrahlung to EB.

radiation from thin sources of ^{32}P surrounded by absorbers covering a wide range of atomic numbers. Her plot of total Bremsstrahlung intensity as a function of atomic number is given in fig. 7.8. We see that the total intensity rises approximately linearly with Z, as one would expect for external Bremsstrahlung production in a thick target, but that the $Z=0$ intercept is finite. This radiation must be produced within the source, and if the source is thin, the predominant mechanism will be the internal Bremsstrahlung process.*

The amount of energy available as IB in ^{32}P was estimated to be, on average, about $0.002\,mc^2$ per β-particle. Fig. 7.7 shows that the amount of energy available when these same β-particles are allowed to produce

* It should be noted that the intensity of radiation with *no* absorber present would depend on several ill-defined factors, such as the atomic number of neighbouring absorbers and radiation shields: it would not by itself yield useful information.

Fig. 7.8. Internal and external Bremsstrahlung production (Wu 1941).

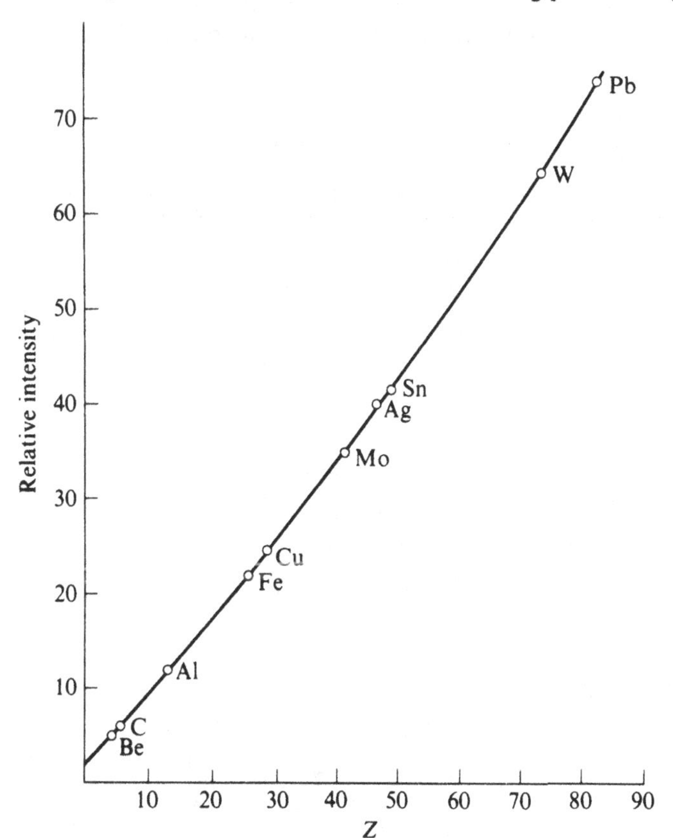

external Bremsstrahlung in aluminium is about four times this amount. The internal process in ^{32}P thus produces about one quarter of the radiation that would occur if the β-particles were allowed to produce EB in an absorber of similar atomic number. The value of 0.002 mc^2 was confirmed by Bolgiano *et al.* (1953) who obtained an average value of 2.32×10^{-3} for ^{32}P. This quantity is expected to depend on W_0, the end-point energy of the β-spectrum, although we shall see below that the amount of energy radiated by external Bremsstrahlung varies with energy in a similar fashion so that the ratio of IB to EB is a function of Z only, and is independent of W_0.

The IB is in the form of a continuous spectrum, and the probability per β-decay for the emission of a quantum of energy between E_γ and $E_\gamma + \mathrm{d}E_\gamma$ has been given by the expression:

$$S(E_\gamma)\mathrm{d}E_\gamma = \int_{E_\gamma}^{E_0} P(E_\beta)\phi(E_\beta, E_\gamma)\mathrm{d}E_\beta \mathrm{d}E_\gamma \tag{7.15}$$

where $P(E_\beta)$ is the energy distribution of the β-spectrum, and $\phi(E_\beta, E_\gamma)$ is the probability of an electron of energy E_β emitting a photon of energy E_γ. The integration extends over the continuous β-spectrum from E_γ (the minimum

Fig. 7.9. IB energy spectrum for $W_0 = 3.6$ and $Z = 0$ (allowed transition). The ordinate of the main curve is the energy per unit energy interval per β-decay radiated as IB. The inset shows the number of photons having energy greater than k (k is the photon energy in units of mc^2). (Knipp and Uhlenbeck, 1936).

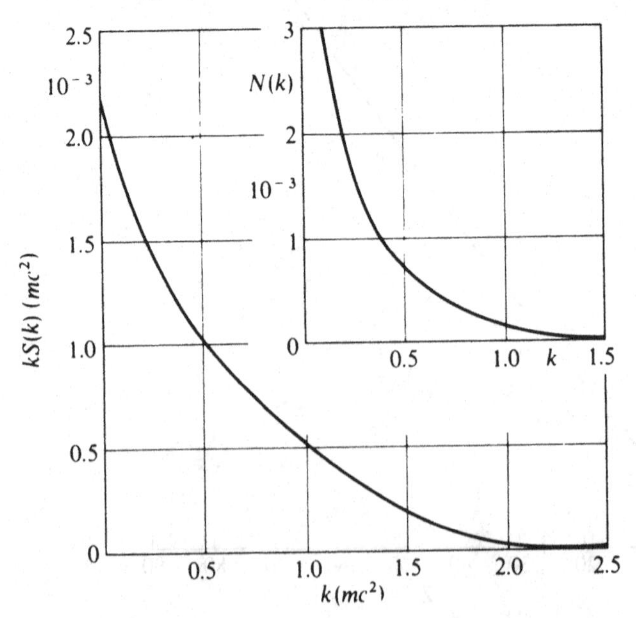

β-particle energy required for the production of a photon of this energy) to E_0, the end-point of the β-spectrum. The form of (7.15) implies that the β-emission and Bremsstrahlung production are independent processes, so that the probability of photon production is given by the product of two mutually independent functions. This is valid (Siegbahn, 1965) for allowed transitions only.

Knipp and Uhlenbeck (1936) have given graphs of the energy emitted per unit energy interval for an allowed transition for which $W_0 = 3.6$ (i.e. $E = 1.3$ MeV), for the $Z = 0$ approximation (fig. 7.9). The inset shows the number per β-decay of photons produced with energies greater than k. An experimental determination of the IB from ^{147}Pm is shown in fig. 7.10(a). Data for ^{35}S are shown in fig. 7.10(b). An additional study of ^{204}Tl and ^{91}Y

Fig. 7.10. IB from (a) ^{147}Pm and (b) ^{35}S (Boehm & Wu, 1954).

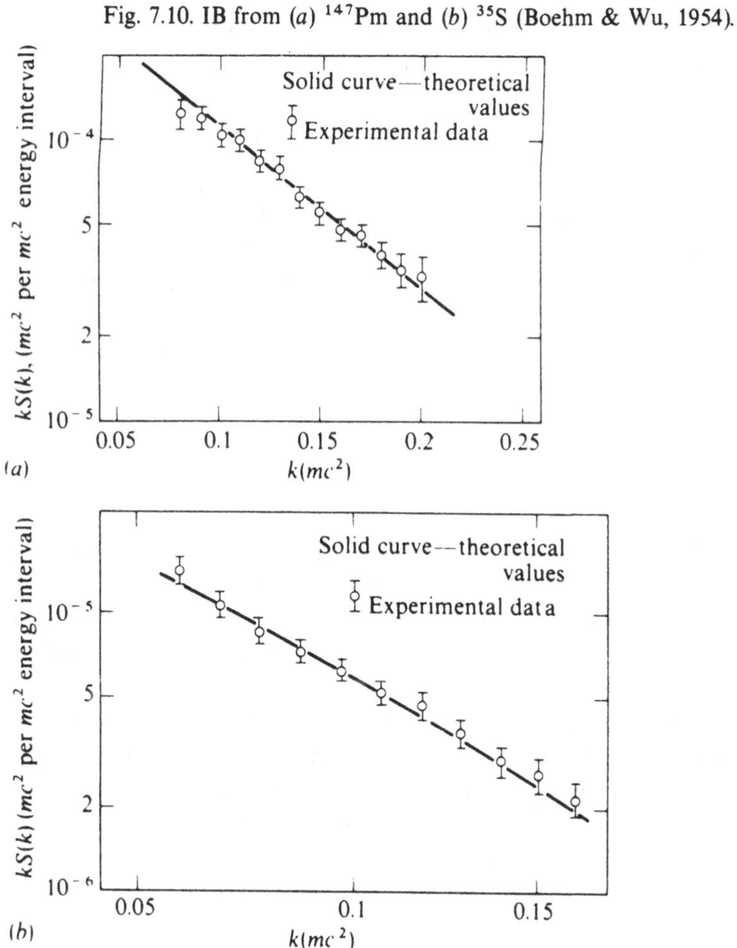

has been reported by Narasimhamarty and Jnanananda (1967). More recent studies of internal Bremsstrahlung associated with β^- emission are reported by Berenyi and Varga (1969) (^{32}P), and Babu *et al.* (1976) (^{45}Ca and ^{35}S). Studies associated with β^+ emission have been described by Berenyi *et al.* (1969) (^{11}C).

Even when the β-decay takes place by electron capture, a weak Bremsstrahlung is emitted on account of the change in electric dipole moment associated with the capture process. Although the intensity is 1 or 2 orders of magnitude lower than the IB produced by charged particle emission, it is sometimes easier to detect because of the complete absence of externally produced Bremsstrahlung. The IB produced in electron capture is a continuous spectrum, and has the important property that its end-point is equal to the disintegration energy of the decaying nucleus, and thus offers an experimental method of determining the Q-value of the decay. The spectral distribution for an allowed transition, neglecting Coulomb effects, has been given by Morrison and Schiff (1940) as

$$N(E_\gamma)\mathrm{d}E_\gamma = P_{\mathrm{EC}} \frac{\alpha}{\pi(mc^2)^2} \frac{E_\gamma}{E_{\mathrm{EC}}{}^2} (E_{\mathrm{EC}} - E_\gamma)^2 \mathrm{d}E_\gamma \tag{7.16}$$

where E_{EC} is the disintegration energy for the EC transition and E_γ the IB photon energy. P_{EC} is the probability per unit time of the EC transition (7.5b). A plot of $[N(E_\gamma)/E_\gamma]^{\frac{1}{2}}$ is proportional to $(E_{\mathrm{EC}} - E_\gamma)$ and is therefore expected to yield a straight line, the intercept of which with the energy axis gives the disintegration energy directly. From (7.16) it is readily seen, by integration, that the number of photons per EC transition is given by

$$\frac{\alpha}{12\pi}\left(\frac{E_{\mathrm{EC}}}{mc^2}\right)^2 \tag{7.17}$$

For iron-55 ($E_{\mathrm{EC}} = 0.43\,mc^2$), it is found experimentally to be approximately 3×10^{-5}, (Bell *et al.*, 1952), which is consistent with (7.17).

A more recent and detailed study of internal Bremsstrahlung, in ^{57}Mn, is described by Kadar *et al.* (1970), in which the IB quanta are studied in coincidence with the 835 keV γ-rays.

By examining the IB produced in coincidence with the K X-rays the IB associated with $1s$ capture processes alone can be obtained, and isolated from the rest of the IB spectrum. When this is done, it is found (Madansky and Rassetti, 1954) that there is a large surplus of low energy photons (<100 keV) which must accordingly be produced by capture from one or more the L sub-shells. A comparison between the $1s$ and total capture spectrum is shown in fig. 7.11.

A further internal process associated with β^- or β^+-decay is the 'internal ionization'. The mechanism for this process is thought to be the 'shake-off' which occurs in the electronic shells, caused by the sudden change in nuclear charge in β-decay, although the possibility of a direct collision process between the β-particle and an orbital electron of the same atom cannot be entirely discounted. The 'shake-off' process has been discussed by Isozumi and Shimizu (1971), who have investigated internal ionization experimentally in ^{147}Pm and ^{63}Ni. The β-particle and ejected K-electron are stopped in a 4π counter, and provide a 'sum' pulse which is then measured in coincidence with the K X-rays. Absolute probabilities for internal ionization are given by Boehm and Wu (1954) for ^{147}Pm ($K: 3.85 \times 10^{-4}$; $L: 2.05 \times 10^{-3}$) and radium E ($K: 1.2 \times 10^{-4}$; $L: 0.6 \times 10^{-3}$). The probabilities are greater in the L shell because of the smaller energy required. This process has also been examined by Mukoyama and Shimizu (1978). Furthermore, it has been found possible to measure internal ionization associated with β^+ decay. If the K X-rays are counted as triple coincidences with the annihilation photons associated with the β^+ particles, the much stronger X-rays associated with the electron capture process invariably occurring as an alternative to β^+-decay may be rejected. Examples of this technique are described by Scott (1980, 1984).

7.5 External Bremsstrahlung in β-decay

The process is readily observable in many β-emitters, and the amount of energy radiated in this way can approach several per cent of the energy of the β-particles radiating it. To examine the intensity of X-

Fig. 7.11. IB following EC in ^{131}Cs (Biavati *et al.*, 1962).

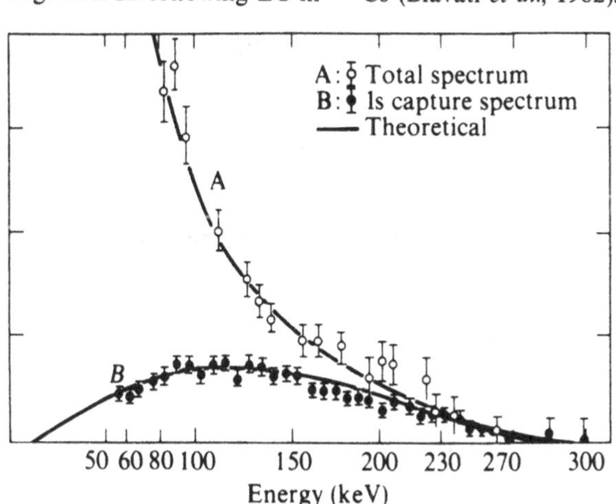

radiation produced in this way we shall use (2.42b), written in a modified form

$$I = KZE^2$$

for the total energy of Bremsstrahlung produced *per electron* in a thick target by monoenergetic electrons of kinetic energy E.

Following the treatment of Evans (1955), the intensity of Bremsstrahlung emitted when the β-radiation from a radioactive source is incident on an absorber of atomic number Z will be proportional to

$$I = (mc^2)^2 KZ \int_1^{W_0} (W-1)^2 N(W) \mathrm{d}W$$

where $W\left(=\dfrac{E}{mc^2}+1\right)$ is the mass-energy of an electron emitted in β-decay and $N(W)$ is the probability distribution of the β-spectrum.

The total number of electrons from the source will be proportional to $\int_1^{W_0} N(W)\mathrm{d}W$ so the average energy radiated per β-particle will be given by

$$E_b = \frac{K(mc^2)^2 Z \int_1^{W_0} (W-1)^2 N(W)\mathrm{d}W}{\int_1^{W_0} N(W)\mathrm{d}W}$$

$$= KZ(E_{rms})^2 \text{ per disintegration.} \tag{7.18}$$

The fraction of radiated energy will be equal to

$$f = \frac{E_b}{E_{av}} \text{ where } E_{av} = \frac{mc^2 \int_1^{W_0} (W-1)N(W)\mathrm{d}W}{\int_1^{W_0} N(W)\mathrm{d}W} \tag{7.19}$$

$N(W)$ is readily obtained from (7.9) by using the relativistic relationships

$$W^2 = 1 + \left(\frac{p}{mc}\right)^2$$

and

$$W\mathrm{d}W = \frac{p}{mc}\mathrm{d}\left(\frac{p}{mc}\right)$$

We obtain

$$N(W)\mathrm{d}W = \text{const.}\ F(Z,p)(W_0 - W)^2(W^2-1)^{\frac{1}{2}} W\mathrm{d}W \tag{7.20}$$

Evans has given graphical data (fig. 7.12) for

$$\frac{E_{\text{rms}}}{E_0} \text{ and } \frac{E_{\text{av}}}{E_0}$$

for the $Z=0$ approximation (in which case $F(Z,p) \sim 1$) from which it is seen that both these quantities are slowly varying functions of E_0, so that both E_b and f are given to useful accuracy by the simple expressions

$$E_b = KZE_0^2 \times (0.45)^2 \text{ and } f = KZE_0 \times 0.51 \tag{7.21}$$

in which the constants are obtained from fig. 7.12.

A difficulty in the use of (7.21) arises from the uncertainty in the value of the thick target continuous efficiency constant K. In (2.42b) written in its original form, with I in SI units ($\mathrm{Js^{-1}A^{-1}}$ or $\mathrm{JC^{-1}}$) and V in volts, K is approximately $1.2 \times 10^{-9}\,\mathrm{V^{-1}}$ for accelerating potentials less than about 100 kV. The experimental and theoretical work supporting this is detailed in section 2.8. The ratio I/V is dimensionless and represents the fractional efficiency of continuous X-ray production in thick targets. In the present context I and I_b are expressed in eV per β-particle, and the particle energies E_0, E_{rms} and E_{av} are in eV. The constant K thus has the same numerical value as before, but with dimensions $\mathrm{eV^{-1}}$ instead of $\mathrm{V^{-1}}$. It is, however, dependent on energy to some extent. Buechner *et al.* (1948) obtained data

Fig. 7.12. Average and RMS energies from β-spectra.

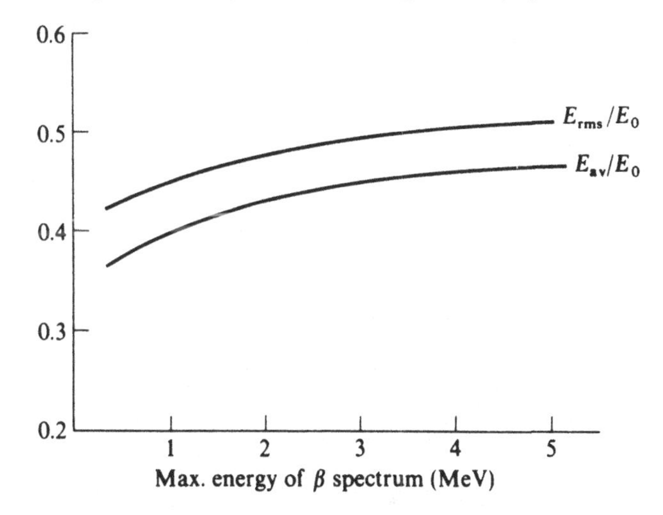

for monoenergetic electrons of 1.25–2.35 MeV, from which a value of 0.39×10^{-9} may be deduced. Rester *et al.* (1970) obtained values in the region of 0.4×10^{-9} for bombardment of thick targets for several targets in the range 0.2–2.8 MeV. A working value of 0.7×10^{-9} is suggested by Evans to cover the energies of β-particles encountered in practice.

The spectral distribution may be predicted from (2.2) (neglecting the small term in Z^2) by writing

$$dI = 2KZd(h\nu)\frac{\int_{W_1}^{W_0}(E-h\nu)N(W)dW}{\int_1^{W_0}N(W)dW} \tag{7.22}$$

where $E = mc^2(W-1)$, and W_1 corresponds to the minimum β-ray energy capable of exciting radiation of photon energy $h\nu$, i.e. $(W_1 - 1)mc^2 = h\nu$. As would be expected, this spectrum (fig. 7.13) is relatively more abundant in

Fig. 7.13. External Bremsstrahlung spectrum from phosphorus-32: β-particles on a brass target (Goodrich *et al.*, 1953).

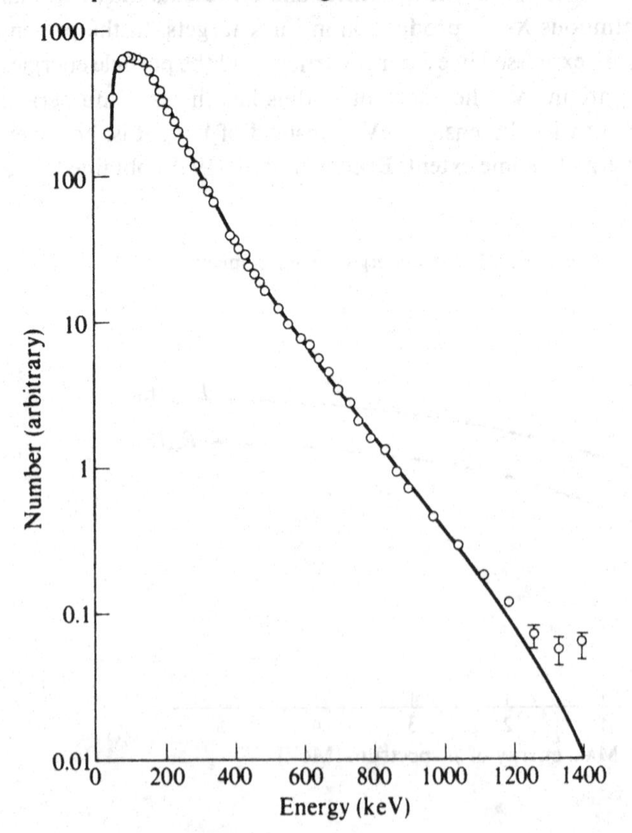

low frequency photons than the spectrum for monoenergetic electrons illustrated in fig. 2.10.

Among the earlier papers reporting studies of external Bremsstrahlung, that of Wu (1941), already referred to, is noteworthy. The main purpose of this investigation was to study the inner Bremsstrahlung produced in ^{32}P. In all studies of this type, the main experimental difficulty is to allow for the effect of EB produced when the emitted β-particles are slowed down and brought to rest in the neighbouring materials, the detector, shielding materials, etc. One approach is to make the source as thin as possible, and to 'sandwich' it between absorbers thick enough to stop all the emitted particles. If the atomic number of the absorber is then varied, the EB will of course vary, and an extrapolation to $Z=0$ enables the intensity of the remaining IB to be obtained. The graph in fig. 7.8 leads to an average radiated energy of $300 \times Z$ eV per β-particle from ^{32}P. Equation (7.21) with $K=0.7 \times 10^{-9}$, leads to a value of $410 \times Z$ eV per β-particle from ^{32}P, establishing its value for approximate calculations of this quantity.

The production of EB has also been investigated by Bustard and Silverman (1967). In this work the emphasis is on the use of very thin absorbers so that the Bremsstrahlung produced in successive layers of absorber material can be measured without the uncertainty associated with the γ-ray absorption corrections which have to be introduced when thick ('electron-opaque') absorbers only are used. These authors used a source of ^{90}Sr–^{90}Y deposited on mylar*.

The source cavity was enclosed on all sides with low Z material (methyl methacrylate). Absorbers of Al, Ni, Nb, Pd, Sn and Ta were used in transmission and in sandwich arrangements. A beryllium plate was used to prevent β-particles from entering the phosphor of the radiation detector. Fig. 7.14 illustrates the variation of total Bremsstrahlung with absorber thickness for absorbers of niobium and tantalum. As expected, the output initially rises to a maximum as the absorber thickness is increased, and then falls off for thicker absorbers due to partial self-absorption of the emitted γ-radiation.

Bustard and Silverman obtained the number of photons per β-particle, and the energy yield in MeV per β-particle, as functions of Z. The expected linearity with Z is well displayed in this data (fig. 7.15).

* Yttrium-90 decays by the emission of energetic β^--particles of 2.26 MeV. Its half-life is, however, only 64 hours, and so it is convenient to use it in the form of a preparation of its long-lived parent (strontium-90, half-life 28 years) which generates the yttrium-90 in equilibrium amounts. The strontium-90 emits a β-particle, the energy of which is relatively low (0.545 MeV) but which would need to be taken account of in accurate work.

Goodrich *et al.* (1953, 1954), in an investigation of the IB and EB associated with ^{32}P in conjunction with a brass target, found that the amount of energy radiated as EB was 9.3 keV per β-particle, and that the ratio of IB to EB was approximately 1:4. This ratio is appreciably higher than the value of 1:6 obtained from the data of Wu shown in fig. 7.7.

The spectral shape of IB and EB are not unlike each other, but the internal spectrum contains a somewhat higher proportion of photons near the high energy limit.

Fig. 7.14. Variation of total (EB+IB) output from targets of tantalum and niobium bombarded by ^{90}Sr-^{90}Y β-particles. The curves marked 'top' relate to a situation in which there is one target between the β-particle source and Bremsstrahlung detector; 'sandwich' relates to the arrangement in which there is an additional target (also variable in thickness) behind the β-particle source. The ordinates are in energy per *emitted* β-particle. (Bustard and Silverman, 1967).

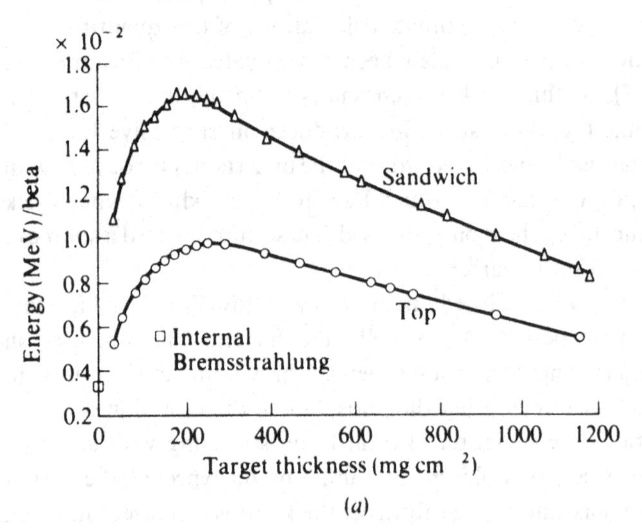

(*a*)

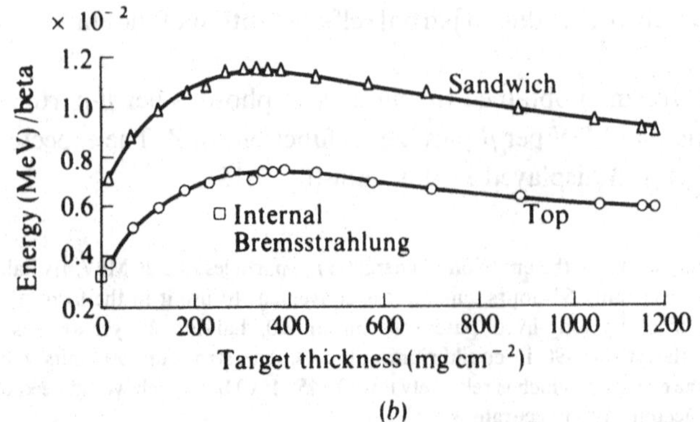

(*b*)

Cameron *et al.* (1963) have examined the EB produced by tritium (β^--emitter; end-point 18 keV) in titanium and zirconium. Here again, the intensity of the radiation was observed to be approximately proportional to atomic number. Cameron and Rhodes (1962), using the calculations of Wyard (1952) obtain the results

$$E_{EB} = 5.8 \times 10^{-4} ZE^2, \text{ with } E_{EB} \text{ and } E \text{ in MeV} \tag{7.23}$$

for monoenergetic electrons, and for a β-spectrum

$$E_{EB} = 5.8 \times 10^{-4} ZE_0^2 \int_0^1 \lambda^2 N(\lambda) \mathrm{d}\lambda, \text{ where } \lambda = E/E_0$$

This may be written

$$E_{EB} = 5.8 \times 10^{-4} ZE_0^2 S, \text{ where } S \text{ is a shape factor.} \tag{7.24}$$

S is in fact $(E_{rms}/E_0)^2$ and can therefore be obtained from the data of fig. 7.12.

Also,

$$E_{EB} = 3.7 \times 10^{-4} ZE_0^2, \text{ with } E_{EB} \text{ and } E_0 \text{ in MeV.} \tag{7.25}$$

These relations show that the ratio of IB : EB is independent of β-particle energy, but inversely proportional to Z. Using (7.24) and (7.25) we obtain

$$\frac{E_{IB}}{E_{EB}} = \frac{0.64}{ZS} \tag{7.26}$$

Fig. 7.15. Photon and energy yields as a function of atomic number. The ordinates are per *incident* β-particle, i.e. the 'top' and 'sandwich' arrangements show the same yield in these units (Bustard and Silverman, 1967).

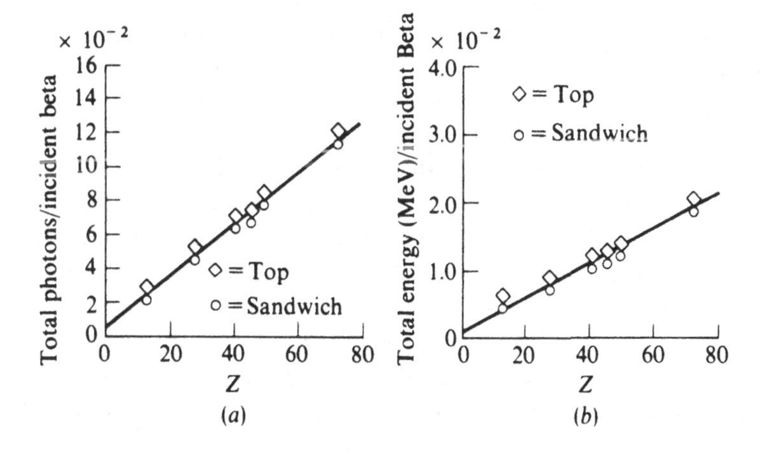

This expression has the correct dependence on Z, but gives ratios lower than those obtained experimentally. Evidently further work is needed to reach agreement between theory and experiment in this area.

More recent studies of EB may be found in the papers by Sama and Morty (1976) where the radiation is studied as a function of target thickness, and by Rudraswamy *et al.* (1984), using β-particles from ^{99}Tc impinging on thick targets. The reader is referred to these papers for details of this work.

7.6 Externally produced characteristic radiation using β-emitters

We turn now to the 'external ionization' produced by β-particles in radioactive decay. Studies of this phenomenon have achieved some practical importance because of the usefulness of small portable sources of characteristic X-radiation consisting of β-emitters mixed with (or otherwise allowed to bombard) the element whose characteristic radiation is required. These sources are very convenient for the calibration of X-ray spectrometers because they have an accurately constant output, or an output which decays at an accurately known rate; they are completely independent of power supplies, and their positioning is made easy by virtue of their small size. Stronger sources have applications in industrial radiography in circumstances where low photon energy and moderately high absorption (as in the examination of light alloys) are desirable.

An early study by Edwards and Pool (1946) reported investigations of characteristic X-rays excited by the β-radiation from ^{32}P. More recent studies have been carried out by Liden and Starfelt (1954), and Starfelt, Cederlund and Liden (1957). In this work, radiators of tin and lead were used in order to excite the K characteristic radiation of these elements, and were mounted on both sides of the β-emitter. The radiators were not designed to be thick enough to stop all the electrons in the case of the higher energy β-emitters, because this would have introduced a large amount of self-absorption of the X-rays produced. It was found that the X-ray peak intensity increased with increasing β-energy and with increasing atomic number of the radiator. Some Bremsstrahlung was detected, with an intensity which increased with these same variables. The yields of characteristic X-rays from these two radiators, using β-emitters ^{35}S, ^{147}Pr, and ^{144}Ce–^{144}Pr (see table 7.2) are listed in table 7.3. These yields include a contribution to the characteristic radiation produced by photoelectric absorption of Bremsstrahlung, which is significant in the case of lead excited by the higher-energy β-emitters. In the case of a 230 mg/cm^2 lead radiator and a ^{32}P source, the 'directly produced' characteristic X-radiation was increased by 25% by this effect. The yield of X-rays fell with

Table 7.2 *β^--emitters used in Bremsstrahlung production*

	E_0 (MeV)	$t_{1/2}$	
Tritium	0.018	12.4 y	No γ-radiation
Sulphur-35	0.167	87 d	No γ-radiation
Promethium-147	0.223	2.64 y	Weak γ-radiation 0.121 MeV
Calcium-45	0.26	160 d	No γ-radiation
Krypton-85	0.672	10.3 y	No γ-radiation
Thallium-204	0.764	4 y	β^-98%; EC 2%
Praseodymium-143	0.93	13.7 d	No γ-radiation
Phosphorus-32	1.707	14.3 d	No γ-radiation
Strontium-90/	0.545	28 y	$^{90}Sr \xrightarrow{-\beta} {}^{90}Y$. No γ-radiation
Ytrtrium 90	2.26	64 h	
Cerium-144/	2.98	285 d	$^{144}Ce \xrightarrow{-\beta} {}^{144}Pr$. No γ-radiation (Cerium-144 βs: 0.31 MeV (70%); 0.16 MeV (20%); 0.25 MeV (~5%))
Praseodymium-144		17 m	

Table 7.3 *Yields of X-rays (photons per* 100 β-particles) *produced by β-particle excitation (Starfelt and co-workers)*

Radiator \ Source	^{35}S 0.167 MeV	^{143}Pr 0.93 MeV	^{32}P 1.71 MeV	^{144}Ce–^{144}Pr 2.98 MeV
Tin	0.15 [37 mg cm^{-2}]	2.8 [68 mg cm^{-2}]	3.6 [74 mg cm^{-2}]	—
Lead	—	4.3 [120 mg cm^{-2}]	11 [220 mg cm^{-2}]	21 [270 mg cm^{-2}]

The radiator thickness is given in square brackets.

decreasing atomic number because of the fall in fluorescence yield. The purity of the X-ray spectrum is better at low β-ray energies and low atomic numbers. The ratio of the number of Bremsstrahlung photons to characteristic X-ray photons is quoted as 1 : 3 for ^{143}Pr β-particles on tin, and 1 : 1 for ^{144}Ce–^{144}Pr β-particles on lead. Somewhat similar studies have been reported by Filosofo *et al.* (1962). In order to produce good spectra from high Z radiators, it was found that high-energy β-emitters were preferable. These could be used as sealed sources with an external radiator. In the case of the medium Z radiators investigated, low energy β-emitters were suitable, and gave good yields of X-rays without an unacceptably high

Table 7.4 *Yields of X-rays (photons per 100 β-particles) produced by β-particle excitation (Filosofo et al. 1962)*

Radiator \ Source	^{147}Pm 0.223 MeV	^{85}Kr 0.672 MeV	^{90}Sr–^{90}Y 2.26 MeV
Tin	0.3	0.7	3.5
Lead	<0.1	1.0	2.0

Table 7.5 *Per cent increase of K X-ray intensity for various backing conditions*

Beta Emitter	Backing Material	Backing Thickness[1] mg/cm^{-2}	Increase in X-ray intensity %	Backing Thickness[2] mg/cm^{-2}	Increase in X-ray intensity %
^{90}Sr – ^{90}Y	Sn	89	118	1326	168
	Ta	219	126	1100	156
	Pb	317	134	1196	168
^{32}P	Sn	102	111	1020	161
	Ta	219	144	877	176
	Pb	317	160	1081	183

[1] Thickness giving maximum K X-ray yields when used as a target
[2] Thickness approaching that which absorbs all β-particles

proportion of Bremsstrahlung. In these cases a mixture of source and target material was found to be preferable to the sealed source and external target configuration. The yields obtained in this work are shown in table 7.4. They are considerably lower than those reported by Linden and co-workers. The effect of absorber 'target' thickness and presence or otherwise of backing material have been investigated by Kereiakes and Krebs (1958). Their experimental arrangement is shown in fig. 7.16. The variation of yield with thickness is illustrated in fig. 7.17 from which it is seen that the optimum yield is obtained for thicknesses which are very much less than the range of the β-particles in this material. The effect of placing backing material behind the source is shown in table 7.5. To obtain maximum intensity in the forward direction, it is clearly advantageous to use a backing foil of thickness much greater than the optimum thickness for the absorber.

Fig. 7.16. Apparatus for determining the intensities of characteristic radiation excited by β-emitters (Kereiakes and Krebs, 1958).

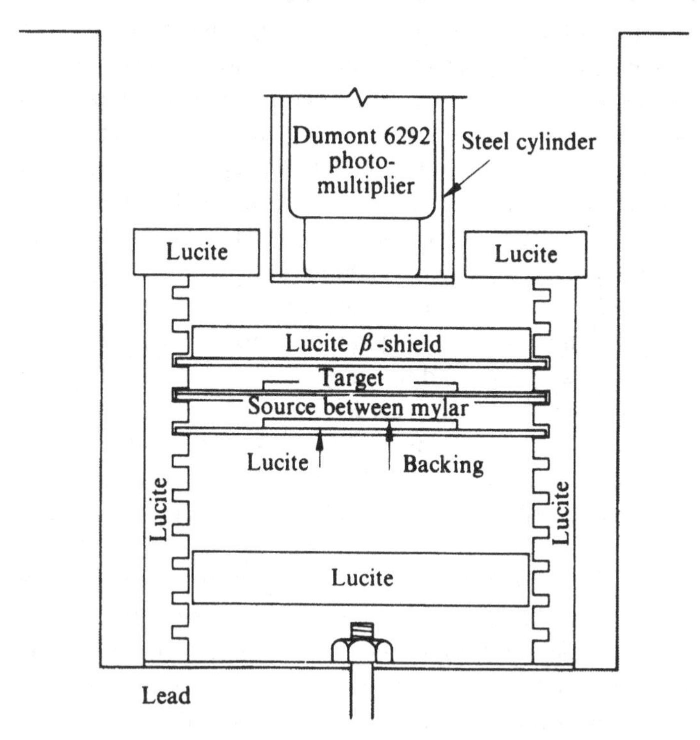

Fig. 7.17. *K* X-ray intensities excited by ^{204}Tl, ^{32}P and ^{90}Sr-^{90}Yβ-particles in various target thicknesses of tin (Kereiakes and Krebs, 1958).

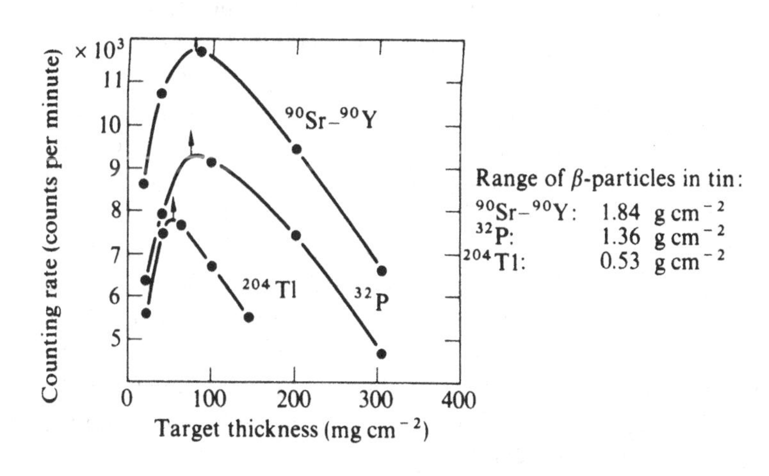

7.7 Practical radioisotope sources of X-rays

We have already referred to the use of EC radionuclides in the calibration of spectrometers, X-ray diffraction apparatus, and so on. A short list of these is given in appendix 3. These radionuclides decay wholly or predominantly by electron capture, and yield an abundant supply of X-ray photons. The presence of small amounts of γ-radiation or Bremsstrahlung is not a serious disadvantage for calibration purposes, as it provides only a low background spectrum, varying only slowly with energy, in the region of interest. This list could be greatly extended, if nuclides showing strong internal conversion were included, but in such cases the γ-radiation may be too strong for the source to be easily used as a calibration source. The presence of large numbers of γ-ray photons of high energy will tend to overload the counter amplifier, and this may cause appreciable 'base-line lowering', affecting the pulse height recorded by a subsequent pulse-height analyser. Of course, the use of fast counters associated with bipolar amplifiers improves the situation considerably, and nuclides which previously were liable to cause errors of this kind (e.g. ^{137}Cs) may now be used with ease.

Sources of ^{51}Cr, ^{55}Fe and ^{65}Zn, may be made by depositing a few drops of solutions on to aluminium or mylar foil, and evaporating to dryness. It is nearly always preferable to precipitate the active material first, for example, as hydroxide, as this will contain less absorbing material than, say, the chloride, sulphate or carbonate radicals in an inorganic salt. Also, the acid nature of e.g. ferric chloride is an obvious disadvantage when preparing the source on many metallic foils. It is desirable to avoid the presence of elements with atomic numbers immediately below that of the X-rays emitted by the source material, which could emit a considerable amount of fluorescent radiation. This would give rise to a spurious peak on the low energy side of the main spectral line, or, in the case of a detector of rather limited resolution, would give rise to a low energy tail and consequent asymmetry of the calibration line. A thickness of up to 1 mg/cm^2 of organic or light inorganic material is acceptable for chromium or iron sources. For X-rays of energy greater than about 15 keV, this condition can be considerably relaxed. Green and Cheek (1958) investigated the use of sources of ^{145}Sm and ^{153}Gd for industrial radiography. These sources were prepared by neutron irradiation of $^{144}Sm_2O_3$ and $^{152}Gd_2O_3$ encapsulated in quartz ampoules with a wall thickness of 0.2 mm (53 mg/cm^2). These were then placed in aluminium containers with wall thicknesses of 0.4 mm. These sources emitted uniformly over an angle of approximately 2.5π steradians, and thus were suitable for 'inside-out' radiography of metallic specimens a few mm in thickness, and were also considered suitable for the determination of liquid levels inside opaque pipes.

Cook *et al.* (1958) have listed ^{55}Fe, ^{131}Cs and ^{181}W, as being pure X-ray emitters suitable for use as the primary sources in X-ray fluorescence analysis.

The excitation of X-rays by β-emitters has been used to provide sources of monoenergetic X-rays for the calibration of scintillation counters by Kereiakes *et al.* (1958), using emitters of ^{204}Tl and ^{90}Sr-^{90}Y. X-ray targets of tantalum, tin, and lead were used, in the form of foils, but in addition a variety of salts, compressed into discs $\frac{1}{2}''$ in diameter and a few mm in thickness were employed, including ammonium molybdate, sodium tungstate, thallous chloride, uranyl nitrate, and several halides.

Cook *et al.* considered that for maximum purity of the X-ray spectrum, the end-point of the β-emitter should not exceed about twice the excitation energy of the desired X-radiation. This implies that for many elements a β-spectrum of considerably lower energy than those so far considered would be useful. Cameron *et al.* (1963) have used sources of tritium absorbed on titanium and zirconium to excite the K spectra of Ca, Ti, Cr, Fe, Ni and Cu, and the L spectra of Ag, Sn, Ta and Au. For industrial and other radiography, the radiation does not need to be monoenergetic and sources of mixed characteristic radiation and Bremsstrahlung have been used for this purpose. Thulium-170 is a γ-emitter (the decay scheme is to be found in appendix 3) yielding ytterbium X-rays ($K_{\alpha}=52.5\,\text{keV}$), some 84 keV γ-radiation, and Bremsstrahlung (end-point 0.97 MeV). Daggs (1956) has described a portable thulium-170 unit which can accommodate up to 40 curies, and which can be used for radiography of limbs, the hand, etc. Coleman *et al.* (1958) have discussed the requirements for a radioisotope source suitable for medical radiography. It must be of high specific activity so that the source can be physically small and so yield radiographs of good definition. Its radiation spectrum should be essentially the same as that from conventional X-ray tubes used for medical radiodiagnosis, and the half-life of the source should be acceptably long (at least 1 year*). If the half-life is too long, however, the maximum specific activity of such a source may be insufficiently high to enable radiographs of good definition to be obtained (requiring a source of small dimensions) in a reasonably short exposure time. These authors have examined the use of thulium-170, and in addition investigated the potential value of sources of promethium tungstate (^{147}Pm), and thallium iodide (^{204}Tl) for radiography. In each case the β-emitter is 'built-in' to a molecule or crystal containing heavy atoms, to give high and reproducible Bremsstrahlung efficiency.

* This requirement seems somewhat severe, and shorter half-lives would probably be acceptable in some circumstances.

7.8 X-rays in association with α-activity

The production of X-rays by heavy charged particles has already been described in chapter 6. From considerations of this kind, it would be expected that α-particles from radioactive sources could be used to excite X-rays in an absorber placed in contact with the source, or mixed intimately with it. This is found to be the case, and studies of this have been reported by Bothe and Franz (1928) and more recently by Buhring and Haxel (1957). 5.3 MeV α-particles from ^{210}Po have been used to produce the characteristic *K* spectrum of nickel for the calibration of X-ray detecting systems (Rothwell and West, 1950), and spectra have been obtained from aluminium and chlorine by Cook *et al.* (1958). These spectra are very pure due to the very low efficiency of Bremsstrahlung production by heavy particles and this may be a major advantage in the use of α-sources for X-ray production.

Many α-active nuclides yield X-rays as a result of internal conversion occurring in the low-lying states which are a feature of heavy nuclei. In many cases (e.g. ^{241}Am, ^{242}Cm) it is found that these transitions have insufficient energy to excite the *K* levels, and that only *L*, *M* etc., X-rays are produced. Barton *et al.* (1951) have pointed to the richness of the *L* spectra in the α-active transuranium elements, and have stressed their value in the study of internal conversion in these nuclides, and in the determination of the multipolarity of the many transitions occurring in them following α-decay.

8

Some additional fields of X-ray study

8.1 X-ray microscopy and microanalysis

The development of X-radiography for medical purposes was one of the early successes in the application of X-rays to practical problems, and the attractions of applying radiography on the microscopic scale, for study of small biological specimens or sections, are obvious. In recent years, methods of viewing small objects by means of a magnifying system using X-rays have been developed and are in use in many biological and metallurgical laboratories.

The first exploratory studies appear to have been made by Sievert (1936). In this work an aperture a few micrometres in diameter was placed in front of an X-ray tube, enabling magnified images to be produced by shadow projection. A system such as this would suffer from the rather small intensity of X-radiation which would be available through a small aperture used in this way, but photographs with a resolution of 5–10 μm were obtained.

However, developments in electron optics were necessary before projection X-ray microscopy, at anything approaching optical resolution, became a practical proposition. Von Ardenne (1939) proposed the use of an electron lens for demagnifying an electron source, thereby enabling X-rays to be generated in a region with a diameter of the order of a few micrometres only. An X-ray tube using this principle was constructed by Cosslett and Nixon (1952), and the method of projection X-ray microscopy has undergone continuous development from that time onwards, and has become an investigational method of considerable importance in biology and metallurgy.

The essential features (fig. 8.1(*a*)) of such an X-ray tube are (a) an electron gun, (b) a lens system for demagnifying the source, and (c) a target in which the X-rays are produced.

The distance from target to photographic plate is limited to a few centimetres, for reasons of exposure time and also because the somewhat soft radiation (0.2–0.4 nm) required to image biological material satisfactorily is absorbed in air rather readily. In order to achieve useful magnifications, of the order of × 100, the specimen has to be not more than 1 mm from the X-ray source. This leads to a target design in the form of a thin transmitting foil forming part of the vacuum wall, and requires that the specimen positioning mechanism be capable of working at this proximity to the X-ray tube.

The electron gun must be capable of producing an electron beam of adequate intensity and brightness from a source which can act as a suitable object in the electron-optical system. Much experience was already to hand through the development of the electron microscope, in which somewhat similar electron source requirements apply. The electron gun normally used in both the electron and X-ray microscopes consists of a thermionic emitter in the form of a tungsten hair-pin filament, a so-called Wehnelt cylinder (or cathode shield) and a circular anode. The electrons are accelerated in the space between the shield and the anode, and then pass through the central aperture of the anode in the form of a divergent beam. The anode is normally at earth potential, the filament is negative with respect to earth and the shield negative with respect to filament. The potential on the shield is variable and may be used to control the beam intensity.

Fig. 8.1. (*a*) Schematic diagram for projection X-ray tube; (*b*) Illustrating the electron-optical path.

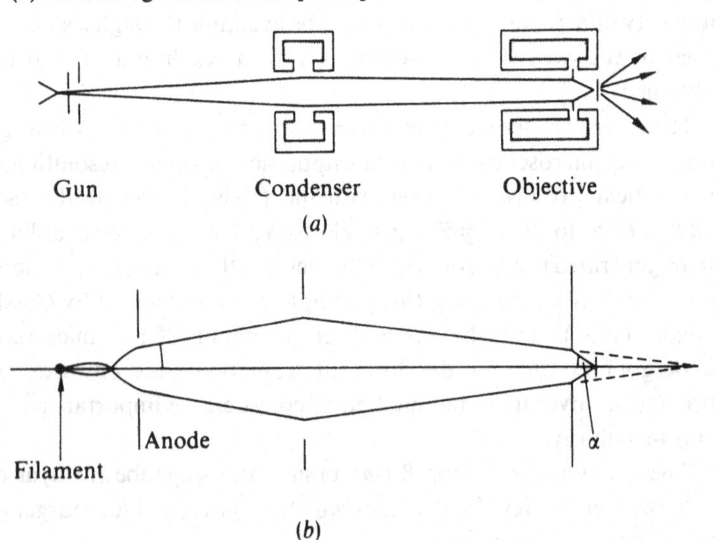

The source of electrons is, from the electron-optical stand-point, located at the so-called 'cross-over' which occurs at a point a few millimetres beyond the tip of the filament, and which is, in a sense, an image of the filament (fig. 8.1b). The cross-over appears to have a diameter in the region of 25 μm, so a demagnification of × 100 is sufficient to yield an image diameter which, if used to excite X-rays in a target, would enable resolutions of optical order to be achieved in the micro-radiographs. The current drawn from the gun may well approach 1 mA, but only a small fraction of this will reach the target; the beam has to be limited to an angular aperture small enough to reduce the spherical aberration and the astigmatism of the objective lens to a tolerable value. Beam currents (at the target) of 5–10 microamps are usual.

To examine quantitatively the amount of current reaching the target, in terms of the lens parameters and the desired spot size in the electron-optical image plane (at which the X-ray producing target is to be placed) we first write down the 'Lagrange' condition for this system in terms of the object and image of the system.

$$r_1 V_1^{\frac{1}{2}} \sin \alpha_1 = r_2 V_2^{\frac{1}{2}} \sin \alpha_2 \tag{8.1}$$

where V_1 and V_2 are the potentials at the source and image positions and $V_1^{\frac{1}{2}}$ and $V_2^{\frac{1}{2}}$ are analogous to the refractive indices μ_s and μ_i of an ordinary optical system.

Equation (8.1) leads to

$$r_1 = r_2 \frac{\sin \alpha_2}{\sin \alpha_1} \left(\frac{V_2}{V_1} \right)^{\frac{1}{2}} \tag{8.2}$$

Now the thermionic emitter can be regarded as a disc of radius r_1, and an emission current density ρ. The emission may be supposed to occur over the whole of the forward hemisphere. Hence

$$i = \rho \pi r_1^2 \text{ and } \sin \alpha_1 = 1$$

$$\therefore i = \rho \pi r_2^2 \sin^2 \alpha_2 \cdot \frac{V_2}{V_1} \tag{8.3}$$

The maximum size of α_2 is limited by the spherical aberration of the lens. This is characterised by a constant C_s such that the radius δ_s of the 'disc of confusion' in the image plane is given by $C_s \alpha_2{}^3$. C_s thus has the dimensions of length. If we impose the condition that $\delta_s \leqslant r_2$, we may write, for the limiting case,

$$C_s \alpha_2{}^3 = r_2 \tag{8.4}$$

This condition restricts α_2 in practice to very small values, so putting $\sin\alpha_2 = \alpha_2$ in (8.3), and substituting (8.4) into (8.3) we obtain

$$i = \rho\pi\left(\frac{V_2}{V_1}\right)C_s^{-\frac{2}{3}}r_2^{\frac{8}{3}} \tag{8.5}$$

This shows that the maximum current falls rapidly as the required value of r_2 is reduced.

Detailed calculations of the spherical aberration constant C_s have been carried out by Liebmann and Grad (1951). The general trend of the data is that to achieve small values of C_s the focal length of the lens must be small. These two parameters are often of the same order of magnitude, and a value for C_s of about 1.5 mm can be achieved with a focal length in the region of 2–3 mm. However, constraints in the design of electron lenses with such short focal lengths are such that the focal plane lies well within the pole pieces of the lens, and this imposes severe restrictions on the design of the specimen holder. Accordingly the present trend is to relax the requirements on C_s somewhat, to allow greater freedom in the design of the objective lens, and a value of 3 mm is quoted by Anderton (1967) as being sufficiently small for good design.

Although single lens X-ray tubes for projection microscopy have been constructed and can give resolutions in the region of 1 μm, it is usual to include a weak condenser lens to give greater flexibility in operation. The overall demagnification of the system can be altered over a wide range whilst maintaining the objective lens at its minimum value of focal length and spherical aberration constant.

The target foil must be thin enough to transmit the X-radiation, including the softer components required for imaging biological material, and must also support the vacuum. A layer of gold evaporated on to a substrate of plastic or other organic material is often used, although thin metallic foils may be used providing they are free from pinholes.

The question of the heat balance in the target has been discussed by Cosslett (1952). For spot sizes of 1 μm or less the conditions are extremely favourable and allow for loading of the order of 10 MW cm^{-2}. The limitation on loading is in fact caused by the operation of (8.5) rather than by inability of the target to conduct the heat away from the point of production. This is true only for a focussed spot, and care has to be taken in operation not to allow the spot to become much defocussed while the gun is operating at maximum brightness.

We have discussed the design of X-ray tubes for projection microscopy in some detail because of the many differences when compared with tubes designed for other purposes. Fig. 8.2 illustrates an X-ray tube described by Anderton (1967).

In order to obtain a high X-ray output from the source, a target of high atomic number is usually chosen, and tungsten or evaporated gold are often used. An accelerating voltage of 10–20 kV is found to be convenient. A higher voltage would reduce the contrast in the image, and a lower voltage would cause insufficient X-ray yield. The voltage also has a direct influence on the effective size of the X-ray source through its effect on electron penetration and diffusion. The range of electrons in gold at 20 kV is in the region of 0.5 μm, and to a good approximation, we may take this to be the diameter of the X-ray producing region in the target. Clearly a high target

Fig. 8.2. An X-ray tube for projection microscopy (Anderton, 1967).

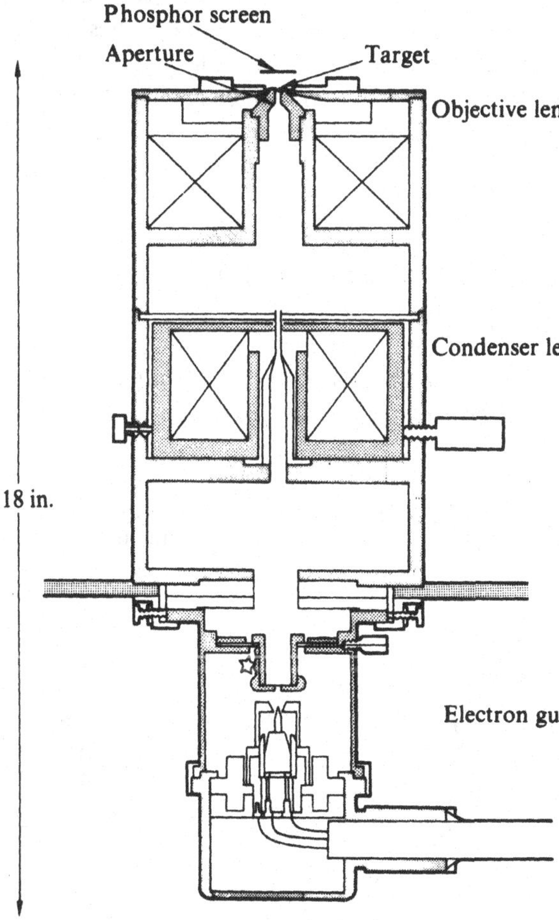

density is an advantage in minimising this diameter, and the accelerating voltage should be kept low: at 20 kV the penetration is already higher than desirable if the best possible resolution is to be obtained.

Referring now to fig. 8.3 the resolution is given by

$$r = d_s(1 - 1/M) \tag{8.6}$$

which is essentially equal to the spot diameter d_s. Equation (8.5) indicates the rapidity with which the current at the target falls as the spot size is reduced and in practice the spot size cannot be reduced below 100 nm without the exposure times becoming impracticably long. The accelerating voltage should not exceed about 10 kV if this resolution is to be achieved, because of the adverse effects of electron penetration at higher voltages. A resolution of this order also places strong demands on the stability of power supplies, and the mechanical stability of the various parts of the apparatus, although experience with the electron microscope has enabled these requirements to be satisfied without undue difficulty.

One further factor which must be considered is Fresnel diffraction of the specimen. For a wavelength λ the width w of the first fringe is given by

$$w = \left[\frac{\lambda a(a-b)}{b}\right]^{\frac{1}{2}} \tag{8.7}$$

Referred to the object plane, and putting $a - b \sim a$, we get

$$w = (b\lambda)^{\frac{1}{2}} \tag{8.8}$$

At a wavelength of 0.2 nm, b must be less than 50 μm if w is not to exceed 100 nm, and this can be achieved by careful axial positioning of the source.

Fig. 8.3. To illustrate the resolution of the projection microscope. a/b is the magnification, and is denoted by M in the text.

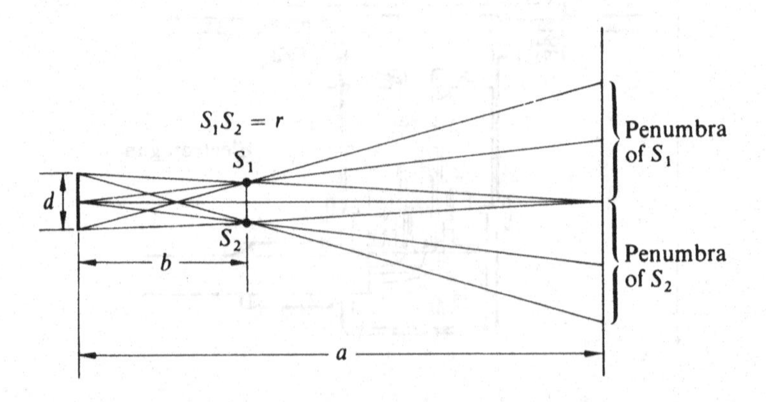

However, the effective wavelength is often rather longer than this, and complete elimination of the diffraction fringe is not always possible. It is much less obtrusive than might be imagined, and this is probably because of the spreading out of the fringe due to the wide range of wavelength (0.1–0.4 nm) contributing to the photographic image.

One of the features of the projection microscope is the unlimited depth of focus caused by the lack of a focal plane. This is a great advantage when setting up the instrument but can cause some degree of confusion when thick specimens are used, because of the large amount of overlapping detail imaged at high definition. Even this can be turned to advantage by the preparation of stereographic pairs (obtained by moving the specimen laterally by a controlled amount between exposures) but in general the biological applications of the projection microscope are best served by specimens only 10–100 μm in thickness, which is of course considerably thinner than the penetrating power of X-rays of this quality. The question of whether sufficient contrast can be obtained to display details of the order, say, of one-tenth of this lower limit of thickness has therefore to be examined.

By calculating the fraction of a continuous X-ray spectrum which would be transmitted by organic specimens of varying thickness, it has been shown that the wavelength range of importance in the imaging of biological material extends quite some distance into the softer parts of the continuous spectrum. The path length between source and plate has an appreciable effect upon the 'effective wavelength' for a particular set of conditions, if we define 'effective wavelength' as the wavelength of monoenergetic radiation which would produce the same contrast at the photographic plate as would the continuous spectrum under consideration. Fig. 8.4 demonstrates that a short path length is advantageous if maximum contrast is to be achieved, and on the basis of absorption calculations, it appears that the thickness detection limit is in the region of 1 μm. It is known, however, that somewhat thinner structures (e.g. fine hairs) may be detectable by virtue of the enhancing effect of the Fresnel diffraction fringe.

Probably a wide spectrum is a highly desirable feature of an X-ray microscope for biological work. This facilitates reasonable contrast in the thinner regions of a specimen whilst avoiding excessive opacity in the denser regions.

The projection microscope has found much application in the field of biology and metallurgy, and we illustrate in figs. 8.5 and 8.6 two examples of such applications.

Projection microscopy is based on an elementary method of producing a magnified image of a specimen, but other imaging methods have been

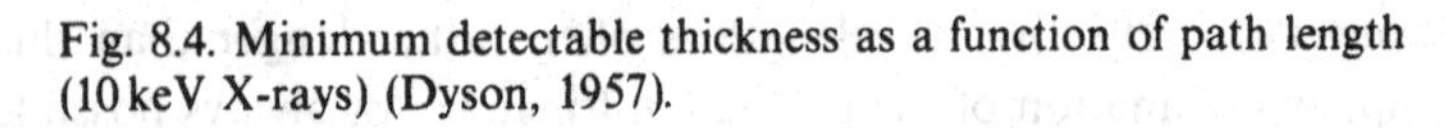

Fig. 8.4. Minimum detectable thickness as a function of path length (10 keV X-rays) (Dyson, 1957).

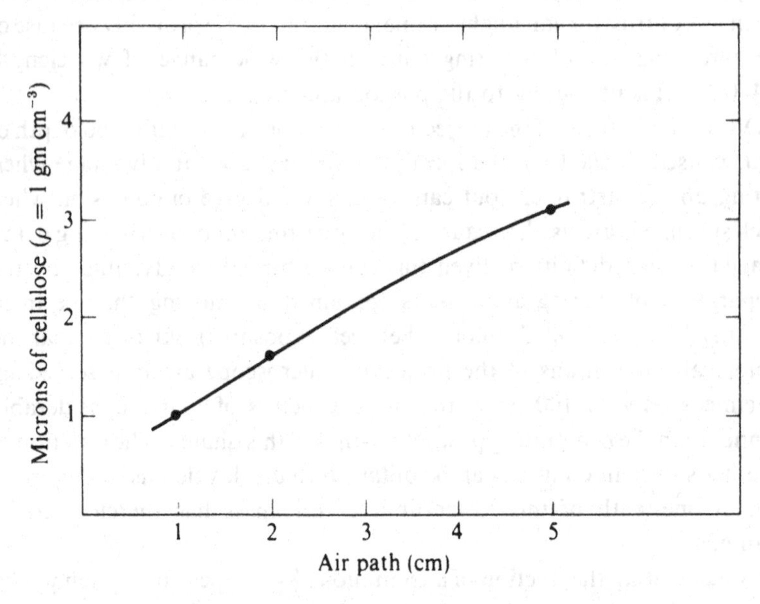

Fig. 8.5. X-ray projection micrograph of a section taken from the base ('hippocampus') of the human brain. The pyramidal nerve-cells with their tapering dendrites are displayed. Nerve fibres can be seen at the base of some of the cells. (By courtesy of Professor R. L. de C. H. Sanders, Dalhousie University).

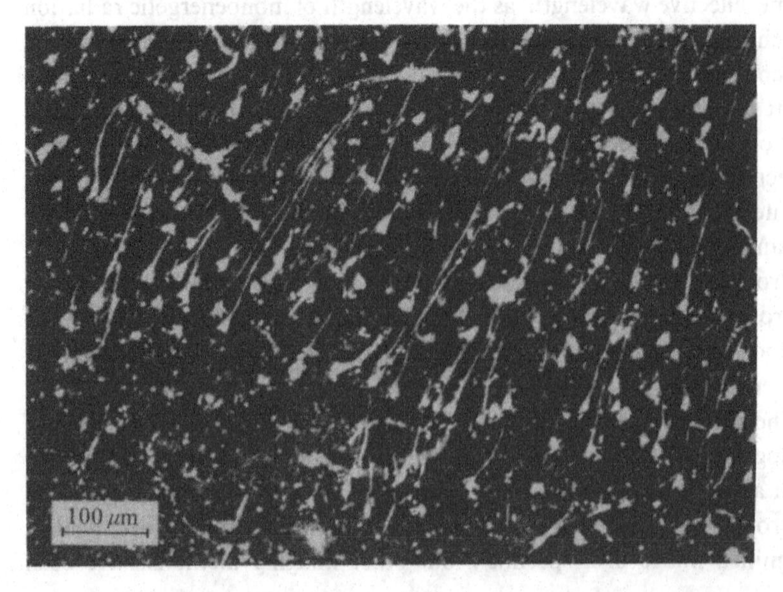

developed, in which a focussing system is used. This encounters considerable technical difficulties because the refractive index of matter for X-rays is close to unity, rendering the construction of a lens system virtually impossible. However, total reflection is feasible, at angles close to grazing incidence, and focussing mirrors have been prepared from which X-ray images can be obtained.

The refractive index for X-radiation is treated in texts on crystallography and optics. For photon energies well above electron binding energies, it is given by

$$\mu = 1 - \frac{ne^2\lambda^2}{2\pi mc^2} \tag{8.9}$$

Fig. 8.6. Graphite-coated nuclear fuel particles. The uniformity of the coating can be gauged from photographs of this type. Discrete contamination spots can be seen on the surface of the graphite. A portion of the standardising mesh (600 lines per inch) can be seen, although the magnification of the particles is somewhat less than that of the grid because of projection effects. (By courtesy of R. S. Sharpe).

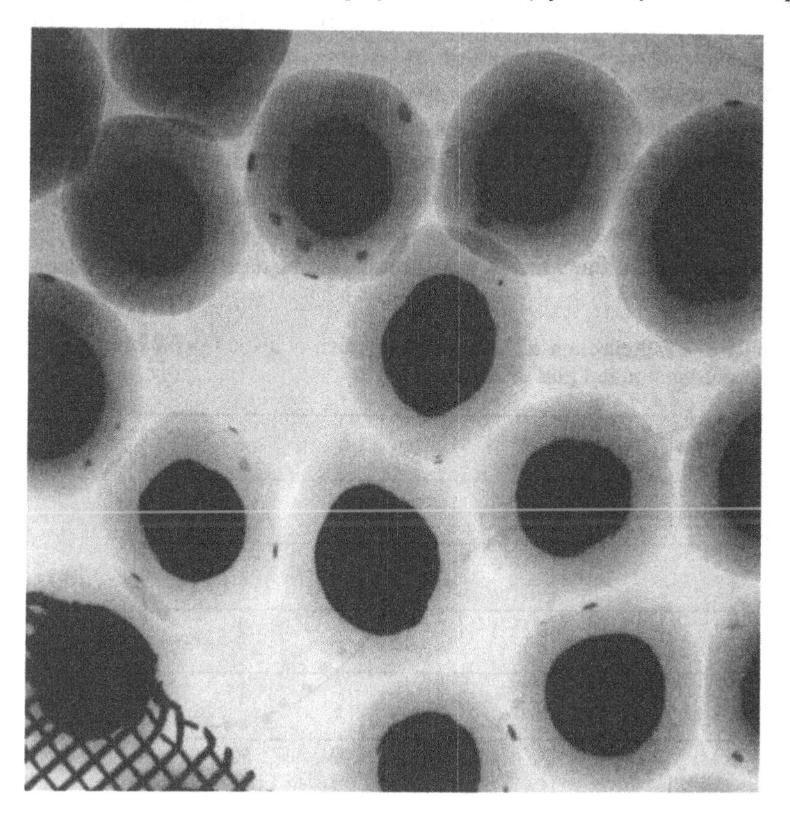

where n is the total number of electrons per unit volume and λ the wavelength. For photon energies below the binding energies of one or more electron shells, (8.9) is still of value (though with a reduced value for n) providing that the photon energy is not close to an absorption edge.

Total external reflection is possible from metallic surfaces at angles of 2° or less, for 0.83 nm radiation, as can be seen from (8.9) and from the experimental data of fig. 8.7.

A spherical or cylindrical surface used under these conditions introduces very large aberrations, but these can be reduced by the use of a second mirror at right-angles to the first (fig. 8.8). This method of microscopy has been pioneered by Kirkpatrick and his co-workers (see Kirkpatrick and Baez (1948), Kirkpatrick and Pattee (1957)), and the resolution over a very restricted area can, under the most favourable conditions, be quite competitive with the projection microscope. It is unlikely, however, that reflection microscopy will become a technique of widespread application mainly because of the high degree of technical difficulty in the setting up of the lenses, the extreme demands placed on the surface polishing, and the very restricted field of view.

Although we have been concerned with the production of enlarged images using X-rays, the method of *contact microradiography* is an alternative process which has been developed to a high degree during the past twenty years or so. In this method, the specimen is placed in contact (via a thin plastic film) with a fine-grain emulsion, a radiograph made, and the result then enlarged photographically. Alternatively, the radiograph may be viewed directly in an optical microscope.

For best results a fine focus X-ray tube should be used and the emulsion

Fig. 8.7. Reflection of X-rays as a function of angle ($\lambda = 0.83$ nm) in aluminium and gold (Rieser, 1957).

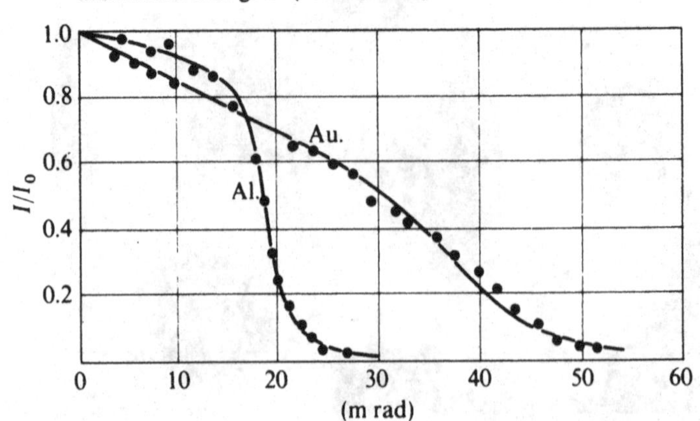

should be thin. The final result obtained depends as much on the properties of the optical microscope or enlarger, as on the primary process, and the use of a thin emulsion (~5 μm) overcomes the difficulties caused by the very limited depth of focus of an optical microscope used at high magnification. Long exposures are unavoidable because of the extreme slowness of the fine grain emulsion used for this work. But grain sizes of the order of 0.1 μm can be achieved, and with their use resolutions of 0.1–0.2 μm can be demonstrated.

Important advances have been made by the development of 'grainless emulsions', which can then be examined in an electron microscope, giving improved resolutions. Certain types of plastic film undergo changes on exposure to X-rays due either to degradation or polymerisation. By placing this material in contact with the object to be examined, and then enlarging in the electron microscope, resolutions of better than 100 nm were demonstrated by Asunmaa (1960) though with long exposure times, of the order of 1 hour. Further details of developments at that time were given by Cosslett (1965). More recently, the availability of synchrotron radiation (see section 8.4) has enabled further developments to be made. The X-ray sensitive material, known as a 'photoresist' is placed in contact with the specimen and is then developed in a solvent which removes radiated and unirradiated material at different rates, is coated with a metal film, and then examined in a scanning electron microscope. The high intensity available in synchrotron irradiation enables much shorter exposure times to be used, and very high resolutions have also been achieved. This work has been reported by Feder *et al.* (1977).

The use of X-rays for microanalysis has developed greatly during the

Fig. 8.8. A stigmatic pair of mirrors for X-ray microscopy (Cosslett, 1965).

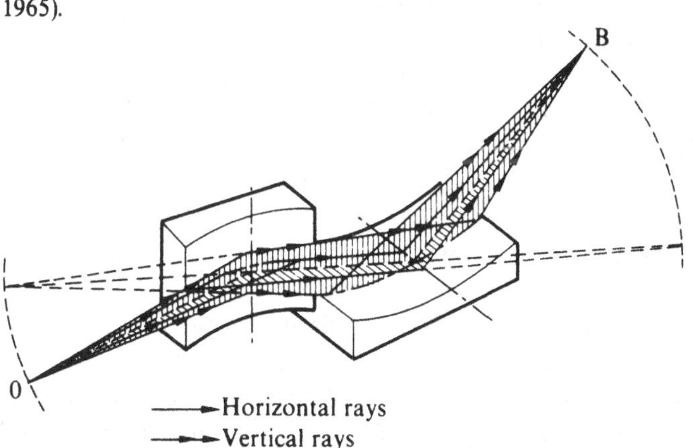

period under discussion. The contact method of microradiography has been refined to enable quantitative studies to be carried out by using the absorption edges of the materials under investigation to provide photographs in which the relative contrast of different parts of the specimen can be correlated with the spectrum of the incident radiation. Accounts of this technique have been given by Engstrom (1963) and Lindstrom (1963). Here again, the availability of synchrotron radiation has made great advances possible (Polack *et al.*, 1977) and chemical microanalysis has been achieved by this method at a resolution of a few micrometres only. However, the study of spectra obtained in emission, rather than absorption, has perhaps attracted more attention, following the introduction by Castaing (1951) of a method in which X-radiation emitted from a specimen undergoing electron bombardment is examined and used for element identification. Initially the specimen had to be examined point by point, but the development of scanning techniques has enabled analysis to be carried out over the whole surface of the specimen, and the data can be displayed on a video display unit.

The first scanning X-ray microscope was constructed and described by Duncumb (1957), and since that time several manufacturers have taken up the principle, and very many are currently in use. In this apparatus, the demagnified electron image is deflected by a set of coils energised by time-base waveforms, and the emitted radiation detected by means of a proportional counter, a curved crystal spectrometer, or a Si(Li) detector. A 'video' signal is derived from the detector output and is then applied to modulate the brightness of the display system. If an 'energy-dispersive' detector is used, the associated electronics can be set so as to be sensitive to characteristic radiation from a selected element. The display is normally on a long persistence tube, because the time required to scan a complete raster may be of the order of one second. The image on the cathode ray tube may then be photographed. If it is desired to examine one particular point in the specimen, the scanning can be stopped and the spot positioned manually against a background of the afterglow of the image. Detailed spectral analysis of the selected point can then be carried out. A third possibility is to scan along a *line* in the specimen and to display the result in the form of a normal oscilloscope trace, with the signal on the y-plates in the usual way.

The proportional counter lends itself well to the analysis of elements of very different atomic numbers present together in the specimen. The detection efficiency is high and this is a great advantage. To analyse specimens containing elements which are adjacent in Z is more difficult by this method, and, in the early days of X-ray microanalysis, special methods were used to 'deconvolute' the unresolved peaks produced by these adjacent elements. This problem has been discussed by Dolby (1959). To analyse such specimens, improved energy resolution is very desirable, and for this reason spectrometers using Bragg reflection are often incorporated in microanalysers. A curved crystal operated in the Dumond mode is used, and the radiation detector (which may still be a proportional counter) is then required simply to deliver an output proportional to intensity. An equipment of this kind has been described by Duncumb (1967). If a Si(Li) detector is used, the arrangements are somewhat simpler, although the K_β line may lie rather close to the K_α line of the element next above it in order of Z, and a correction procedure is then necessary to allow for this.

In order to interpret quantitatively the results of intensity measurements of this kind, calibration samples of various elements (usually in pure form) are introduced into the electron beam, and provide a set of standard readings for each element. However, although the counting rate at a given wavelength (photon energy) is proportional to the amount of element present to a first approximation, several corrections have to be applied if the results of analysis are to have quantitative significance. The atomic number of the other constituents in the specimen affects counting rates because the backscattering coefficient and the stopping power are dependent on this. The *atomic number effect* is therefore important and has to be corrected for. Further, the emitted radiation is partially absorbed on the way out, depending on the composition of the specimen, and so the *absorption correction* has to be investigated. Thirdly, some characteristic radiation can be produced by *fluorescence*, and this may be caused either by continuous radiation generated in other parts of the specimen, or by characteristic radiation of higher energy from other constituents. The mode of applying these corrections (collectively known as Z-A-F corrections) is described in specialised texts dealing with the subject, for example those of Reed (1975) and Heinrich (1981).

Some applications of X-ray microanalysis in the field of metallurgy have been reviewed by Long (1963), Shinoda (1963), Goldstein (1969) and Reed (1975). The technique has been found to have important applications in mineralology, for example in the study of fine-grained materials, or in the examination of the different phases in intergrowths, and this can be carried out on features with dimensions of only 1–2 μm. A review of the applications of the electron microprobe in geology has been given by Keil (1973). Reviews of biological applications have been compiled by Anderton (1967) and Robertson (1968).

The microanalysis of light elements in the region $4 \leqslant Z \leqslant 13$ presents particular problems in view of their soft characteristic radiation, but by the use of vacuum spectrometers and suitable curved crystals or multi-layer crystals this has been achieved. The whole range of elements from $Z=4$ upwards is therefore amenable to analysis down to quantities of the order of 10^{-15} g.

8.2 Chemical influences in absorption and emission; the isotope effect

(a) *Emission line-shifts*

We have seen that in special circumstances broadenings and displacements of X-ray emission lines can occur. This was noted in connection with X-ray 'flash' tubes, in section 3.8, and an account was given of the main phenomena observed in the emission from these high current density devices. Much of what is observed in these circumstances appears to be due to vaporization of the target material, with consequent disappearance of the electron energy bands which are associated only with the solid. A second consequence of the vaporization is that the lifetimes of atoms which are ionized in their outer shells begin to approach values which could allow of multiple ionization by two independent electrons, thereby introducing current-dependence shifts and broadenings into the existing line and satellite structures. We have also seen that multiple ionization, with a consequent shift of the centroid of a line is a regular feature of X-rays produced by heavy-ion bombardment.

The existence of chemical shifts was first examined systematically in 1924 when Lindh and Lundquist (1924) measured the relative intensities and wavelengths of lines in the K_β group from rhombic sulphur, and also the sulphides and sulphates of several metals. Valasek (1933) adopted a different experimental arrangement in which the substances under examination were excited by fluorescence rather than by electron bombardment. In this way the chemical transformations induced by the electron beam were avoided, and results were obtained which were more readily interpreted.

As a general principle we may assert that chemical combination modifies X-ray emission spectra through the effects of the *outer* (valence) electrons upon the energy levels of *inner* shells. If a valence electron (especially an *s* electron) is removed from an atom, a reduction in screening occurs, and all the remaining electrons become bound at somewhat deeper levels within the atom. This effect is discernible even at the innermost shell of the electronic structure, and in fact, the K spectra have been examined in much of the recent work on chemical displacements. Often the absolute magnitude of the shift in levels is greatest for the K shell, and least for electrons in the outermost shell. This means that the highest member of a series (i.e. transitions between an outer and an inner shell, rather than transitions between adjacent shells) will show the greatest effect. In chemical combination, the valency electrons may be displaced but not completely removed from their atom, in which case the extent of the line displacement can be related to the degree of displacement of the electrons, known as the *ionicity* of the bond.

Chemical effects are greatest in the case of light elements for which the valence electrons are in the L shell and are thus able to exert a relatively large influence on the K binding energies.

As an example of a detailed study of chemical shifts, we look at the chemical shifts in compounds of tin and other metals, which have been reported in a series of papers by Sumbaev and co-workers. The first paper in the series is by Sumbaev and Mezentzev (1965), in which the experimental method is described and some early results given. A detailed discussion of the whole series, with a full bibliography, is given by Sumbaev (1970). A Cauchois bent quartz crystal spectrometer was used, and the X-radiation was excited by fluorescence using primary radiation generated at 200 kV. The essential nature of the phenomena is well-illustrated by their measurements on Sn and SnO_2. SnO_2 is ionic in character and the levels are bound more strongly as shown in fig. 8.9. It was found that the shifts relative to β-tin amounted to 0.18 ± 0.02 eV, 0.192 ± 0.010 eV and 0.22 ± 0.02 eV for the K_{α_2}, K_{α_1} and K_{β_1} lines respectively, the *lines* from tin oxide being of *higher*

Fig. 8.9. Electronic energy levels in Sn and SnO_2.

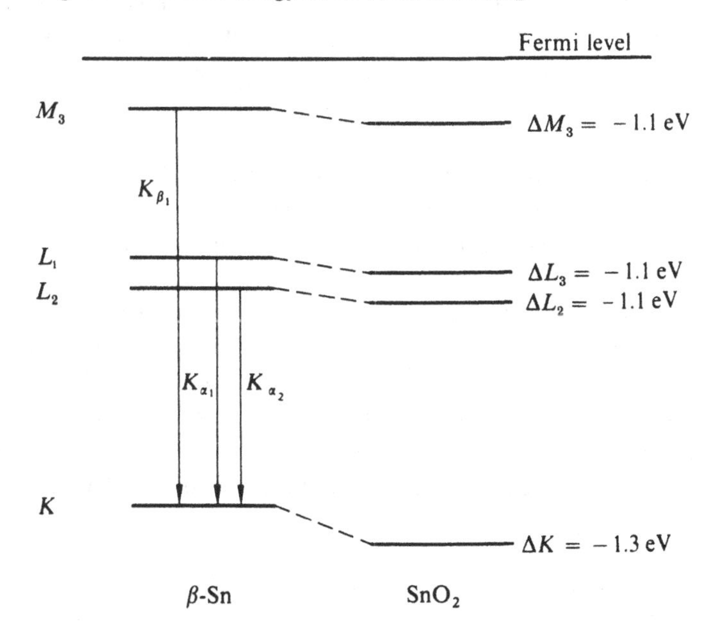

energy in each case. In order to determine the displacements of the individual levels, the displacement of any one level needs to be known from independent data. This is supplied by the work of Nordling (1959) who had measured the shift in the energy of photoelectrons ejected from the L_3 sub-level of tin, thus enabling the shifts of the other levels to be calculated. In each case the levels are nearer the Fermi level in tin than in tin oxide. In a later paper (Sumbaev *et al.*, 1966) the shifts between Mo–MoO_3, Sn–SnO_2 and W–WO_3 were observed, and amount to $+192 \pm 7$ meV, -152 ± 5 meV, and $+110 \pm 33$ meV respectively*, these values being $K_{\alpha 1}$ (metal) $-K_{\alpha 1}$ (oxide).

We note that in tin oxide, the X-ray lines have a slightly *increased* energy, because SnO_2 is moderately ionic, i.e. with a net loss of electronic charge around the tin atom (reduced screening). In molybdenum and tungsten, the oxides show a *decreased* transition energy, because these compounds tend towards a covalent structure, with the metallic atoms being surrounded with a slightly increased electronic charge derived from the oxygen atoms (increased screening).

The energy shift may be expressed in the form

$$\Delta E_{K_\alpha} = im\{p_K \Delta E_{K(Z:Z-1)} - p_L \Delta E_{L(Z:Z-1)}\} \tag{8.10}$$

where the ΔEs refer to the shift in the K and L levels which would result if 1 outer electron were wholly removed from the atom, i is the ionicity and m is the valency. p is the fraction of time for which each valence electron is within the K or L electron shell. The dependence of the shift upon redistribution of the valence electron density, and the approximate proportionality between shift and ionicity of the bond appear to be well-established.

The linear relationship between 'shift per unit ionicity' and valency has been made clear by a further investigation in this series in which a series of metallic oxides of progressively increasing valency has been studied. The results of this investigation are given in fig. 8.10.

Sumbaev (1970) has summarized this work and table 8.1 is reproduced from his paper.

The chemical X-ray shift is closely related to the *nuclear isomer shift*, which has also been studied recently. Nuclear energy levels are modified (raised) to a slight degree by the finite probability of an orbital electron existing within the nuclear volume. If the nuclear radius, or the nuclear charge distribution, changes during a nuclear transition the energy of the transition will be modified to an extent depending on the change in radius

* meV = millielectron-volts, not to be confused with MeV = 10^6 eV.

Table 8.1 *Emission line energy in the K_{α_1} line from the oxides of elements with $Z>32$ (Sumbaev, 1970)*

No.	Z	Investigated pair A	B	$\Delta E = E_B - E_A$ (meV)	Coordination number N	Pauling ionicity i
1	32	Ge	GeO_2	$+244\pm20$	6	0.62
2	32		GeS_2	$+123\pm13$	6	0.39
3	32		GeS	$+110\pm11$	6	0.70
4	33	As	As_2O_3	$+151\pm6$	3	0.43
5	38	Sr	SrO	-30 ± 4	6	0.93
6	39	Y	Y_2O_3	-146 ± 10	6	0.87
7	40	Zr	ZrO_2	-229 ± 15	8	0.82
8	41	Nb	Nb_2O_3	-260 ± 5	6	0.63
9	42	Mo	MoO_3	-199 ± 5	6	0.39
10	47	Ag	Ag_2S	$+51\pm4$	2	0.56
11	48	Cd	CdO	$+115\pm6$	6	0.85
12	48		CdSe	$+82\pm13$	4	0.56
13	49	In	In_2O_3	$+112\pm8$	6	0.78
14	50	Sn_α	Sn_β	$+37\pm10$	4	0
15	50		SnO	$+108\pm12$	6	0.85
16	50		SnO	$+131\pm10$	6	0.85
17	50		SnS	$+113\pm14$	6	0.72
18	50		SnSe	$+79\pm11$	6	0.70
19	50		SnTe	$+103\pm12$	6	0.68
20	50		SnO_2	$+229\pm10$	6	0.64
21	50		SnO_2	$+204\pm11$	6	0.64
22	50		SnO_2	$+210\pm10$	6	0.64
23	50		SnS_2	$+149\pm12$	6	0.40
24	50		$SnSe_2$	$+113\pm13$	6	0.37
25	51	Sb	Sb_2O_2	$+121\pm17$	3	0.51
26	51		Sb_2O_4	$+172\pm10$	6	0.59
27	51		Sb_2O_5	$+200\pm15$	6	0.49
28	52	Te	TeO_2	$+176\pm5$	6	0.59
29	52		TeO_3	$+269\pm5$	?	?
30	56	Ba	BaO	$+42\pm20$	6	0.94
31	57	La	La_2O_3 (*A*)	-3 ± 10	6	0.89
32	58	Ce	CeO_2	-457 ± 15	8	—
33	59	Pr	Pr_2O_3 (*A*)	-20 ± 15	7	0.89
34	59		$PrO_{1.82}$	-263 ± 9	—	—
35	60	Nd	Nd_2O_3 (*A*)	$+14\pm11$	7	0.87
36	60		Nd_2O_3 (*C*)	$+50\pm10$	6	0.85
37	62	Sm	Sm_2O_3 (*B*)	$+32\pm11$	6	0.85
38	63	Eu	Eu_2O_3 (*C*)	-644 ± 10	6	0.87
39	64	Gd	Gd_2O_3 (*C*)	$+36\pm12$	6	0.85
40	65	Tb	Tb_2O_3 (*C*)	$+19\pm15$	6	0.85

Table 8.1 (*continued*)

No.	Z	Investigated pair A	B	$\Delta E = E_B - E_A$ (meV)	Coordination number N	Pauling ionicity i
41	65		$TbO_{1,66}$	-266 ± 13	—	—
42	65		$TbO_{1,72}$	-394 ± 18	—	—
43	66	Dy	Dy_2O_3 (C)	$+10 \pm 14$	6	0.85
44	67	Ho	Ho_2O_3 (C)	$+1 \pm 17$	6	0.85
45	68	Er	Er_2O_3 (C)	$+18 \pm 43$	6	0.85
46	69	Tm	Tm_2O_3 (C)	$+46 \pm 16$	6	0.85
47	70	Yb	Yb_2O_3 (C)	-582 ± 30	6	0.87
48	70		Yb_2S_3	-520 ± 25	6	0.69
49	71	Lu	Lu_2O_3 (C)	$+3 \pm 20$	6	0.85
50	72	Hf	HfO_2	-6 ± 30	8	0.82
51	73	Ta	Ta_2O_5	-113 ± 30	6	0.63
52	74	W	WO_2	-110 ± 33	6	0.39

The letters A, B, C for entries from No. 31 onwards refer to different crystallographic structures. See, for example, Wells (1962) p. 464 ff.

Fig. 8.10. Chemical shift as a function of valency (Sumbaev *et al.*, 1968). The energy shift divided by the ionicity is plotted in the full line. The dotted curve illustrates the uncorrected values of ΔE.

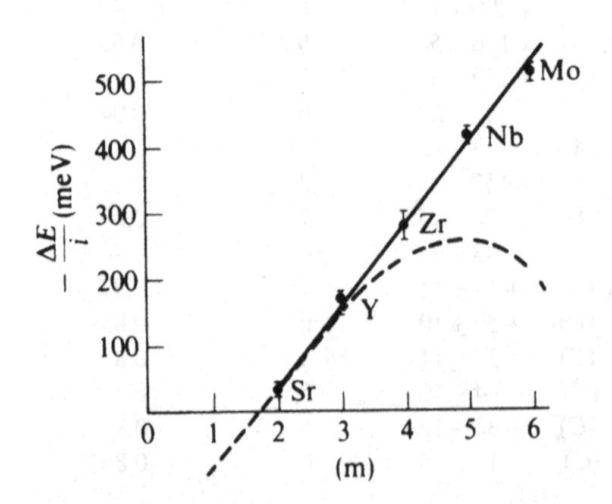

and the electronic charge density at the nucleus. The change in transition energy is given by

$$\Delta E_\gamma = \text{const.}\, \psi^2(0)[R_e{}^2 - R_g{}^2] \tag{8.11}$$

$\psi^2(0)$ is determined by the amount of electronic charge within the nuclear volume and so is closely related to the quantities i, p_K, m and p_L in (8.10).

By Mössbauer techniques (section 5.8) small differences of E_γ between source and absorber of different chemical composition can be measured and are proportional to differences in ionicity between source and absorber. This shift may be written, from (8.11), as

$$\Delta E_\gamma = \text{const.}\frac{\Delta R}{R}(\psi_1^2(0) - \psi_2^2(0)) \tag{8.12}$$

where the subscripts 1 and 2 refer to the two compounds being compared.

The directions of the several shifts involved are illustrated in fig. 8.11. The

Fig. 8.11. Illustrating chemical and nuclear isomer shifts.

Nuclear energy levels:

E_γ

$\psi^2(0)$ Increasing

High ionicity (e.g. SnO_2) — Low ionicity (e.g. α-or β-tin)

Electronic energy levels:

Fermi level

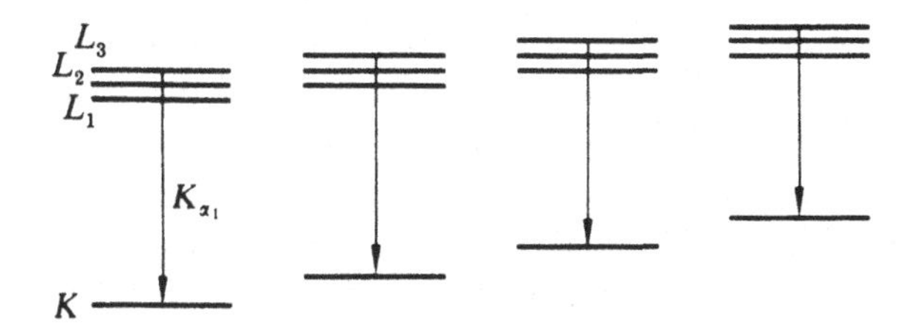

shifts in nuclear levels shown here are appropriate for a case where the *excited* nuclear state has the *larger* radius. This is not always the case, but is so in, for example, the Mössbauer nuclide of tin, ^{119m}Sn.

A detailed study of the light elements oxygen and fluorine has been carried out by Fischer (1965) and changes in energy, line width, and line structure were observed. A further factor which can modify line energies is the possibility of the excitation probabilities of electrons being modified by chemical combination, and there is the further possibility that changes of bond character might cause forbidden transitions to become allowed or vice versa (Baun, 1969). Clearly the possibilities of relating X-ray emission spectra to chemical structure are very rich.

(b) *Absorption and isochromat structure*

We now turn to the influence of chemical combination upon X-ray absorption spectra. In general the effects are much more marked than in studies of emission spectra. This is understandable, because line shifts are normally not as great as the shifts of the levels between which the transition occurs, and absorption spectroscopy is primarily concerned with the liberation of electrons from a single inner level, to quasi-free states in the valence or conduction band. For example, the shift of the K absorption edge of copper in compounds has been studied by Verma and Agarwal (1968). The edge is shifted by amounts varying between 2 and 10 eV, usually in the direction of higher energies, corresponding to a lowering of the K level by this amount, as would be expected when comparing copper in an ionized state with the metallic element. An exception appears to be Cu_2O in which a net lowering of the edge is observed, presumably caused by approximately equal lowerings of K shells and valence levels.

Cobalt has been alloyed with a series of rare earth metals and the absorption edges studied by Sarode and Chetal (1977). The L_3 discontinuity of the rare earth constituent moves to the higher energy side compared with the pure metal. The magnitude of the shift decreases as the atomic number increases and this is interpreted as being associated with the decreasing trend of the ionic character of rare earth-cobalt bonds as Z increases. The cobalt absorption discontinuity shifts towards lower energies compared with the pure metal, the shift again diminishing as the atomic number of the rare earth constituent increases. This study correlates well with the calculated ionicity of the bonds in this series of alloys.

A study of rare earth metals and their oxides is reported by Kushwaha *et al.* (1975), in which the so-called 'white line' is investigated in the L absorption spectra. The white line – a well-established feature of absorption spectra – appears on the high energy side of the absorption edge and is a maximum of increased absorption, so-called because it corresponds to reduced exposure (i.e. reduced blackening) of a photographic plate. It is caused by the raising of an inner-shell electron to vacant states in the outer regions of the atom, falling short of complete ejection of the electron. In this work white lines associated with the L_3 and L_2 edges are observed (those in Ce and Pr, and the L_2 white line in La, for the first time) and are attributed to the raising of $2p$ electrons to vacant $5d$ states. No white line is associated with the L_1 shell, because the $2s$–$5d$ transition would be forbidden. However, Kawata and Maeda (1977) have reported a white line associated with the L_1 edge in molybdenum, and attribute this to the transition of an L_1 ($2s$ state) electron to lattice levels of p symmetry or to a $5p$-like state above the Fermi level.

On the high energy side of the K edge a pronounced structure is often found which may extend to 100 eV or so from the edge itself. The 'near' fine structure is thought to be due to the ejected K electron taking up unfilled states in the valency shells of the atoms. The extended fine structure (or 'Kronig' structure) is influenced principally by the energy bands in the crystal which are available to the ejected electrons. Preferred absorption occurs at those energies for which the ejected photoelectron can enter an energy band within the crystal. Calculations of the energy states available have been carried out by Chivate *et al.* (1968) and have been used to explain the shifts in the K edge of manganese in the oxides MnO, Mn_3O_4, Mn_2O_3 and MnO_2. The extended fine structure is thus determined primarily by structural considerations. A detailed study of the extended fine structure in manganese metal has been reported by Vainganker *et al.* (1979). Data are given in graphical and in tabular form in their paper, and are compared with previous data and with a theoretical treatment. A study of the K absorption edge of cobalt in the pure metal and in a series of Co–Al alloys has been described by Khasbardar *et al.* (1980). We illustrate (fig. 8.12) the extended fine structure near the K edge of copper (Tsustumi *et al.*, 1958), from which it can be seen that the fluctuations are sufficiently large to have an appreciable effect on the absorption coefficients of suitably monochromatic radiation and that the fluctuations can amount to $\sim 10\%$ of the mean value.

A third method of investigating the effects of chemical or structural factors is by the study of the isochromats (q.v. section 2.10) which are obtained when the continuous spectrum is examined within a narrow photon energy band as a function of incident electron energy. For electrons 0–50 eV above the threshold, isochromats result which contain a pronounced structure. We have already noted the presence of forbidden energy bands causes dips to appear in the isochromats, giving rise to pronounced differences between metals and semi-conductors which can be correlated with the energy gap between valence and conduction bands. Again, a more extended structure appears in most materials and is determined by the same factors which operate in the extended fine structure of *K* absorption edges. Detailed studies of Cu and its oxides have been reported by Fujimoto (1965 a,b). A further possibility is that the incident electron loses energy of preferential amounts *before* the emission of the Bremsstrahlung photon. This would give rise to pronounced maxima which can in fact be correlated with the characteristic energy losses which are known to occur when electrons enter metallic foils or targets. Sugawara *et al.* (1967) have obtained good agreement between the positions of their maxima (relative to the first maximum) and the characteristic energy losses in tungsten, tantalum and molybdenum.

(c) *Electronic band structure*

Although absorption spectra can provide information regarding available energy bands in crystals, the study of soft X-ray emission spectra is perhaps more useful for examining valence and conduction bands in metals and insulators. The *K* and *L* spectra of many of the lighter elements have been studied in detail, and appear as bands rather than sharp lines, and bands which involve an electronic transition from the conduction band are often characterised by a sharp upper limit and a low energy tail. The upper limit represents the Fermi surface of the electrons, and this can be correlated with the absorption edge, because energy levels lower than this

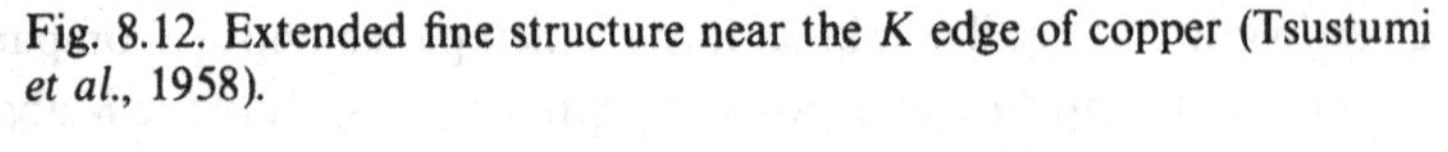
Fig. 8.12. Extended fine structure near the *K* edge of copper (Tsustumi *et al.*, 1958).

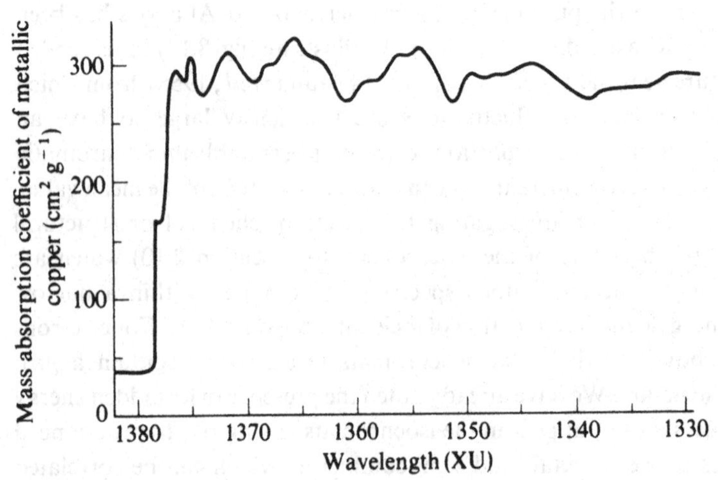

are filled, and are therefore not available as vacancies for ejected electrons. The L_3 emission and absorption spectra of Li, Na and Al illustrate this. The sharp upper limit to the K bands are in practice 'smoothed out' to a slight extent because of the finite width of the K level, which in moderately light elements is of the order of a few electron volts in magnitude.

The origin of the low energy tail is not so clear. When an electron from the lower part of the conduction band 'falls in' to fill the vacancy in the K or L shell, a vacancy is created in the conduction band, which is then filled by a radiationless transition from higher up in the conduction band. This has the effect of increasing the width of the levels at the bottom of the conduction band, and therefore produces a low energy tail on any X-ray band originating from the conduction band. An alternative explanation proposes the existence of occupied excitation orbitals immediately below the conduction band, from which electrons may fall into the K or L shells. This would clearly contribute to the low energy side of the X-ray band.

The shape of the band is proportional to $\{N(E)\cdot T(E)\}$, where N is the density of states of given symmetry (s, p, d, etc.) and T the transition probability. Each level in the band structure contains a mixture of electrons of different symmetry, and information about this can be obtained by studying spectra of more than one series in the same metal. K spectra provide information on p states, L_2 and L_3 spectra on s and d states, etc. because of the operation of the selection rules for dipole radiation.

The first attempt to interpret soft X-ray spectra in terms of electronic band structure appears to have been made by Houston (1931), but much of the experimental development of the subject has been due to Skinner (e.g. 1938, 1940). The techniques and experimental results are described in detail by Tomboulian (1957) and there are reviews by Parratt (1959), Appleton (1964) and Fabian *et al.* (1971).

A completely different approach to the study of electrons in metals is being actively pursued at present, and uses the fact that the radiation scattered by Compton effect (normally monoenergetic to an extent which is limited only by the energy width of the incident photons and the instrumental broadening) is broadened if the scattering electrons are not at rest. Early studies of this effect were made soon after the discovery of the Compton effect itself, but its use in connection with the electron momentum distribution in solids has not been pursued in detail until relatively recently. It may be shown that the difference in wavelength as a result of Compton scattering can be written in two parts as

$$\lambda' - \lambda = \frac{2h}{mc}\sin^2\frac{\phi}{2} - \frac{2\lambda \sin\phi/2}{mc}p_z \tag{8.13}$$

The first term represents the usual Compton shift for a scattering angle ϕ and the second term represents the broadening due to the z component of the electron momentum, the z axis being taken in the direction of the scattering vector $\mathbf{k}'-\mathbf{k}$ ($\mathbf{k}$ and $\mathbf{k}'$ are the wave vectors of the incident and scattered photon respectively).

At any position in the line profile, the intensity $J(p_z)$ is proportional to the probability of the scattering electron having momentum p_z in the z direction,

$$J(p)=\iint_{p_x,p_y}^{P}(p_x,p_y,p_z)\mathrm{d}p_x\mathrm{d}p_y \tag{8.14}$$

and

$$p_z=\frac{mc\Delta\lambda}{2\lambda\sin\phi/2}$$

where $\Delta\lambda$ is the shift from the centre of the profile.

The electron momentum distribution can thus be investigated by this method. Cooper (1977, 1985) has reviewed the present state of development of this technique and gives an account of experimental investigations in this field.

(d) *The isotope shift*

The *X-ray isotope shift* may conveniently be discussed here because its origins relate it closely to the chemical and nuclear isomer shifts. The addition of neutrons to a nucleus will increase the nuclear volume, and reduces the effective nuclear charge experienced by an electron within the nucleus. The effect of this is to reduce the binding of s electrons, particularly $1s$ electrons, thereby introducing a negative shift into the energy of the K_α lines. The effect was first reported by Brockmeier *et al.* in 1965, using targets of ^{233}U and ^{238}U. The characteristic radiation was excited by fluorescence, and the K_{α_1} energy of the heavier isotopes was the lesser of the two by 1.8 ± 0.02 eV. This was observed using a Dumond curved crystal spectrometer. Sumbaev and Mezentsev (1966) report a similar effect in molybdenum and obtain the result

$$^{92}\mathrm{MoO_3}-{}^{100}\mathrm{MoO_3}=0.3\,\mathrm{eV};\ {}^{94}\mathrm{MoO_3}-{}^{100}\mathrm{MoO_3}=0.27\,\mathrm{eV};$$

and

$$^{92}\mathrm{MoO_3}-{}^{94}\mathrm{MoO_3}=0.001\pm0.005\,\mathrm{eV}$$

for the K_{α_1} line in each case. Similar data are available for isotopes of tungsten, samarium, and tin (Chesler *et al.*, 1967) and more detailed discussions, together with new data, have been given by Boehm and Chesler (1968), Bhattacherjee *et al.* (1969), Seltzer (1969) and Lee and Boehm (1971).

These effects are much smaller than the isotope shifts observed in muonic X-ray spectra discussed in section 8.3 because of the much smaller fraction of time occupied by the electron within the nuclear volume.

8.3 Mesonic X-rays

Our treatment of characteristic X-rays so far has related to normal atoms, but we now turn to new categories which became amenable to study as the result of the availability of accelerators with energies in excess of 100 MeV. The π- and μ-mesons are readily produced by proton bombardment in such machines, and negative pions decay in flight to produce negative μ-mesons (muons). The essential properties of the muon for our purpose are that it possesses a negative charge and is weakly interacting. It can therefore enter into bound states with a nucleus. These states can exist independently of the electronic structure of the atom, and transitions of muons between bound states can take place. These transitions result in the production of muonic X-rays.

The Bohr radius is less than that for a hydrogen atom by a factor of $m_\mu/m_e = 207$, and the binding energies are increased by this order. All states are unoccupied, and spectra appropriate to a hydrogenic atom are therefore produced.

The μ-mesons are slowed down in matter by the normal processes of ionization by collision until they reach energies of the order of 2 keV. They are then captured by the atom and enter into orbits of high energy and angular momentum. The time for the slowing down process is of the order of 10^{-13} s for a solid but considerably longer for a gas, then being of the order of 10^{-9} s.

The muon then begins the process of cascading to lower energy states and is either absorbed by the nucleus from the 1*s* state, or decays with its natural life time of 2×10^{-6} s.

The orbit which has the same radius as the hydrogen atom has the principal quantum number $n = 207^{\frac{1}{2}}$ or 14 approximately, and will cascade by allowed transitions usually of the form n, $\ell(=n, n-1) \rightarrow n-1, \ell-1$ emitting X-ray photons or Auger electrons. For transitions between outer states the probability is that Auger emission will take place, but in the later stages of the cascade, the transitions become more energetic $\left[E \frac{Rhc}{n^2},\right.$ therefore $\left.\Delta E = \left(-\frac{Rhc}{n^3}\right)\Delta n\right]$ and the radiative transition probabilities increase accordingly. Complete Lyman ($np \rightarrow 1s$) and Balmer ($nd \rightarrow 2p$) series have been observed from $n = 14$ in titanium by Kessler *et al.* (1967) and in the heavier elements the K_{α_1} and K_{α_2} [$2p_{\frac{3}{2}} \rightarrow 1s$ and $2p_{\frac{1}{2}} \rightarrow 1s$] are resolvable. The data for manganese are illustrated in fig. 8.13.

Radiation from mesonic atoms was first observed by Chang (1949) and the first observations of muonic X-rays were made by Fitch and Rainwater (1953), using a sodium iodide crystal as detector. The introduction of the Ge(Li) detector in 1964 enabled great improvements in energy resolution to be obtained, and a considerable body of data now exists.

The radius of the 1s Bohr orbit may be seen to be $a_H/207Z$ which for a $Z = 20$ atom is approximately 13 fm, and for $Z = 80$ approximately 3.5 fm.

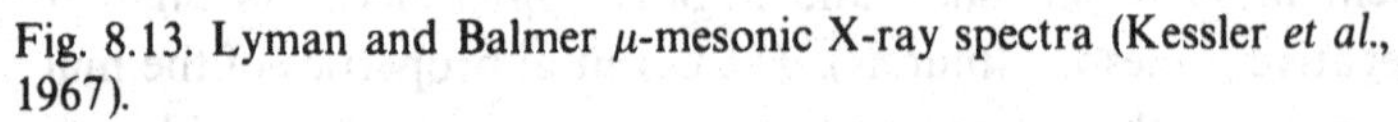

Fig. 8.13. Lyman and Balmer μ-mesonic X-ray spectra (Kessler *et al.*, 1967).

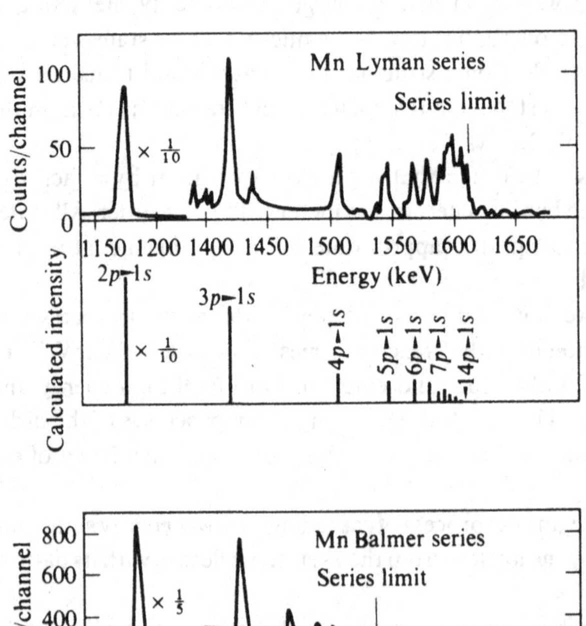

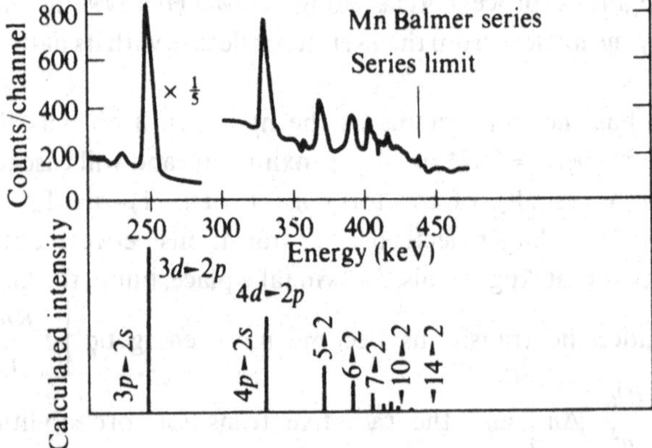

Even in moderately light elements the probability of finding the muon within the nucleus is considerably greater than for an electron with the same (n,ℓ) values, and for a heavy element the muon is almost wholly contained within the nucleus. We therefore expect large deviations between observed energies and those calculated from, e.g., (3.1). For example, the observed value μK_{α_2} in lead equals 5788.33 keV, to be compared with about 15 MeV calculated from the formula for a hydrogenic atom. A further expectation is that the isotope shifts will be large, on account of the change in nuclear volume as extra neutrons are added. We note, however, that the effects of electron screening will in general be small because of the very small radius of the muonic orbits. The isotope shift is illustrated in fig. 8.14 for several

Fig. 8.14. Muonic K X-ray spectra showing the isotope shift in neodymium. (Wu and Wilets, 1969; Macagno *et al.*, 1967).

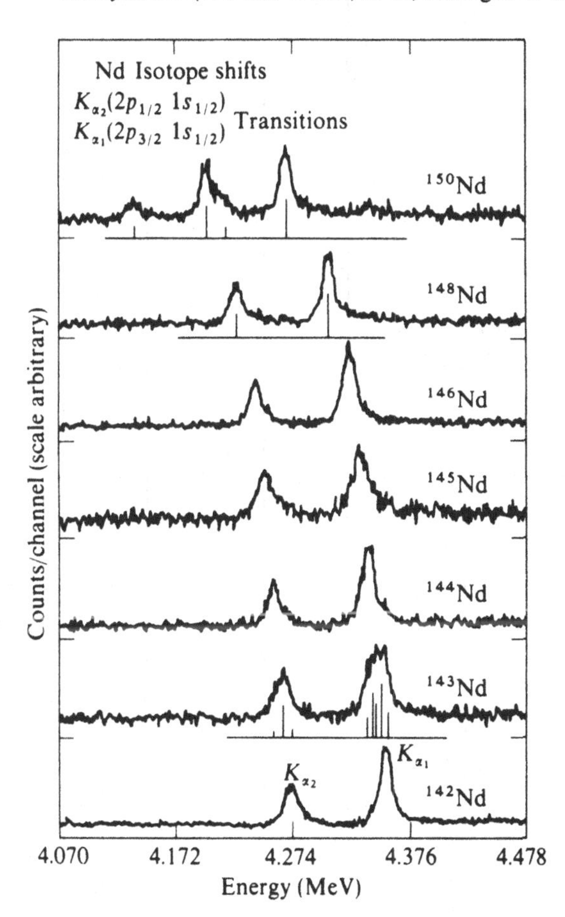

isotopes of neodymium; the progressive reduction in energy as the number of neutrons is increased is well displayed.

Tabulations of the energies of muonic transitions are given by Wu and Wilets (1969) for μK_{α_1} and μK_{α_2} radiation. We have plotted in fig. 8.15 some determinations of the $2p \rightarrow 1s$ transition [i.e. unresolved μK_{α_1} and μK_{α_2} line] from a number of elements as reported by Quitmann *et al.* (1964).

Fig. 8.15. $2p \rightarrow 1s$ transition in μ-mesonic atoms.

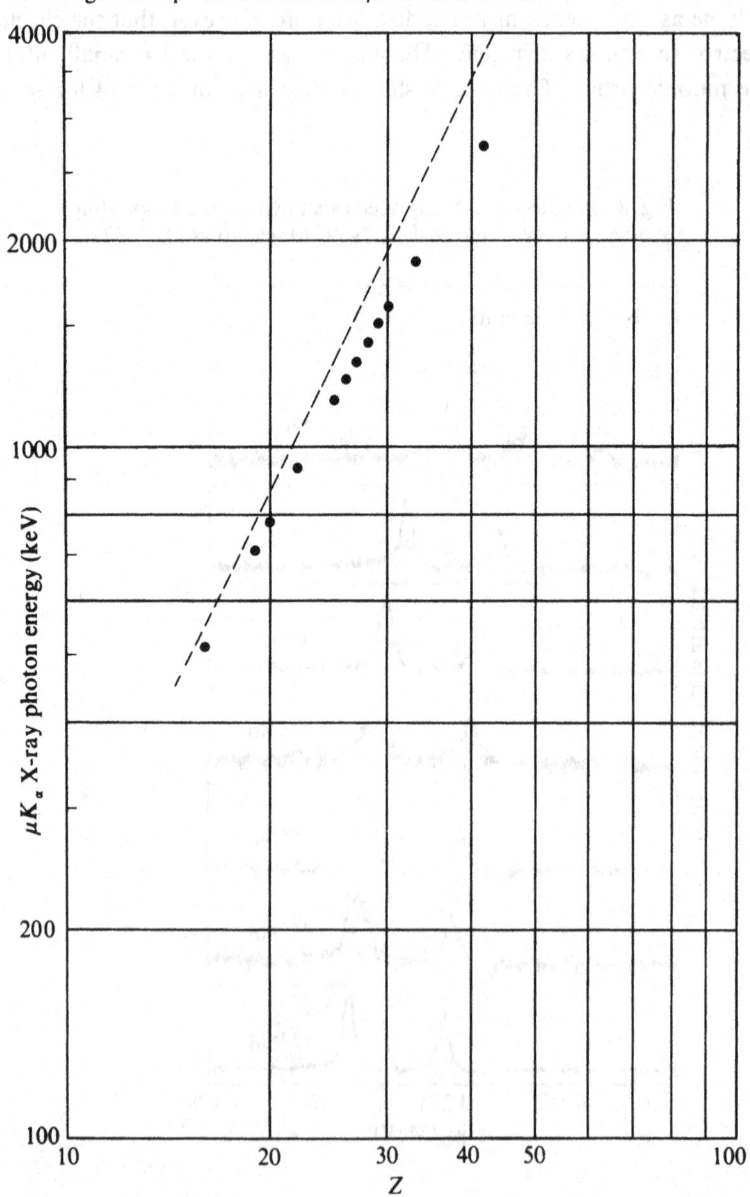

An important product of muonic X-ray studies follows from the very close proximity of the electron orbit and the nucleus – the transition energies are very sensitive to nuclear radius, and enable values for the nuclear radius to be obtained with greater reliability than has hitherto been possible. A value of $1.20\,A^{\frac{1}{3}}$ fm is obtained from muonic X-ray measurements.

Further species of mesonic X-rays exist – π-mesonic X-rays are radiated by pionic atoms in a manner very similar to the processes already described. Transition energies are somewhat higher than for muonic X-rays (e.g. lithium $\mu K_\alpha = 18.7$ keV, $\pi K_\alpha = 24.2$ keV, magnesium $\mu K_\alpha = 296.5$ keV, $\pi K_\alpha = 330$ keV) but the general pattern of behaviour is rather similar to the production of muonic X-rays. Differences do arise, however, because of the strong pion–nucleon interaction which, for example modifies the transition energies, particularly in the heavy elements. The most striking difference between muonic and pionic X-ray spectra, however, lies in the relative intensities of X-ray lines. Because absorption of the pion into the nucleus takes place with rather high probability, the yield of X-ray transitions gradually falls with increasing atomic number, finally disappearing altogether. The K lines are observed only for Z less than 11, the L lines for Z less than 30 and the M lines for Z less than 59. The pion absorption rate increases more strongly than the transition rate, and eventually reduces the probability of transitions taking place to unobservably low values. The properties of pionic atoms have been reviewed by Backenstross (1970).

K-mesonic X-rays are known, and were reported for the first time by Burleson *et al.* (1965), who observed the $n = 3$ to $n = 2$ transition (L_α line) in helium at 6.7 ± 0.2 keV. Most of the K-mesons cascade to low-lying levels before being captured by the nucleus, but it is estimated (from the predominance of the L radiation) that about 80% are captured from the $2p$ state and that only about 20% cascade down to the $1s$ state before capture takes place. A subsequent study by Wiegand and Mack (1967) detected $_KL$ X-rays ($3d \rightarrow 2p$) from Li, Be, B and $_KM$ X-rays ($4f, 5f, 6f \rightarrow 3d$) from boron and carbon. No L X-rays were detected from carbon although a search was made.

Chemical effects are strongly marked with muonic atoms. The ratio of intensities $np \rightarrow 1s/2p \rightarrow 1s$ is very different in oxides from the ratios found in metals. This suggests that the details of the cascade process must differ in the two cases. The slowing down process may be different, leading to different conditions at the time of capture. Auger transitions will in all probability be affected by the availability or otherwise of outer electrons and this will clearly affect the relative probability of the radiative process.

Molecular structure appears to have an effect on the meson capture process. During the process of slowing down and stopping, mesons may be captured into molecular orbits rather than into orbits associated with individual atoms. Theories of this process together with experimental data have been reviewed by Ponomarev (1973). Chemical effects associated with capture of muons in nitrogen have been studied by Dubler *et al.* (1976), in which the relative intensities of the Lyman series ($n \rightarrow 1$) in nitrogen were measured in $NaNO_3$, $NaNO_2$ and boron nitride. The different valencies of nitrogen in these three compounds has been invoked as an explanation of the observed effects.

In recent years the study of muonic and pionic X-rays has been carried over into the biomedical field, and so has become relevant to applied physics. X-ray fluorescence analysis has been used increasingly in recent years for the identification and quantitative measurements of elements *in vivo*. In general this technique excludes the lighter elements because the relatively low photon energy of their characteristic radiation prevents ready transmission outwards through the overlying tissue, so detection is difficult. However, if the body is irradiated with muons, the muonic X-rays are of much higher energy, and can be detected without difficulty. Taylor *et al.* (1973) have described an experiment in which bone has been irradiated in this way, and the muonic X-rays from calcium (784 keV) detected. Although this experiment was carried out *in vitro*, it was demonstrated that the radiation dose to a living subject would be acceptably small, enabling the calcium content of parts of the skeleton to be determined by this method.

A further example from the biomedical field is to be found in the study of pions for radiotherapy, where the particular depth-dose distribution of these particles has certain advantages. In order to understand the mechanisms by which pions deposit their energy in biological tissue, their relative capture probability in carbon and oxygen needs to be known. This has been studied (Lewis *et al.*, 1982) by observing the relative intensities of pionic X-rays in the Balmer series when the pions are stopped in tissue. Detailed X-ray spectra are given by these authors, and information is obtained which has important implications for pion dosimetry.

8.4 Synchrotron radiation

In chapter 2 we have discussed in detail the continuous radiation, or Bremsstrahlung, produced when electrons are slowed down in matter. A simple treatment of this would suppose that the electrons are rapidly decelerated during their slowing down period, and a more sophisticated treatment takes into account the fact that the acceleration vector will have a direction which varies with time during the trajectory of the electron past the nucleus with whose Coulomb field it is interacting. This vector is, however, generally opposed in direction to the velocity vector and it is this fact which determines the form of the angular distribution of the continuous spectrum, as given by (2.8), *et seq.*

We now turn to a different situation – that of circular motion – in which the electrons experience a centripetal acceleration perpendicular to the direction of motion. For many years it has been known that such a system would produce electromagnetic radiation, and early experiments showed that emission in the visible and ultra-violet region took place. The work of Tomboulian and Hartmann (1956), using the 320 MeV electron synchrotron at Cornell University demonstrated that emission occurred in the 8–30 nm region, thereby bringing this phenomenon into the soft X-ray field. This radiation, known as synchrotron radiation, has in recent years become a very important source of continuous X-radiation. The acceleration is maintained by a strong uniform magnetic field perpendicular to the direction of motion and the radiation is occasionally known as magnetic Bremsstrahlung.

The emission would follow the pattern of familiar dipole radiation when viewed in the frame of the electrons. In the laboratory frame of reference this is strongly concentrated in the forward direction, and may be shown to be concentrated in a cone of order m_0c^2/E radians where m_0 is the electron rest mass and E the electron energy, i.e. the direction of *minimum* emission may be written as

$$\theta = (1-\beta^2)^{\frac{1}{2}} = \frac{m_0c^2}{E} \tag{8.15}$$

This is illustrated in fig. 8.16.

These electrons are highly relativistic; the cone of radiation is a few minutes of arc only.

To examine the frequency of the emitted radiation we note that, when viewed from a particular direction tangential to the beam, the observer will receive a pulse of electromagnetic radiation at each revolution of the electron in its orbit, and the electron will radiate in that direction for a time which is of the order of $2(R/c)(1-\beta^2)^{\frac{1}{2}}$, where R is the orbit radius and $v \sim c$. However, the end of the pulse, as seen by the observer, will be emitted from a point considerably nearer the observer than the beginning of the pulse. This will result in a foreshortening of the pulse by a factor

$$\left(1-\frac{v}{c}\right).$$

This may be written as

$$\frac{1-\beta^2}{2}$$

when velocity approximates to c, so the duration of the pulse as seen by a stationary observer will thus be of the order of

$$\frac{R}{c}(1-\beta^2)^{\frac{3}{2}} \text{ or } \frac{R}{c}\left(\frac{mc^2}{E}\right)^3 \tag{8.16}$$

The pulses will recur with a frequency equal to $c/2\pi R$, and so the radiation will consist of a fundamental at this frequency and a series of harmonics up to frequencies of the order of

$$\frac{c}{R}\left(\frac{E}{mc^2}\right)^3 \tag{8.17a}$$

or down to wavelengths of the order of

$$R\left(\frac{mc^2}{E}\right)^3 \tag{8.17b}$$

Fig. 8.16. Angular distribution of synchrotron radiation (Ansaldo, 1977).

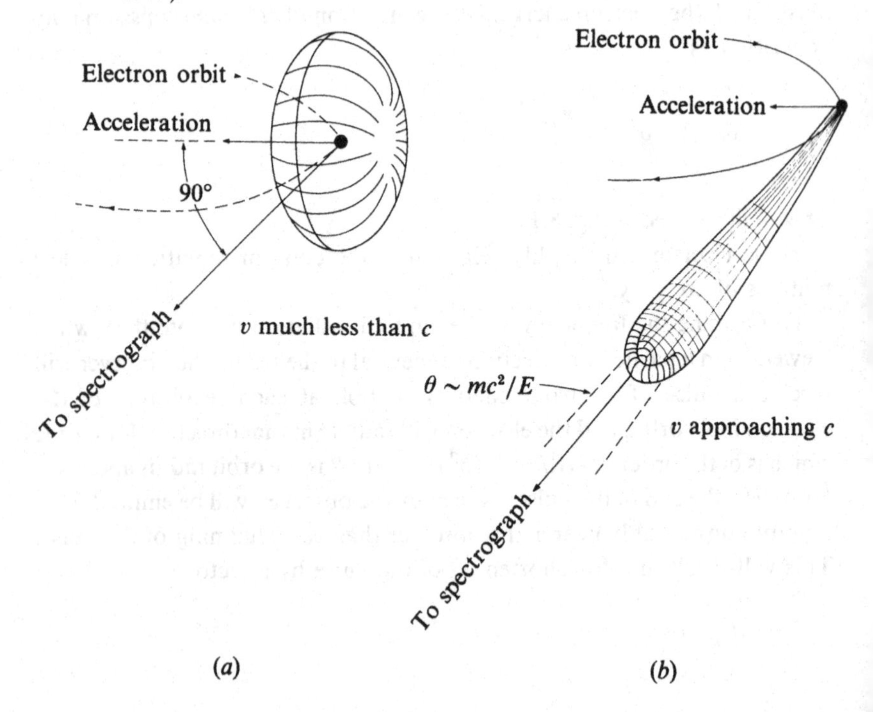

This is in broad conformity with experimental observations. A more detailed theory (Schwinger, 1949; Tomboulian and Hartmann, 1956) expresses this result in terms of a characteristic wavelength λ_c, where

$$\lambda_c = \tfrac{4}{3}\pi R\left(\frac{mc^2}{E}\right)^3. \tag{8.18}$$

λ_c is such that half the power is radiated at longer wavelengths than this, and half at shorter wavelengths. An example of spectral distributions calculated from the detailed theory is given in fig. 8.17.

Detailed theory shows that the total radiated power varies strongly with electron energy, and has been given, for a single electron in circular motion (Winick and Bienenstock, 1978) as

$$P = \tfrac{2}{3}\frac{e^2 c}{R^2}\beta^4\left(\frac{E}{mc^2}\right)^4 \tag{8.19}$$

For highly relativistic electrons, for which $\beta \sim 1$, this may be multiplied by the circulating current I to give the total radiated power $P = 88.47\, E^4 I/R$, (P in kW, E in GeV, I in amperes, R in metres). Equations (8.18) and (8.19) show the advantages of pushing to higher electron energies, and several machines are currently in use or under construction. The bending radius is

Fig. 8.17. Spectral distribution of synchrotron radiation (from the storage ring SPEAR, Stanford; bending radius 12.7 m). (Winick and Bienenstock (1978)).

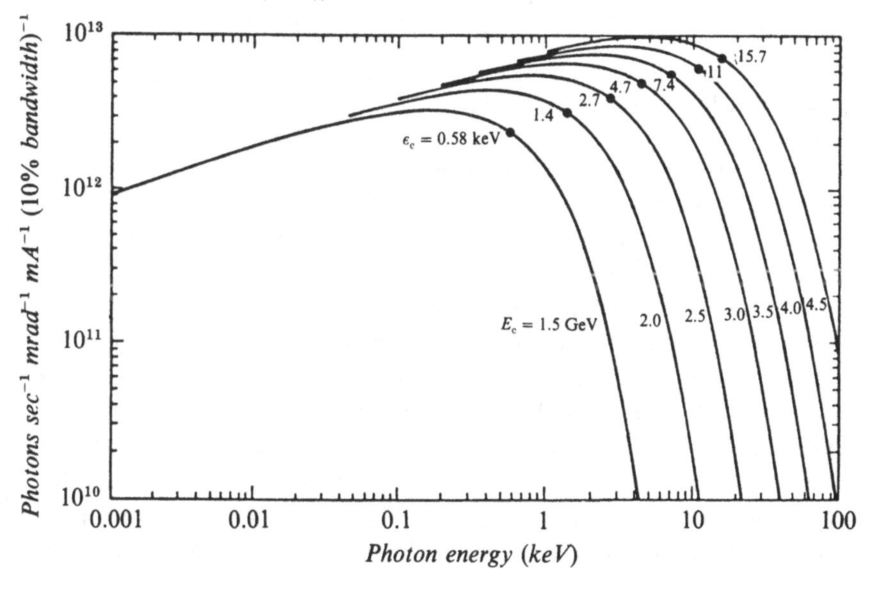

several metres, but, as most synchrotrons consist of arcs alternating with field-free straight sections, the effective orbit radius is normally several times this. The electrons are maintained in orbit by high-frequency generators, and several hundred kilowatts may be required. If this amount of radiated power in the X-ray region is compared with 10 kW radiated by a rotating-anode conventional X-ray generator, we see that a synchrotron of 2–5 GeV energy and a current of several hundred mA represents an unsurpassed source of X-ray power of very high brightness which is extremely attractive for much work in X-ray spectroscopy, crystallography (particularly in the biomedical field) and related areas such as X-ray microscopy.

An important development took place with the building of electron storage rings, into which electrons are injected and then maintained in orbit by the provision of RF power. Lifetimes of several hours can be achieved and currents built up to the order of 1 ampere. As an example we may take the synchrotron radiation source facility at Daresbury (UK). The electron beam is initially produced by a commercially available linac operating in the region 10–15 MeV. This is injected into a 600 MeV booster synchrotron the accelerated beam of which is in turn injected into the storage ring, operating at 2 GeV. The RF power for the storage ring is provided by a 250 kW klystron supply. The storage ring has a bending radius in its curved sections of 5.56 m, and a mean orbit radius of 15.28 m. The facility produces at its exit ports a beam strongly peaked in the forward direction, highly polarized in the plane of the electron orbit, and with a well-defined continuous spectrum reaching from the infra-red to X-rays. An account of the constructional features of this machine is given by Suller and Thompson (1978). Other facilities (either synchrotrons or storage rings) exist at Frascati (Italy), Tokyo, Hamburg, the USA, USSR and elsewhere.

8.5 Plasma physics and astrophysics

(a) *Introduction*

Two of the most important developments in X-ray physics in recent years can be grouped together, principally because they both relate to the production of X-rays at extremely high temperatures. The study of fusion (thermonuclear) reactions has stimulated intensive study of highly ionized gases, including those in which the atoms are stripped of a substantial proportion of their orbital electrons. In such plasmas many of the ejected electrons remain free and are capable of emitting Bremsstrahlung by virtue of their thermal energy, or by acceleration in electric fields associated with the discharge initiating plasma formation. The astrophysical developments have been opened by the inclusion of X-ray detecting equipment in high altitude rockets (since 1962) and satellites (1970), and more recently using the Mir space station and the space shuttle, by the use of which many discrete X-ray sources have been detected, together with a diffuse background. The Sun is also a (relatively weak) X-ray source. X-rays from celestial sources have a range of photon energies comparable with those from laboratory plasmas, and the physical processes of X-ray production have some aspects in common.

(b) *X-rays in plasma physics*

A plasma established in a pulsed discharge tube, using magnetic confinement, will yield optical radiation, characteristic X-rays and an X-ray continuum. These radiations represent an important mechanism for energy loss from the plasma, but also provide a valuable diagnostic tool for study of the conditions prevailing in the plasma. *Bremsstrahlung* is produced predominantly by electron–ion interactions and for a temperature of 10^6–10^7 K, Bremsstrahlung emitted by electrons in thermal equilibrium with the gas atoms will have photon energies in the range 10^2–10^4 eV. Electron–electron collisions make an appreciable contribution for $kT > 10$ keV (Maxon and Corman, 1967). The ratio of electron–electron to electron–ion emission is in the region of 5–10% for a photon energy region of 2–20 keV, and is proportional to electron temperature*. *Characteristic radiation* also appears. The line energies will in general be displaced from their normal values because of the presence of multiply ionized atoms and modified (reduced) screening consequent upon this. Characteristic radiation will often appear from vaporized metallic impurities (particularly from heavy metals) derived, for example, from electrodes in the gas. *Recombination radiation* is produced when a free electron recombines with an atom to form a bound state. Because the free electron is in a continuum, recombination radiation forms a continuum also, though because of the discrete nature of the final bound state some structure is to be expected. Recombination radiation and Bremsstrahlung together form the continuum which is observable from these gaseous discharges.

Experimental studies of the soft X-radiation from pulsed discharges consist principally of studies of the variation with time of X-ray output during the discharge, and measurements of the spectral distribution. Bremsstrahlung produced by a Maxwellian distribution of electrons has the form

$$I(E)\mathrm{d}E = \text{const.}\, \mathrm{e}^{-E/kT}\, \mathrm{d}E \tag{8.20}$$

* The term 'electron temperature' is used in situations when the kinetic energy of the electrons, though Maxwellian, is not necessarily equal to that of the ions.

for photons of energy E and an electron temperature T. Measurement of the slope of the distribution against E therefore yields values for electron temperature, and the absolute intensity gives information on electron and ion densities.

The pulsed nature of discharges used for these studies means that photon counting rates during the pulses are very large. Pulse-height analysis can be difficult under these conditions, especially as the energy resolution of the scintillation and proportional counters used as detectors is rather poor in this energy region. Accordingly rather simple absorption methods have been used to study the spectral distribution of the radiation. Boyer *et al.* (1959) and Griem *et al.* (1959) have reported measurements on the 'Scylla' fusion device. In this apparatus, deuterium at pressures of 0.01–1 torr is ionized by the compression and intense heating which takes place during the discharge as a result of the rising magnetic field, and X-rays and neutrons are observed. The X-ray spectra were studied by plotting absorption curves in beryllium, aluminium and nickel. From the slopes of these curves at the point where the softer radiations have already been removed the end-point of the spectrum can be found, and was in the region of 1 keV. From the total yield the electron and ion densities were in the region of $10^{17}\,\text{cm}^{-3}$.

Jahoda *et al.* (1960) have described similar data. They report in addition the appearance of harder X-rays ($\sim$200 keV). These appear only at pressures less than 0.05 torr at certain stages of the discharge, when $\mathrm{d}B/\mathrm{d}t$ is at a maximum. These hard X-rays are due to electrons being accelerated out of the thermal domain into higher energies, subsequently producing Bremsstrahlung. Beckner (1967) has investigated spectra from a pulse source in deuterium (4–5 torr). This very intense source (0.1–5 MW, for pulses with duration of a few tenths of a microsecond) emitted an X-ray continuum, but with contributions from copper contamination.

(c) *X-ray astronomy*

The use of sounding rockets above the X-ray absorbing part of the Earth's atmosphere led to the discovery of a large number of discrete X-ray emitters. Up to 1970 when the first X-ray satellite UHURU was launched, about 50 discrete X-ray sources had been discovered, and by the end of 1971 this number had increased to about 100. At the time of writing (1988) several hundred such sources are known. Many of these sources have been identified with visible objects. The majority of these sources are members of our own galaxy, and lie rather close to the galactic plane, but several sources are extra-galactic, and include some of the strongest X-ray emitters yet discovered, when the inverse square law is taken into account.

The strongest X-ray emitters have an output which is extremely large, and they are clearly objects of an unusual nature. For example, the X-ray emitter Sco X-1 has an output in the region of 10^{29}–10^{30}W, which is of the order of 100 times its output of visible light. The strong source in the Crab nebula and the X-ray source Cas A have X-ray luminosities (i.e. energy outputs) of the same order. The extra galactic source M87 has an output of $\sim 10^{36}$ W compared with an estimated total of $\sim 2\times10^{32}$W from our own galaxy, and its X-ray output is of the order of 100 times larger than its output at radio frequencies. Two main mechanisms appear to be responsible for the X-radiation from discrete sources – the Bremsstrahlung and synchrotron mechanisms. The first yields an exponential spectrum of the form of (8.20), and the synchrotron mechanism has a form which is represented by a 'power law' spectrum

$$N(E)\mathrm{d}E=\text{const.}\,E^{-\alpha}\mathrm{d}E \tag{8.21a}$$

$$(\text{or } I(E)\mathrm{d}E=\text{const.}\,E^{-(\alpha-1)}\mathrm{d}E) \tag{8.21b}$$

A third mechanism, that of black-body radiation, is possible. This would give a spectral distribution of the form

$$I(E)\mathrm{d}E=\text{const.}\,\frac{E^3\mathrm{d}E}{\mathrm{e}^{E/kT-1}} \tag{8.22}$$

It should be noted, however, that spectral distributions may be strongly influenced by secondary processes, e.g., absorption by gas, or by the presence of stellar winds. Deviations from these basic forms are therefore to be expected.

Much data in the spectral distribution from discrete X-ray sources has been reported. The earlier data was collated by Adams (1971), and two of these spectra are illustrated in figs. 8.18 and 8.19. A later review (Adams, 1980) summarises the data available at that time, and discusses some of the possible mechanisms for X-ray production in detail. A detailed review has also been given by Willmore (1978).

Most X-ray astrophysical work has been carried out with proportional counters of large area, which combine high detection efficiency with an ability to work at high counting rates and a useful degree of energy resolution. A high degree of directional sensitivity has been achieved by the use of special mesh or honeycomb structures placed in front of the detector. If two or more such meshes are separated axially or a single honeycomb placed in front of the counter, radiation which arrives from the forward direction will be unobstructed, but radiation from other directions may encounter mesh wires in one or other of the honeycombs which will impede passage. Clearly some sort of polar-diagram pattern will ensue, and the width of the central maximum can be of the order of a few minutes of arc only. This enables a distinction to be made between objects which are 'star-like', and those which have a more extended structure.

Focussing collimators, making use of reflection from curved metallic mirrors at grazing incidence, can be constructed, and provide an alternative to the 'honeycomb' type of structure.

The source Sco X-1 is the brightest source (in X-radiation) yet observed, and has been identified with a star of visible magnitude 12.5. Its X-ray spectrum is shown in fig. 8.18. Analysis of this establishes that it is exponential in form, which shows it to be a Bremsstrahlung emitter with a surface temperature between 4 and 9×10^7 K. It may be a neutron star or a white dwarf in association with another, larger, star, in the form of a close binary system. The continual transfer of matter from the larger star to the collapsed star, by gravitational attraction, would give rise to large amounts of energy as it falls into the potential well, becoming raised thereby to a temperature of the order required to explain the observed Bremsstrahlung spectrum.

Fig. 8.18. X-ray spectrum from Sco X-1 (Adams, 1971).

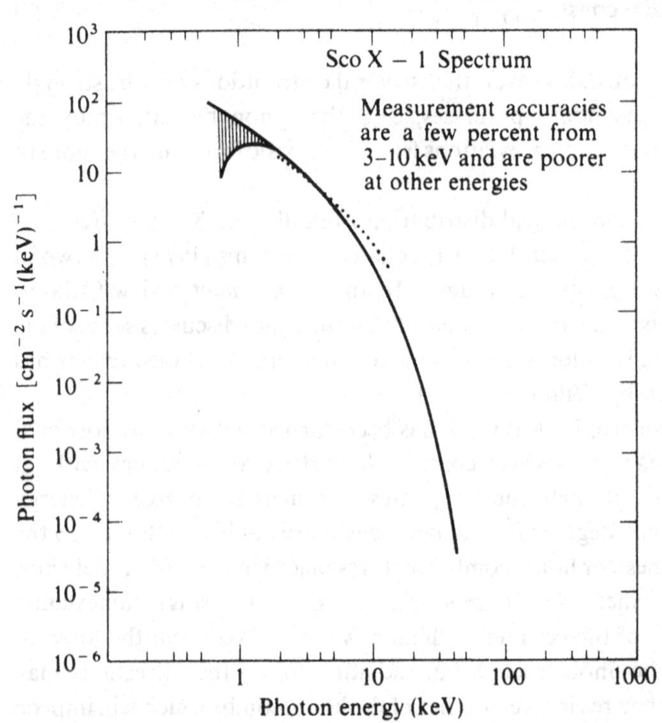

Detailed examination shows that the temperature is not uniform across the surface of the emitter and that it varies between the limits quoted (Grader *et al.*, 1970).

A very different X-ray source is Tau X-1 in the Crab nebula. The latter is a diffused source in the visible, (2–4 arc min.) and is also a radio source with a rather greater extension. The visible spectrum is strongly polarized, suggesting a synchrotron origin for the electromagnetic radiation, and the X-ray spectrum follows a power-law (fig. 8.19) in agreement with this. The visible and radio spectra are thought to be produced by one and the same synchrotron mechanism, but to account for the observed X-ray emission, electrons of much greater energy (up to 10^{13} eV) are required.

A remarkable feature of many X-ray sources is that the output often varies greatly with time. Sco X-1 fluctuates in visible brightness by a factor of 10 or more in the time of a few hours, and the intensity at the low end of the spectrum (fig. 8.18) is highly variable. This may be explained by absorption in cooler clouds of matter around the star, which may be of uneven and variable distribution. More remarkable still is the pulsating nature of certain X-ray emitters. In the radiation output from the Crab

Fig. 8.19. X-ray spectrum from the Crab nebula, Tau X-1, (Adams, 1971).

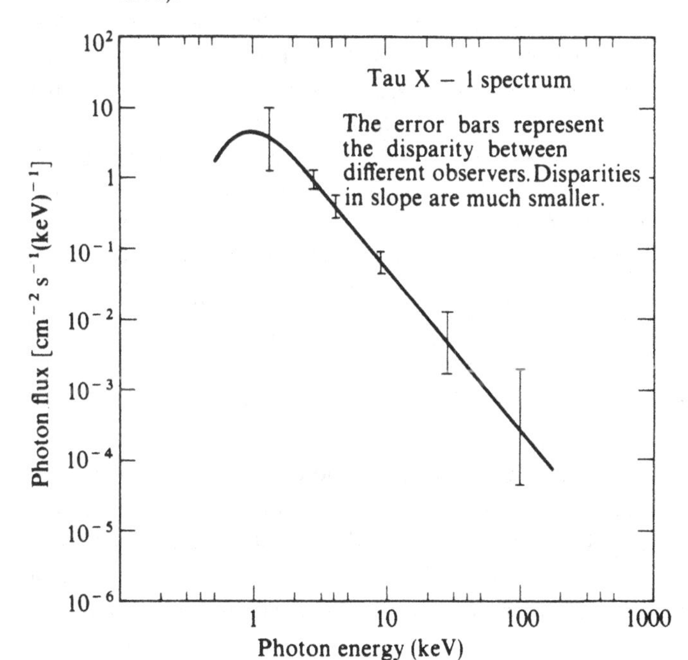

there is an X-ray component pulsating with a period of 33.1 ms (Bradt *et al.*, 1969), in synchronism with the previously discovered pulsation in the visible output. The period of the pulsation is accurately maintained but is subject to a slow rise at the rate of 0.5% per year (Rappaport *et al.*, 1971). This pulsatance is thought to be caused by a rotating neutron star, and the radiated energy can be accounted for by the gradual loss of kinetic energy during its slowing down. This spinning neutron star will also accelerate electrons to high energies because of its associated magnetic field, which will then produce synchrotron radiation as already described.

Further effects have been observed in pulsating stars – the X-ray pulsator Cen X-3 has been found to pulsate with a period of 4.8 s (Giacconi *et al.*, 1971) but the frequency shows sudden changes, either increasing or decreasing, and the rate has a slow change superimposed on it.

Cen X-3 is an example of an X-ray emitting binary system, and in recent years much attention has been paid to such systems, of which a considerable number are known to exist. There are the X-ray pulsators already referred to, and the two members of a binary system appear to be interacting in the sense that matter is being continuously transferred from the companion star to the X-ray emitting star. A review of these has been given by Fabian (1985). A feature of many X-ray emitting binary systems is that one member of the pair is a neutron star. Such stars are formed by gravitational collapse of a normal star following the exhaustion of its nuclear fuel, and much evidence for the existence of neutron stars is obtained from observation of X-ray binaries. For instance, the pulse periods of X-ray emission are often $\ll 1$s, and this small period of rotation is not possible for a white dwarf which might otherwise be an alternative hypothesis. These pulsing sources may have 'off-states' of longer periods superimposed upon them, examples being Cen X-3, and also the X-ray binary Her X-1 for which the 'off-states' occur with a period of 1.70 days. This is the orbital period of the binary system, which can be measured accurately by such observations. The X-ray emitting component of the Cen X-3 binary already referred to is eclipsed with a period of 2.087 days due to orbital motion about its more normal companion star. In the case of some X-ray binaries, other, larger periodicities also occur, and the X-ray emission may disappear at irregular intervals.

If a collapsed stellar remnant which might otherwise turn into a neutron star has more than a certain mass (in the region of 10 solar masses), the escape velocity from the surface is theoretically equal to the velocity of light. No matter or radiation can escape from such an object, which is known as a 'black hole'. There are strong candidates for these amongst the X-ray binaries, notably the massive binary Cyg X-1, though several others have been listed (White, 1983).

Data on X-ray sources is thus accumulating at a great rate, and the periodical literature must be consulted for latest information in this rapidly developing area. Sufficient has been said, however, to illustrate the great expansion which has taken place in the field of X-ray astrophysics following the early observations, and it is clear that further projects will add a great amount of additional data to that which is already published.

APPENDIX 1

Range–energy relations, etc., for electrons

Extrapolated range: light elements ($Z \leqslant 13$)

In chapter 4 (section 4.6(a)), the penetration of electrons into matter was discussed and it was seen that the range in matter is often defined in terms of 'extrapolated range', obtained by extrapolation to zero intensity of data of the type illustrated in fig. 4.21. Katz and Penfold (1952) summarised the experimental data available at that time and also proposed an empirical relation

$$R = 412\, E_0^{(1.265 - 0.0964 \ln E_0)} \tag{A1.1}$$

(R in mg cm^{-2}, E_0 in MeV) which is a good fit to the experimental data from 10 keV (the lowest energy shown in their data) up to 2.5 Mev. The data are derived from studies by a number of workers using monoenergetic electrons, and also from work in which the range of β-particles from radioactive sources is determined. Good agreement between these two methods is found. The experimental data and empirical relation are shown in fig. A1.1 but for full details and bibliography their paper should be referred to.

For higher energies (up to about 10 MeV) data by Hereford and Swann (1950) is used by Katz and Penfold, and an empirical relation

$$R = 530\, E_0 - 106 \tag{A1.2}$$

(R in mg cm^{-2}, E_0 in MeV) is found to be a good fit above 2.5 MeV. This linear form is the so-called 'Feather' plot, and has been used both for monoenergetic electrons and β-decay spectra by several authors, with slightly differing values for the numerical coefficient and constant term. The article by Knop and Paul (1965) may be consulted for details of these variations.

In the region 10–100 keV the range varies approximately as the square of the energy, and this justifies the use of the Thomson–Whiddington Law over a restricted energy range. As the energy increases, the rate of variation becomes less, approaching a linear law above about 1 MeV.

For energies below 10 keV the data has been reviewed by Kanter (1961) and is shown in fig. A1.2(*a*). Data for the lighter elements cluster closely round that of Young (1956a,b) but show a less rapid variation than E^2, and are probably closer to $E^{1.5}$.

The range–energy relationship for light elements in the energy region 1–40 keV has been studied by Lane and Zaffarano (1954) (fig. A1.2(*b*)). Here again the E^2 law is obeyed approximately for E above 10 keV, but below this the variation with E is less rapid. Everhart and Hoff (1971) have given the relation

$$R = 4.0\, E_0^{\,1.75}$$

(R in $\mu g\, cm^{-2}$, E_0 in keV) for the range in thin wafers of silicon. This expression lies very close to the experimental data for other light elements shown in fig. A.1.2(*b*).

Integrated path length: lighter elements ($Z \leqslant 13$)

The *integrated path length* may be defined as the total distance travelled by an electron during the slowing-down process, measured along the electron trajectory. Equation (3.30) gives the Bethe–Bloch expression for the *linear stopping power* for electrons, and the integrated path length is obtained by integration of the reciprocal of this quantity.

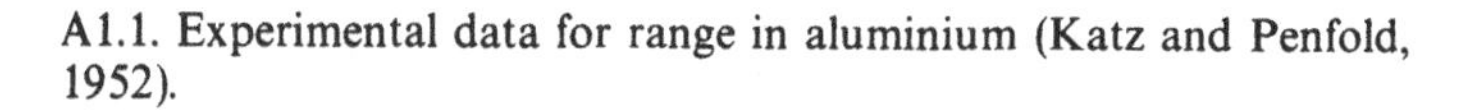

A1.1. Experimental data for range in aluminium (Katz and Penfold, 1952).

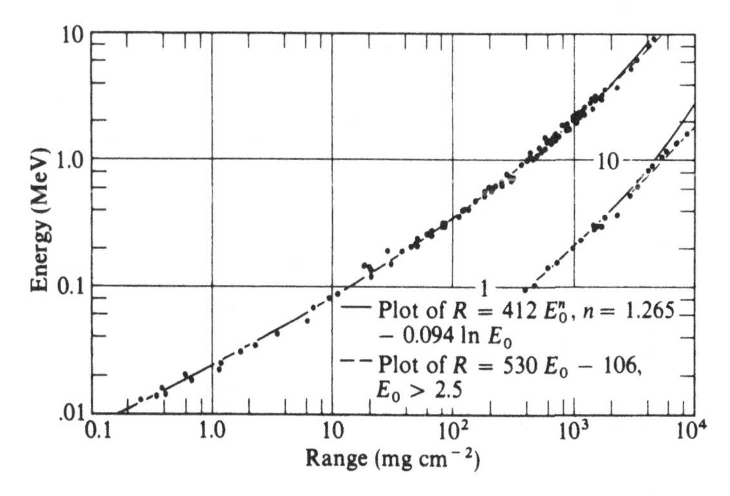

Fig. A1.2(*a*) Practical range-energy relations below 10 keV (after Kanter, 1961).

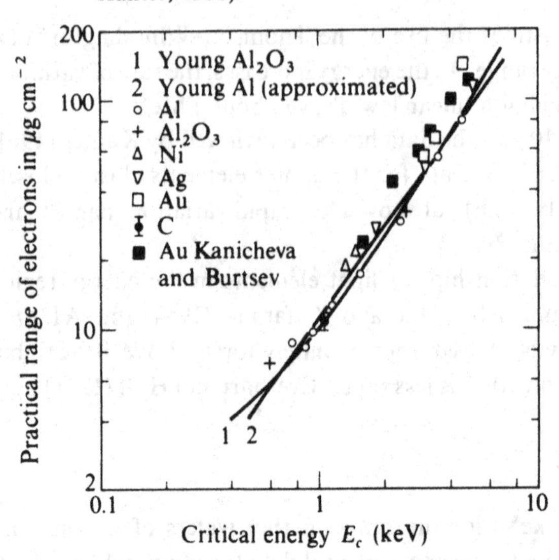

Fig. A1.2(*b*). Practical range-energy relations, 1–40 keV (Lane and Zaffarano, 1954). (Curve *A*: (A1.1); curves *B,C,D*: Bethe-Bloch formula for aluminium, collodion, and formvar respectively).

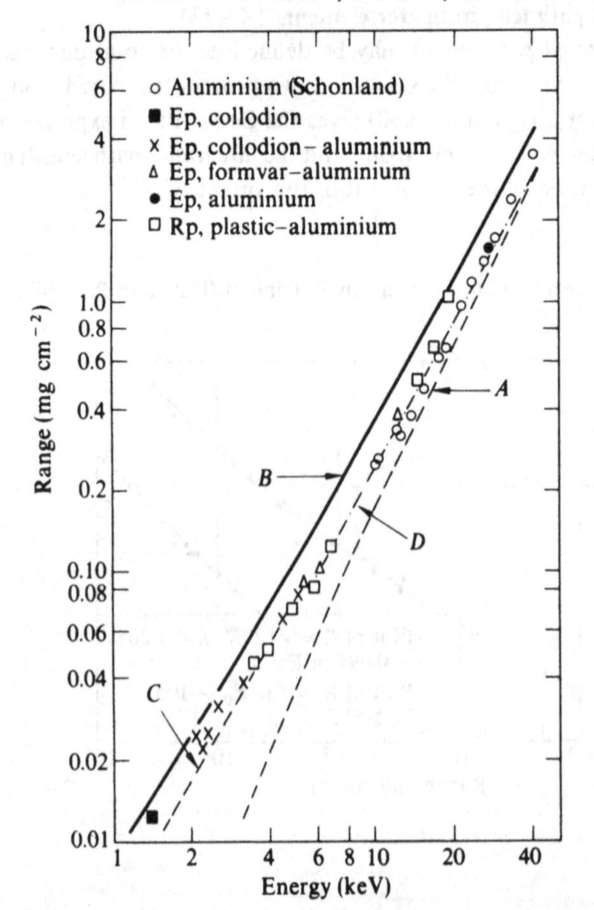

The calculations of Nelms (1956) provide this information for elements covering a wide range of Z, and for electron energies between 10 keV and several MeV (the upper limit is different for different elements). It is difficult to summarize this data. We show (fig. A1.3a) the integrated path length for aluminium, taken from this data, and for comparison the empirical relation for extrapolated range of Katz and Penfold. The integrated path length is to be seen to exceed the extrapolated range by a factor which falls from 1.9 at 10 keV to 1.4 at 70 keV. Evidently, these paths become more nearly equal as the electron energy rises. It is expected that the extrapolated range used would be less than the integrated path length because of the finite average angle between the path of the electron and the direction of the incident beam, the deviation being caused by scattering within the material. This difference is illustrated particularly clearly by the cloud chamber data of Williams (fig. 4.24).

Extrapolated range and integrated path length for heavier elements:

Much less experimental data is available for heavier elements – even the general trend of variation with Z is not clear. Referring to (A1.3), the parameter (b/p) has been determined for several different elements, and is tabulated here in table A1.1. These values suggest a distinct rise in range as Z is increased. In (A1.3),

$$V_0{}^2 - V_x{}^2 = \frac{b}{\rho}(\rho x), \tag{A1.3}$$

V_0 and V_x are the incident and emergent energies (in eV) and (ρx) is the range, in gm cm^{-2}.

The data shown in fig. A1.2(a) indicates that for energies between 1 and 10 keV, the points for gold, silver and nickel lie above the points for lighter elements by an amount which varies between 15 and 45%.

The integrated path length (Nelms) rises steadily with Z (fig. A1.3(b)), the ratio of path in gold to that in aluminium being 1.78 at 40 keV and 1.67 at 100 keV. The general trend therefore seems to be that the extrapolated range and integrated path length both rise to some extent with increase in Z, but that the rise in extrapolated range tends to be less marked than the rise in integrated path length. This is in conformity with the general expectation that for high Z the electrons reach a condition of 'full diffusion' (in which they are moving at large average angles relative to the forward direction) at an earlier stage in the slowing down process than is the case in the lighter elements.

The ratio between extrapolated range and integrated path length has

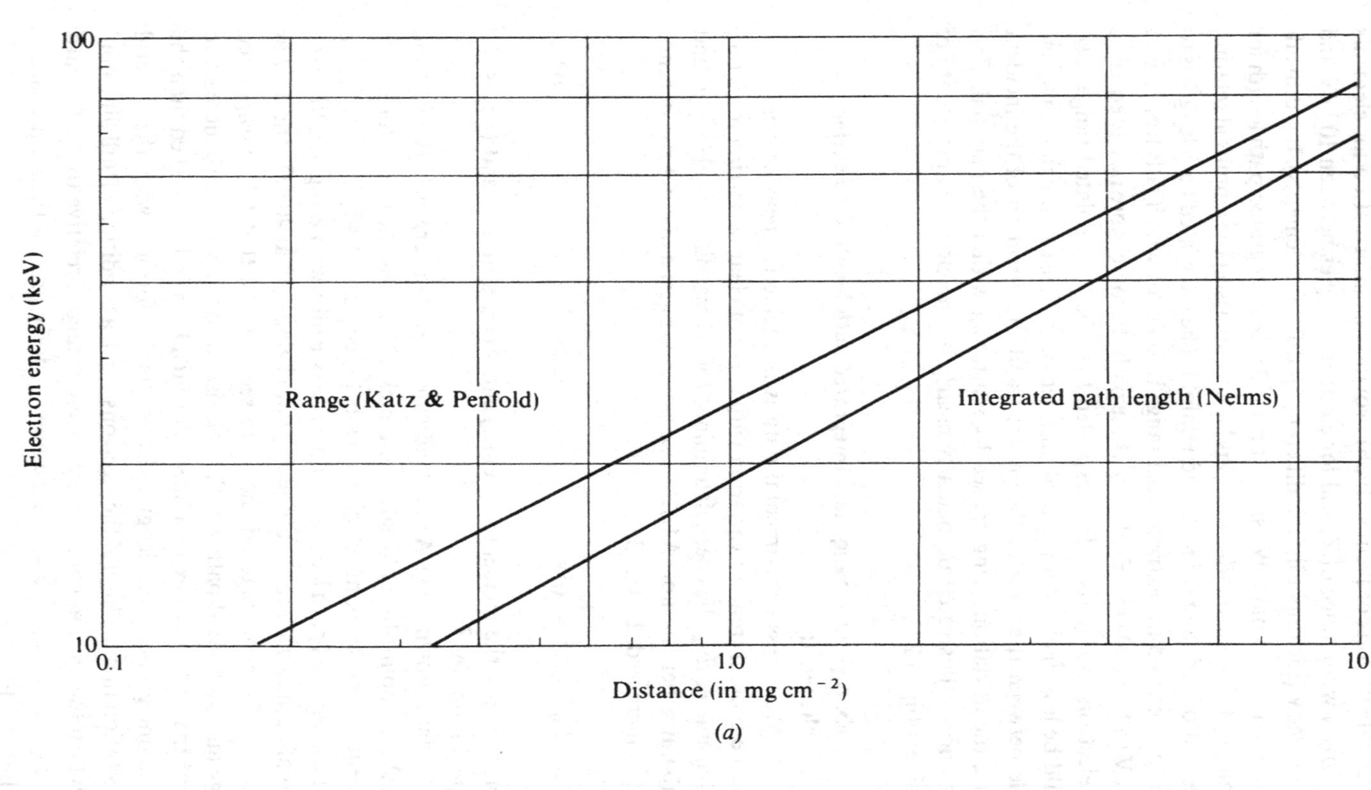

Fig. A1.3(*a*). Range in comparison with integrated path length in aluminium.

Table A.1.1 *Values of* b/ρ *(Cosslett, 1966) for equation A.1.3*
(The units of (b/ρ) *are* $(\text{eV})^2\ \text{g}^{-1}\ \text{cm}^2$*)*

E_0(keV)	9	15	18
Al	0.40×10^{12}	0.46×10^{12}	0.52×10^{12}
Cu	0.34	0.39	0.49
Ag	0.30	0.38	0.46
An	0.27	0.31	0.38

been examined by Ehrenberg and King (1963). The data they review show a steady reduction in the ratio as Z is increased. Experimental data on the penetration of electrons into phosphors are presented by them, but the interpretation of this data is complex because of the non-elemental nature of the slowing down medium. For further details their paper should be consulted.

When discussing heavier elements, it is important to realize that the transmitted intensity varies with depth in a manner which is markedly different from the trend noted in the lighter elements. In particular, the linear part of the curve becomes much less marked. For the heaviest elements the extrapolated range is no longer a realistic concept. This is illustrated very clearly in some early work by Eddy (1929) as shown in fig.

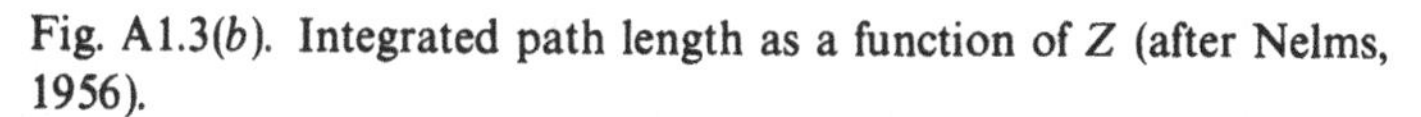

Fig. A1.3(*b*). Integrated path length as a function of Z (after Nelms, 1956).

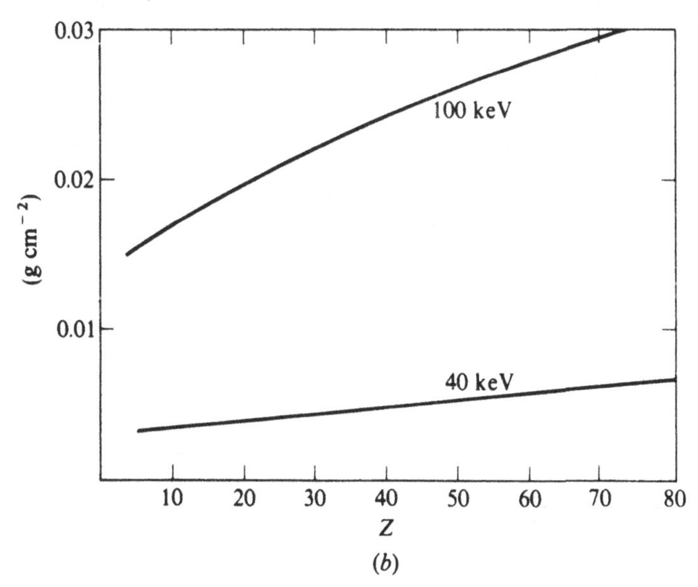

A1.4. It may be understood by noting that elastic (i.e. nuclear) scattering becomes much more important as Z increases, as discussed briefly below. A single scattering event of this type may be sufficient to remove an electron from a collimated system, in which case we might expect a quasi-exponential decrease of intensity with increasing depth; the behaviour of the beam might better be described in terms of a mean-free-path and an associated attenuation coefficient, and this approach is occasionally used. The concept of integrated path length, however, remains sound.

Diffusion depth

This is the depth within the target at which the most probable angle of total deflection (relative to the incident direction) attains its maximum value. Cosslett (1964) has given the data of table A1.2. We add for comparison the 'mean energy range' (Cosslett, 1966) and the integrated path length (Nelms), from which it is clear that the ratio of diffusion depth to mean energy range falls with increasing atomic number.

Fig. A1.4. Transmission of monochromatic electrons (energy 146 keV) through several elements (Eddy, 1929).

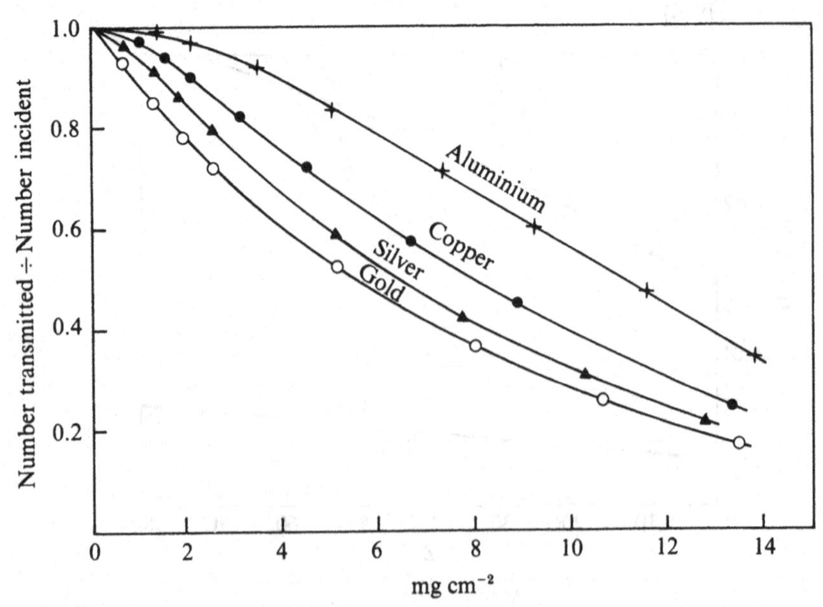

Table A.1.2 *Diffusion depth at 20 keV*

	Z	Diffusion depth (g cm^{-2})	Mean energy range (g cm^{-2})	Integrated path length (g cm^{-2})
Al	13	455×10^{-6}	760×10^{-6}	1120×10^{-6}
Cu	29	390	820	1410
Ag	47	275	870	1670
Au	79	220	1080	2130

Elastic scattering, and the ratio of elastic to inelastic scattering cross-sections

The elastic scattering cross-section is discussed by Lenz (1954). From his expression, the following simple relation may be deduced:

$$\sigma_{\mathrm{el}} = \frac{Z^{\frac{4}{3}}}{\pi}\lambda^2 \tag{A1.4}$$

where λ is the de Broglie wavelength for the electron

$$\lambda = hc\{E(E+2mc^2)\}^{-\frac{1}{2}} \tag{A1.5a}$$

where E is the electron energy, or

$$\lambda = h(2mE)^{-\frac{1}{2}} \tag{A1.5b}$$

at low energies.

The thickness in which, on average, one elastic collision occurs is given by $\sigma N/A(\rho t) = 1$. We shall call (ρt) the *single elastic scattering thickness*. The quantity $Z^{\frac{4}{3}}/A$ is approximately the same for all elements. Lenz's graph of (ρt) as a function of electron energy is reproduced in fig. A1.5.

The single scattering thickness does not vary greatly with Z, and at non-relativistic energies is proportional to electron energy. It is a useful parameter in experimental X-ray work, because it represents a 'criterion of adequate thinness', if the modifying effects of electron scattering are to be avoided. The angles through which the *inelastic* scattering events deflect the electrons are, in general, smaller, and so do not contribute so much to the deviation of the electrons.

In fig. A1.6 we show n the ratio of inelastic to elastic scattering cross-sections as a function of Z, for $E = 50$ keV. Lenz shows that this ratio varies only weakly with E. A variation in the region of $\pm 10\%$ would be expected for a variation in E by a factor of 4 in each direction.

The ratio n is a measure of the relative probability of slowing-down versus scattering processes, and as such is relevant to X-ray production in thin targets.

The number of elastic scattering events in a given thickness has been further investigated by Cosslett and Thomas (1964a,b), and experimental data is presented by them.

Fig. A1.5. Single elastic scattering thickness for carbon, chromium and gold (Lenz, 1954).

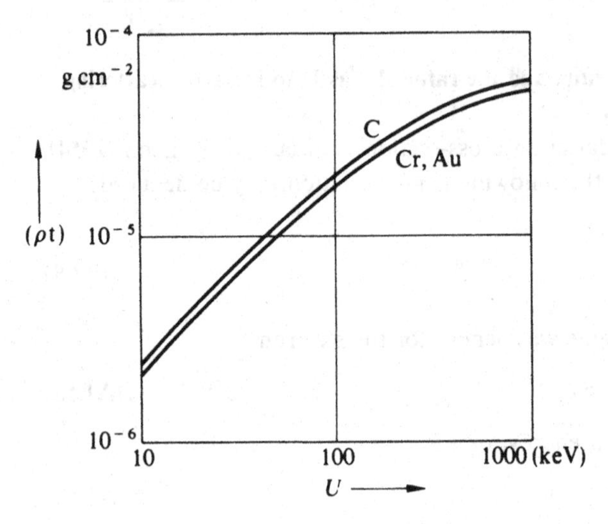

Fig. A1.6. Ratio of inelastic to elastic scattering cross-sections at 50 keV (Lenz, 1954).

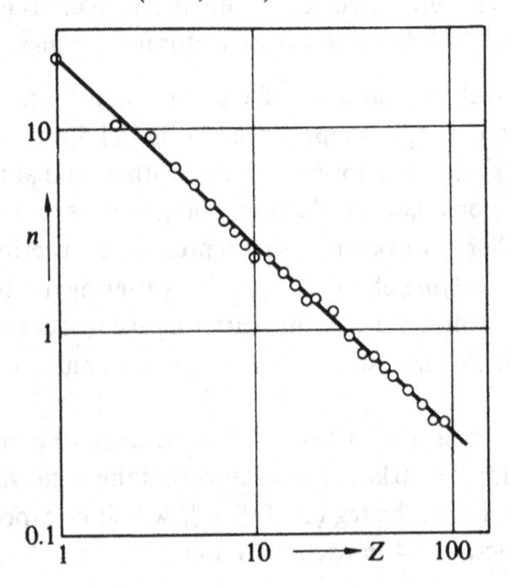

APPENDIX 2

Experimentally determined mass attenuation coefficients

Data from five tables of experimental values of mass attenuation coefficients are included in the information presented here. The data of Cooke and Stewardson (1964) extend from 730 eV to 1.74 keV and are given in their entirety (table A2.1). Hughes and Woodhouse (1966) reported data in the energy region 1.3–22 keV and made a comparison between their data and theoretical and experimental data reported elsewhere in the literature. A few additional values at low photon energies were included in Hughes *et al.* (1968). The whole of their data is included here (table A2.2).

Hopkins (1959) has reported data in the energy range 6–40 keV for aluminium and copper, and in the range 6–14 keV for four other elements. This is presented in table A2.3. To fill out the information in the range 10–25 keV we have drawn on the earlier data of Laubert (1941) (table A2.4).

A detailed set of experimental data, with estimated errors, was published by McCrary *et al.* (1967) extending from 24 to 131 keV, and is reproduced in table A2.5.

The full range of data given in this appendix thus extends from 730 eV to 131 keV. A great deal of additional experimental data, both within and outside this energy range, exists in the literature. A survey of measurements extending from 10 eV to 100 GeV has been given by Hubbell (1971), covering the period 1909 to June 1971, including a complete bibliography. This survey also includes graphs of 17 selected elements, showing the experimental data. For further information, particularly at higher energies, Grodstein (1957) or Davisson (1965) may be consulted. Other compilations have been given by Heinrich (1966) and Henke and Elgin (1970).

Figure A2.1 illustrates the mass absorption coefficient of an element for its own K_α characteristic radiation, as a function of atomic number. There is some discussion of this in chapter 3, section 3.5(d).

Fig. A2.1. Mass attenuation coefficients for the K_α radiation of the same element. (μ/ρ in $cm^2\ g^{-1}$)

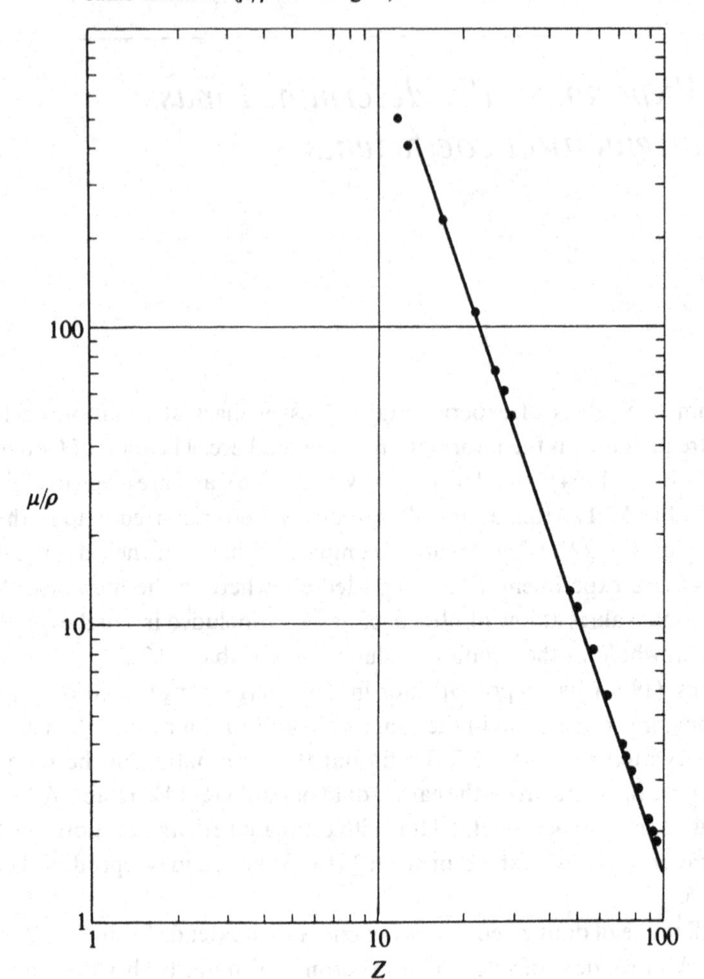

Table A.2.1 *Mass attenuation coefficients* ($cm^2\ g^{-1}$) 0.729–1.743 keV

E (keV)	λ (Å)	Be ($Z=4$)	Mg ($Z=12$)	Al ($Z=13$)	Cu ($Z=29$)	Ag ($Z=47$)
0.729	17.0	—	—	2840	3350	—
0.775	16.0	—			2940	11 500
0.800	15.5	1050	1830	2200		
0.824	15.05				2520	10 900
0.882	14.05	792	1470	1700		
0.886	14.0				2090	9 680
0.954	13.0		1100		L_2——	
0.957	12.95				L_3——	8 260
0.992	12.5	579		1200		
1.033	12.0					6 930
1.078	11.5		767			
1.127	11.0	379		862	L_1——	5 810
1.170	10.6				11000	
1.234	10.05				9190	4 630
1.240	10.0		510			
1.312	9.45	252		556		
1.360	9.12				6830	
1.385	8.95					3 600
1.445	8.6	189		430		
1.540	8.05				4760	2 810
1.579	7.85	148		K——		
1.653	7.5			3890		
1.743	7.13	110		3370	3390	2 080

Table A2.2 *Mass attenuation coefficients* ($cm^2\ g^{-1}$)

Aluminium: $Z=13$

Energy (keV)	Wave-length (Å)	μ/ρ
1.253	9.889	643
1.486	8.338	408
		K——
1.710	7.251	3640
1.741	7.126	3450
1.775	6.983	3299
1.923	6.449	2700
2.051	6.046	2206
2.293	5.406	1644
2.423	5.118	1410
2.839	4.368	936
3.287	3.752	639
3.444	3.600	567
3.769	3.290	430
4.510	2.749	267
5.415	2.290	160.3
6.404	1.936	99.7
7.478	1.658	63.1
8.041	1.542	50.6
9.442	1.313	31.8
10.84	1.144	20.5

Titanium: $Z=22$

Energy (keV)	Wave-length (Å)	μ/ρ
1.486	8.338	2280
1.740	7.128	1515
1.923	6.449	1156
2.051	6.046	981
2.293	5.406	722
2.423	5.118	613
2.839	4.368	402
3.287	3.752	277
4.510	2.749	114
4.952	2.503	89.3
		K——
5.415	2.290	593
5.899	2.102	470
6.404	1.936	377
7.478	1.658	251
8.041	1.542	206
9.442	1.313	133.7
10.55	1.175	99.1
10.84	1.144	91.3

Nickel: $Z=28$

Energy (keV)	Wave-length (Å)	μ/ρ
1.740	7.128	2990
1.923	6.449	2310
2.042	6.070	1980
2.166	5.725	1733
2.293	5.406	1447
2.839	4.368	831
2.984	4.154	720
3.287	3.752	565
3.444	3.600	510
4.510	2.749	240
5.415	2.290	145.5
6.404	1.936	92.0
7.478	1.658	61.3
8.041	1.542	49.6
		——
9.44	1.313	246
10.29	1.205	195.0
10.84	1.144	168.7

Copper: $Z=29$

Energy (keV)	Wave-length (Å)	μ/ρ
1.487	8.338	4680
1.739	7.128	3190
1.923	6.449	2520
2.042	6.070	2170
2.166	5.725	1872
2.293	5.406	1607
2.839	4.368	896
3.287	3.752	631
3.444	3.600	553
4.088	3.032	342
4.510	2.749	261
4.952	2.503	203

Zirconium: $Z=40$

Energy (keV)	Wave-length (Å)	μ/ρ
1.486	8.338	1746
1.741	7.126	1185
1.923	6.449	912
2.042	6.070	784
2.166	5.725	656
3.444	3.600	1329
4.088	3.032	842
4.500	2.754	650
5.415	2.290	403
6.404	1.936	257
7.478	1.658	168.3
8.041	1.542	137.8

Palladium: $Z=46$

Energy (keV)	Wave-length (Å)	μ/ρ
2.423	5.118	810
2.839	4.368	544
2.984	4.154	473
3.134	3.956	413
5.415	2.290	581
5.895	2.103	469
6.400	1.937	379
6.925	1.790	308
7.478	1.659	252
8.041	1.542	208
8.635	1.436	172.8
9.886	1.255	119.8

Copper: $Z=29$

Energy (keV)	Wavelength (Å)	μ/ρ
5.415	2.290	156.0
5.899	2.102	125.4
6.404	1.936	99.7
6.930	1.789	80.1
7.478	1.658	64.6
8.041	1.542	52.2
8.639	1.435	42.7
	K	
9.442	1.313	270
9.713	1.277	246

Silver: $Z=47$

Energy (keV)	Wavelength (Å)	μ/ρ
2.839	4.368	606
2.984	4.154	525
3.134	3.956	460
3.287	3.772	398
	L	
5.415	2.290	606
5.895	2.103	498
6.400	1.937	402
6.925	1.790	327
7.470	1.659	266
8.041	1.542	220
8.635	1.436	182.5
9.886	1.255	127.8

Tantalum: $Z=73$

Energy (keV)	Wavelength (Å)	μ/ρ
4.952	2.503	549
5.415	2.290	458
5.899	2.102	366
6.404	1.936	299
7.478	1.658	200
8.146	1.522	157.1
9.251	1.341	114.5
9.442	1.313	107.2
9.713	1.277	98.5
	L_3	
10.55	1.175	208
10.84	1.144	192.9
11.07	1.120	178.8
	L_2	
11.44	1.083	237
	L_1	
12.21	1.015	226
17.48	0.71	90.2

Gold: $Z=79$

Energy (keV)	Wavelength (Å)	μ/ρ
4.510	2.749	870
4.952	2.503	704
5.415	2.290	570
5.899	2.102	465
6.404	1.936	376
6.930	1.789	308
7.478	1.658	254
8.041	1.542	210
8.639	1.435	175.9
9.713	1.277	129.2
10.55	1.175	103.9
	L_3	
12.97	0.956	150.6
13.38	0.927	131.4
13.61	0.911	131.2
	L_2	
14.17	0.875	169.1
	L_1	
14.93	0.831	169.6
17.48	0.710	117
22.16	0.560	59.7

Table A2.3 *Mass attenuation coefficients* ($cm^2\ g^{-1}$)

Energy (keV)	Aluminium (Z=13)	Chromium (Z=24)	Cobalt (Z=27)	Nickel (Z=28)	Copper (Z=29)	Zinc (Z=30)
6.00					131.0	
6.41		474.0	101.0	90.0		105.0
7.00		380.0	74.0		73.6	88.0
7.50	68.7			61.0		
7.60		309.0	49.0			
7.80			381.0			
8.00					50.3	
8.05		280.0	362.0			62.0
8.22				48.0		
8.40				335.0		
8.50	45.6	231.0	322.0	327.0		
8.59					40.7	
8.75					40.3	
8.83					39.5	
8.92		212.0	276.0	277.0		47.8
8.95					306.0	
9.00					290.0	
9.12					278.0	
9.30					262.0	
9.57						38.3
9.77						270.0
10.00					206.0	
10.04	29.8	149.0	203.0	198.0		247.0
11.00				162.0	159.0	193.0
12.40	14.7	91.0	124.0			149.0
13.20	11.2				98.5	
14.00	10.4					
14.08		71.0	87.0	89.0		104.0
14.23	8.75				83.8	
15.40	7.10				67.6	
16.62	5.47				53.4	
17.44	4.74					
18.20	4.28				41.9	
19.52	3.44					
20.39	3.05				30.5	
22.34	2.35				24.1	
24.65	1.84					
24.89					17.6	
27.99	1.24				12.8	
31.23	1.02					
34.92	0.67				6.3	
39.86	0.50				5.6	

Table A2.4 *Mass attenuation coefficients* ($cm^2\ g^{-1}$) 9–24 keV (*Laubert, 1941*)

Energy (keV)	Wavelength (Å)	Nickel ($Z=28$)	Copper ($Z=29$)	Silver ($Z=47$)	Cadmium ($Z=48$)	Tin ($Z=50$)	Tantalum ($Z=73$)	Tungsten ($Z=74$)	Gold ($Z=79$)
9.87	1.253					146.0			119.8
10.53	1.1747					122.8			100.4
11.21	1.1037	154.3	159.1			103.8			84.75
11.91	1.0389	130.6	145.2	74.22	80.39	89.01			
13.37	0.9250	96.61	99.68	54.00	59.24	64.71			L_3 133.2
									L_2
14.14	0.8758	81.96	87.07	46.72	51.01	55.87			159.6
									L_1
14.93	0.8294	67.16	69.98	38.79	42.18	46.22			
15.75	0.7857	60.83	63.81	34.75	38.15	41.75			140.8
16.58	0.7460	51.78	55.49	30.10	32.67	36.01			121.4
17.43	0.7092	45.57	48.44	26.56	28.55	31.30	90.80		109.0
19.23	0.6433	34.60	36.90	20.12	21.77	23.88	69.03		83.59
20.17	0.6135	30.97	31.96	17.82	19.34	21.33	62.96		74.90
21.12	0.5857	27.06	29.08	15.92	16.92	18.69	55.58		66.77
22.10	0.5597	23.64	25.18	13.14	15.00	16.66	48.54		58.31
23.10	0.5353	20.93	22.21	11.96	13.09	14.66	43.51	45.60	53.01
24.14	0.5125	18.73	19.70	10.61	11.52	13.00	38.71	40.96	46.80

Table A2.5 *X-ray attenuation coefficients* 24.00–131.41 keV ($cm^2\ g^{-1}$) (*McCrary et al., 1967*)

Energy (keV)	Beryllium (Z=4)	Carbon (Z=6)	Magnesium (Z=12)	Aluminium (Z=13)	Sulphur (Z=16)	Titanium (Z=22)	Iron (Z=26)	Nickel (Z=28)
24.00								
25.00	0.1780	0.2968	1.430	1.811	3.109	8.541	13.72	16.93
26.01								
28.00								
30.04	0.1697	0.2466	0.8966	1.085	2.035	4.992	8.167	10.21
35.05								
35.06						3.246		
40.04	0.1572	0.2032	0.4734	0.5561	0.9764	2.239	3.637	4.597
45.01								
50.08	0.1520	0.1841	0.3200	0.3517	0.5711	1.199	1.967	2.469
55.04								
60.03	0.1467	0.175	0.2519	0.2748	0.3975	0.7515	1.207	1.534
61.82								
70.04	0.1421	0.1649	0.2139	0.2219	0.3085	0.5252	0.8110	1.004
70.20								
79.96				0.1970	0.2544	0.3997	0.5920	
80.16								
84.99	0.1366	0.1549	0.1864					0.6319
89.92				0.1812	0.2253	0.3187	0.4542	
90.18								
94.00								
100.06	0.1306	0.1512	0.1658	0.1665	0.2012	0.2670	0.3674	0.4304
100.38								
109.99						0.2297	0.3085	

110.39								
114.87			0.1567	0.1565	0.1773			0.3244
120.21						0.2050	0.2661	
120.67								
123.09								
127.11								
130.31	0.1229	0.1414	0.1439	0.1463	0.1653	0.1910	0.2374	0.2658
131.41								

Energy (keV)	Copper ($Z=29$)	Zinc ($Z=30$)	Zirconium ($Z=40$)	Niobium ($Z=41$)	Molybdenum ($Z=42$)	Silver ($Z=47$)	Tin ($Z=50$)	Lanthanum ($Z=57$)
24.00						10.94		17.25
25.00	18.01	19.79	40.83	43.17	44.35	9.818	11.49	
26.01						53.95		
28.00							8.251	
30.04	10.84	11.95	25.00	25.92	27.30	36.36	41.71	10.25
35.05								
35.06								6.835
40.04	4.857	5.326	11.49	12.13	12.91	17.09	19.13	27.06
45.01								
50.08	2.592	2.880	6.156	6.551	7.034	9.231	10.53	14.44
55.04								
60.03	1.580	1.761	3.779	4.072	4.320	5.708	6.522	8.963
61.82								
70.04	1.055	1.163	2.447	2.688	2.804	3.786	4.271	5.961

Table A2.5 *cont.*

Energy (keV)	Beryllium (Z=4)	Carbon (Z=6)	Magnesium (Z=12)	Aluminium (Z=13)	Sulphur (Z=16)	Titanium (Z=22)	Iron (Z=26)	Nickel (Z=28)
70.20								
79.96						2.636		4.161
80.16								
84.99	0.6604	0.7057	1.442	1.582	1.666		2.525	
89.92						1.934		3.071
90.18								
94.00								
100.06	0.4550	0.4852	0.9456	1.042	1.097	1.454	1.676	2.319
100.38								
109.99								
110.39								
114.87	0.3410	0.3716	0.6798	0.7454	0.7840	1.036	1.172	1.638
120.21								
120.67								
123.09								
127.11								
130.31	0.2802	0.2963	0.5179	0.5762	0.6299	0.7524	0.8476	1.184
131.41								

Energy (keV)	Gadolinium (Z=64)	Hafnium (Z=72)	Tungsten (Z=74)	Gold (Z=79)	Lead (Z=82)	Thorium (Z=90)	Uranium (Z=92)	Plutonium (Z=94)
24.00								
25.00	23.83	34.51	35.96	43.48	50.78	60.62	63.75	67.5
26.01								
28.00								

30.04	15.35	21.37	22.61	26.86	31.60	38.04	41.08	42.98
35.05							28.32	
35.06	9.926	13.96						
40.04	6.949	9.85	10.45	12.79	14.90	18.03	19.60	21.77
45.01	5.107							
50.08		5.378	5.746	6.996	7.948	10.12	11.15	11.70
55.04	15.43							
60.03	12.41	3.341	3.623	4.365	4.875	6.487		7.40
61.82							6.476	
70.04	8.180	10.45		2.892	3.251	4.298		4.953
70.20							4.670	
79.96	5.687	7.449	7.729		2.318	3.101		3.543
80.16							3.397	
84.99				7.602				
89.92	4.165	5.424	5.816	6.592		2.284		2.594
90.18							2.446	
94.00					6.462			
100.06	3.127	4.203	4.362	5.032		1.761		1.956
100.38							1.918	
109.99		3.271			4.327			1.553
110.39							1.507	
114.87	2.229		3.166	3.588		4.725		
120.21		2.557				4.266		
120.67							4.277	
123.09							4.245	
127.11							3.870	
130.31	1.575	2.176	2.322	2.635	2.878	3.516	3.84	3.718
131.41							3.618	

APPENDIX 3

Decay schemes of some radionuclides, giving modes of decay, and intensities of principal γ-radiations

In this appendix we list a number of radionuclides which have been referred to in the text in connection with the X-radiation emitted during decay. Many decay schemes are now understood to a high degree of complexity, and we have attempted to present the data in a simplified manner, so as to stress the emitted X- and γ-radiation, and including only those γ-rays which are relatively readily observable by γ-ray and X-ray spectroscopy. The data are based for the most part on the compilation by Lederer *et al.* (1966).

In table A3.1 are listed 14 radionuclides which decay wholly by electron capture and which yield little or no γ-radiation. If the decay goes directly to the ground state of the daughter nuclide, the only γ-radiation emitted will be the very weak internal Bremsstrahlung associated with the electron capture decay. If the decay proceeds to an excited state, some γ-radiation will be emitted, to an extent depending upon the fraction of electron capture events proceeding to that level, and upon the degree of internal conversion. The latter process becomes important for low transition energies and high multipolarities (large changes of angular momentum during the γ-transition). When internal conversion occurs, the conversion electrons are able to produce external Bremsstrahlung in the source material.

The Q-values for the electron capture transitions are given to enable the free-recoil energy of the nucleus to be calculated ($E_r = E_v^2/2Mc^2$), and to facilitate the calculation of the ratio of $K:L$ capture (7.14). This is normally in the region of 8 : 1 in cases where the Q-value is much higher than the binding energy of the K electrons. When this is not so, the ratio of K:L capture becomes less (table 7.1), and when the Q-value is less than the K electron binding energy, only L, M, etc., capture is possible.

The energy of the K_{α_1} X-ray is given for all cases where capture takes place from the K shell. But it should be remembered that some L-radiation will always be present, partly as a secondary consequence of K characteristic X-ray emission, and partly as a result of electron capture taking place directly from the *L* shell.

Decay schemes are given in all cases except those for which electron capture decay takes place to the ground state only. All the radionuclides listed in table A3.1 emit X-rays strongly, and are, in principle, very suitable for the calibration of γ-spectrometers in the low energy region. In no case is the associated γ-radiation so intense (or so energetic) as to cause difficulties due to amplifier overloading or base-line fluctuations.

The γ-ray intensities given in the decay schemes are numbers of *photons* emitted per 100 disintegrations, the effect of internal conversion having been taken into account in the calculation.

To calculate the intensities of *K* X-ray emission, the percentage decay by electron capture (100% in all the nuclides listed in table A3.1) has to be multiplied by the fluorescence yield (appendix 5) of the daughter atom. Any additional X-rays produced as a consequence of internal conversion in the *K*-shell have to be added to the total.

In table A3.2 ten radionuclides are listed which yield intense X-rays but which also emit γ-radiation in appreciable amounts. This γ-radiation is normally not so strong as to make the X-rays difficult to observe. In the case of radionuclides decaying simultaneously by electron capture and β-particle emission, there will be some Bremsstrahlung produced, and also some X-radiation which is characteristic of the *parent* nucleus (and any other elements present in the source material) produced as a result of electron bombardment. We have also included three α-emitters in this list. In addition to producing *L* X-rays of the *daughter* nucleus following internal conversion in the *L* shell, the *K* and *L* X-rays of the *parent* nucleus are also, in principle, observable as a result of bombardment by the α-particles.

In table A3.3 five additional radionuclides are given, which have been referred to in the text in connection with the X-radiation.

Other information relating to radionuclides is given in table 7.1 (ratios of $L:K$ capture probability) and table 7.2 (β^--emitters used in Bremsstrahlung production).

Table A3.1 *Radionuclides decaying wholly* by electron capture, and yielding little or no γ-radiation*

Nuclide	Half-life	X-rays Element	X-rays $K_{\alpha 1}$ (keV)	Q_{EC} (keV)	Remarks
^{37}A	35 d	Cl	2.622	815	—
^{51}Cr	27.8 d	V	4.952	752	γ at 320 keV (see decay scheme, fig. A3.1)
^{55}Fe	2.6 y	Mn	5.898	217	—
^{71}Ge	11.4 d	Ga	9.251	233	—
^{103}Pd	17 d	Rh	20.214	560	Several γs, all weak (see decay scheme, fig. A3.2)
^{109}Cd	453 d	Ag	22.16	170	γ at 88 keV (see decay scheme, fig. A3.3)
^{125}I	60 d	Te	27.47	149	γ at 35.4 keV (see decay scheme, fig. A3.4)
^{131}Cs	10 d	Xe	29.80	360	—
^{145}Pm	17.7 y	Nd	37.36	140	γs at 67 and 72 KeV (see decay scheme, fig. A3.5)
^{145}Sm	340 d	Pm	38.65	650	γ at 61 keV; weak γ at 485 keV (see decay scheme, fig. A3.6)
^{179}Ta	600 d	Hf	55.76	119	—
^{181}W	140 d	Ta	57.72	190	γ at 6.5 keV; weak γs at 136, 153 keV (see decay scheme, fig. A3.7)
^{205}Pb	5×10^7 y	Tl	L only ($L_{\alpha 1} = 10.27$ keV)	30	—
^{235}Np	410 d	U	98.43 (very little K; mainly L X-rays; $L_{\alpha 1} =$ 13.61 keV)	123	1.6×10^{-3}% α-decay Weak γs (associated with α-decay) at 25.7, 84.2 keV (see decay scheme, fig. A3.8)

* Except ^{235}Np.

Table A3.2 *Radionuclides yielding X-rays and low or medium energy γ-radiation*

Nuclide	Half-life	Mode of decay	X-rays Element	X-rays $K_{\alpha 1}$ (keV)	Principal γ-radiation
^{57}Co	270 d	EC	Fe	6.403	14.39, 122 keV (fig. A3.9)
^{75}Se	120.4 d	EC	As	10.543	265, 136, 280, 131 keV and weaker γs (fig. A3.10)
^{79}Kr	34.9 h	EC(92%) β^+(8%)	Br	11.923	606, 398, 261 keV, annihilation radiation, and weaker γs (fig. A3.11)
^{113}Sn	120 d	EC	In	24.21	393 keV (fig. A3.12)
^{119m}Sn	250 d	IT	Sn	25.27	23.8 keV (fig. A3.13)
^{153}Gd	242 d	EC	Eu	41.53	100, 97, 70 keV and weaker γs (fig. A3.14)
^{170}Tm	130 d	β^- (EC 0.2%)	Yb	52,36	84 keV + Brems. (also some Tm X-rays $K_{\alpha 1} = 50.73$ keV) (fig. A3.15)
^{238}Pu	86 y	α	U *L* X-rays		43.5 keV. Other weak γs, all highly converted (fig. A3.16)
^{241}Am	458 y	α	Np *L* X-rays ($L_{\alpha 1} = 13.95$ keV)		59.6, 26.3 keV and weaker γs (fig. A3.17)
^{242}Cm	163 d	α	Pu *L* X-rays ($L_{\alpha 1} = 14.28$ keV)		Weak, highly converted γs (fig. A3.18)

Table A3.3 *Other X-ray and γ-emitters referred to in the text*

Nuclide	Half-life	Mode of decay	X-rays Element	$K_{\alpha 1}$ (keV)	Principal γ-radiation
^{65}Zn	245 d	EC 98% β^+2%	Cu	8.047	1,115 keV, annihilation radiation (fig. A3.19)
^{67}Ga	78 h	EC	Zn	8.638	93, 185, 296, 389 keV and weaker γs (fig. A3.20)
^{74}As	17.9 d	β^- 32% β^+ 29% EC 39%	Ge	9.885	596, 635 keV, annihilation radiation, and weaker γs (fig. A3.21)
^{126}I	13 d	β^- 44% β^+ 1.3% EC 55%	Te	27.47	386, 667, 484, 753 keV annihilation radiation and weaker γs (fig. A3.22)
^{137}Cs	30.0 y	β^-	Ba	32.19	662 keV (fig. A3.23)

Fig. A.3.1.

Chromium − 51
[EC only: Q_{EC}=752 keV]

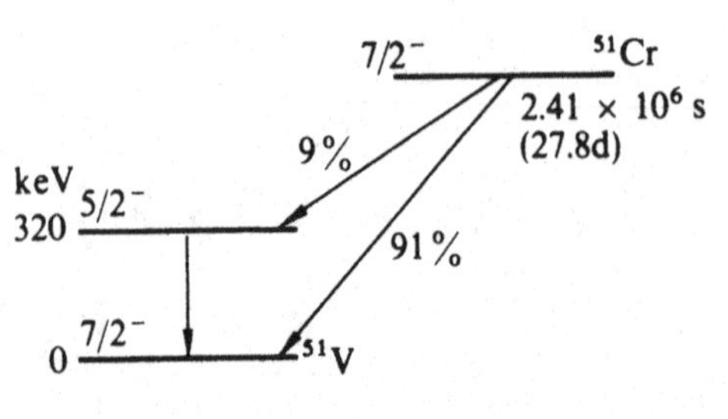

320 keV: 9%

Fig. A.3.2.

Paliadium − 103
[EC only: $Q_{EC} = 560$ keV]

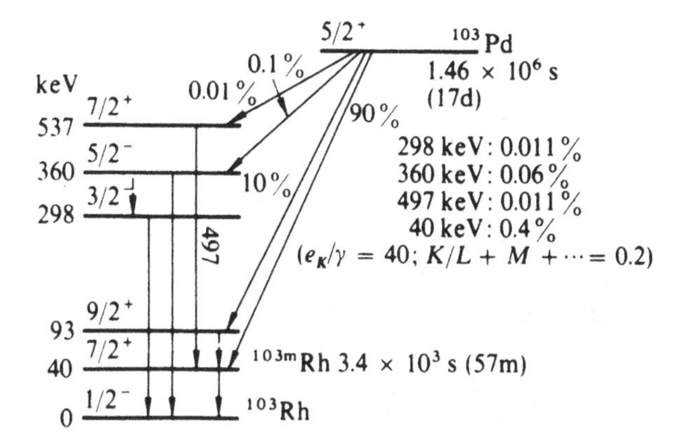

Fig. A.3.3.

Cadmium − 109
[EC only: $Q_{EC} = 170$ keV]

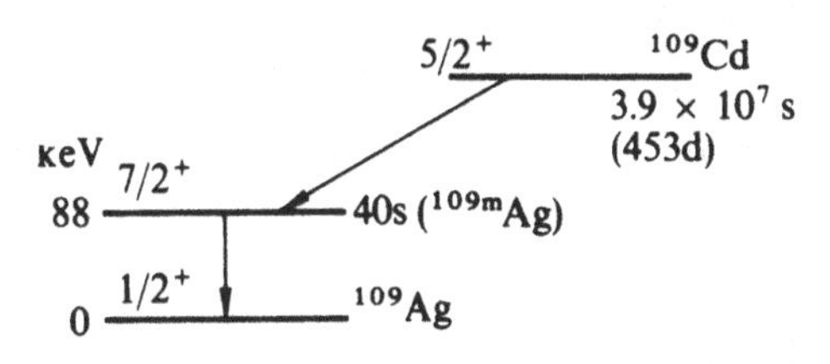

88 keV: 5%($e_K/\gamma = 11$; $K/L + M + \cdots = 0.87$)

Fig. A.3.4.

Iodine − 125
[EC only: $Q_{EC} = 149$ keV]

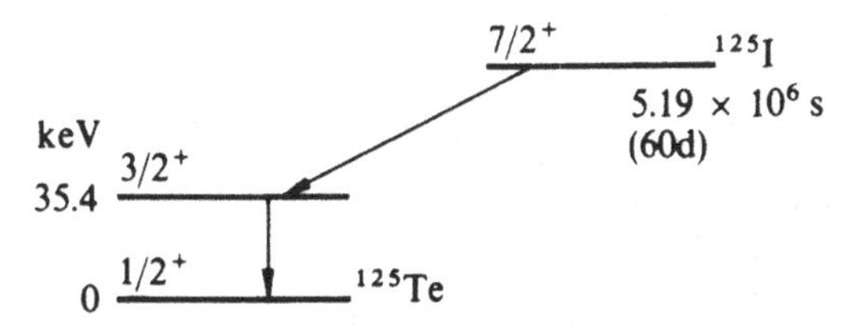

35.4 keV: 7%($e_K/\gamma = 11$; $K/L + M + \cdots = 4$)

Fig. A.3.5.

Promethium – 145
[EC only; Q_{EC} = 140 keV]

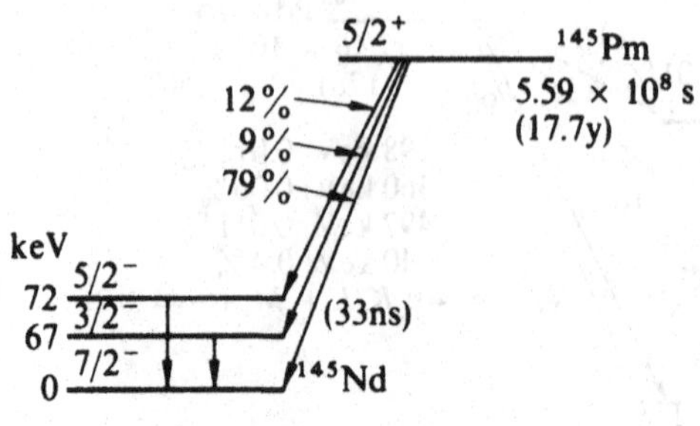

72 keV: 2.3% (e_K/γ = 3.3)
67 keV: 1% (e_K/γ = 3.3: K/L = 1.1)

Fig. A.3.6.

Samarium – 145
[EC only: Q_{EC} = 650 keV]

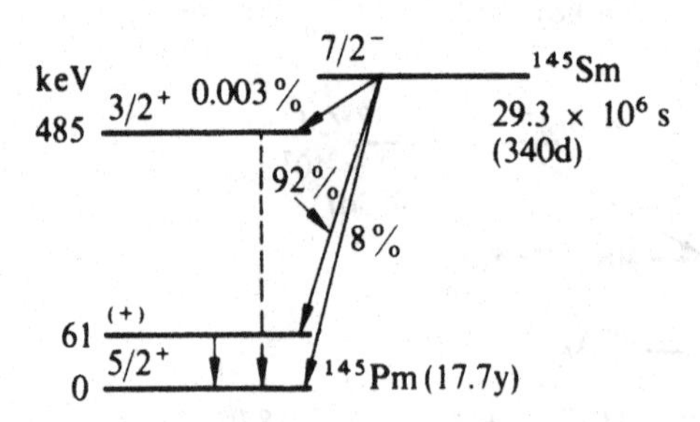

61 keV: 13%(e_K/γ = 5.3; K/L = 6.5)

Fig. A.3.7.

Tungsten – 181
[EC only: Q_{EC} = 190 keV]

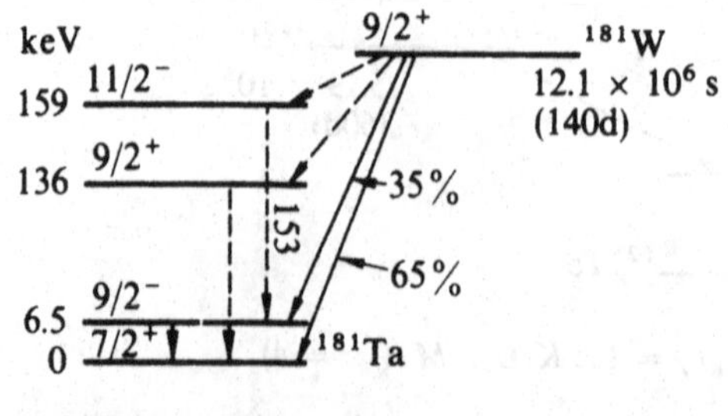

6.5 keV: ≈ 1%(e_K/γ = 46)
153 keV: 0.1%(e_K/γ = 1)
136 keV: 0.1%

Fig. A.3.8.

Neptunium – 235
[EC and α-decay: Q_{EC} = 123 keV]
[$EC_{(M)}$: $EC_{(L)}$: $EC_{(K)}$ = 17:37:1]

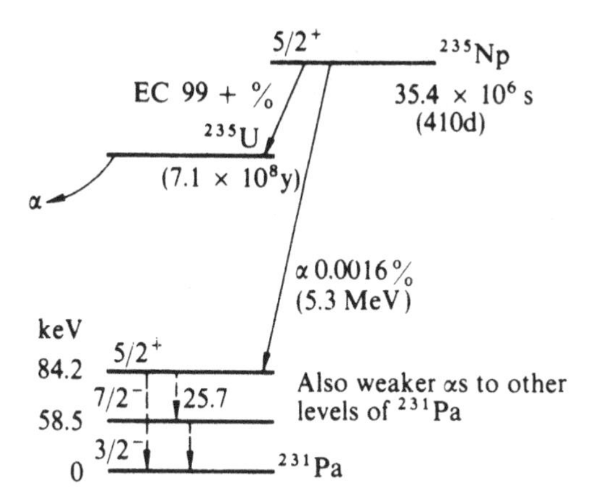

Fig. A.3.9.

Cobalt – 57
[EC only; Q_{EC} = 837 keV]

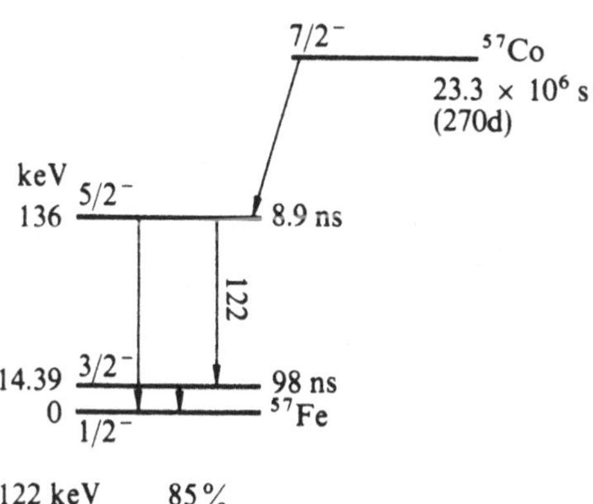

122 keV 85%
136 keV 11%
14.39 keV 8.5% ($e_K/\gamma = 9$)

Fig. A.3.10.

Selenium – 75
[EC only; Q_{EC} = 865 keV]

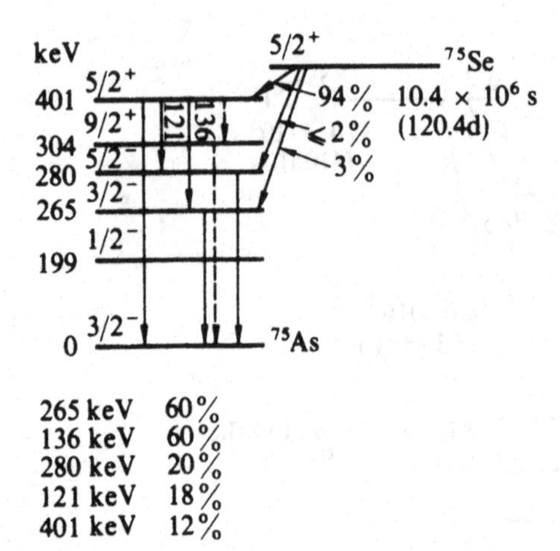

Fig. A.3.11.

Krypton – 79
[EC and β^+; Q_{EC} = 1620 keV]

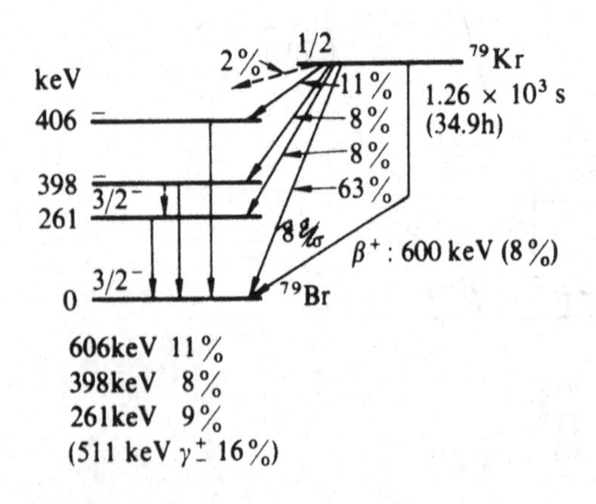

Fig. A.3.12.

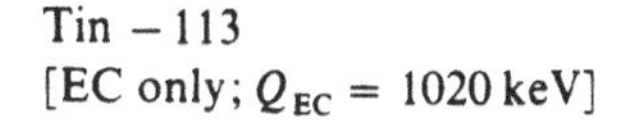

Tin – 113
[EC only; Q_{EC} = 1020 keV]

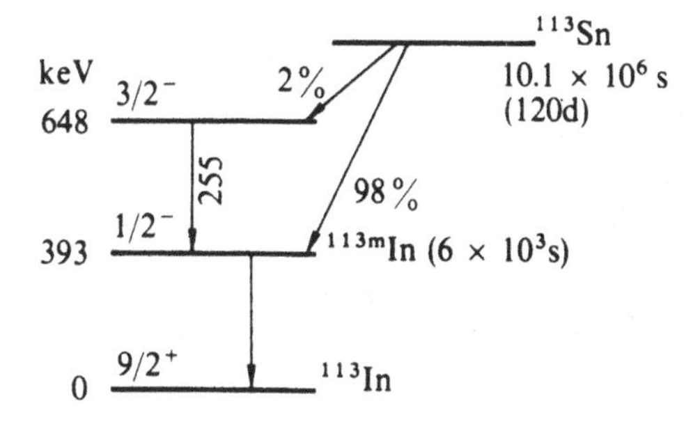

393 keV 64% ($e_K/\gamma \sim 0.5$)
255 keV 2%

Fig. A.3.13.

Tin – 119m
[IT only]

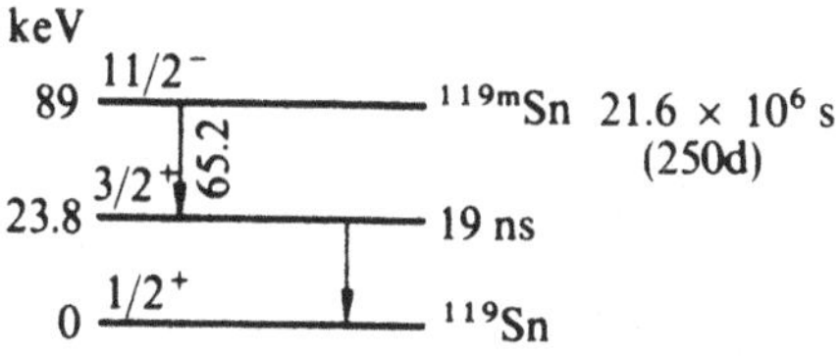

keV
11/2$^-$
89
^{119m}Sn 21.6 × 10^6 s
(250d)
65.2
3/2$^+$
23.8
19 ns
1/2$^+$
0
^{119}Sn

23.8 keV 16% (e_K/γ = 5.2)
(65 keV highly converted)

Fig. A.3.14.

Gadolinium – 153
[EC only; Q_{EC} = 243 keV]

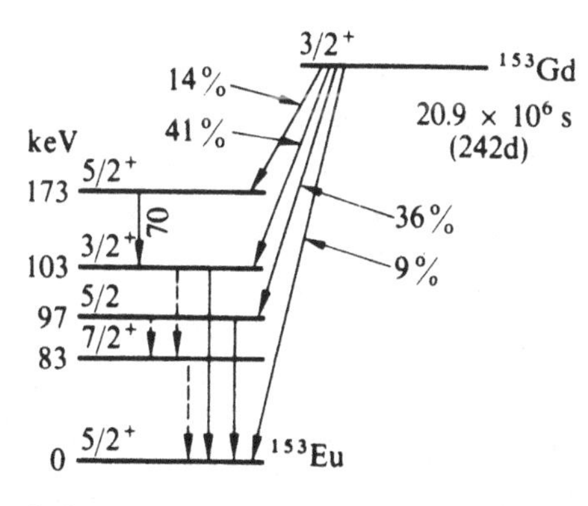

97 keV 28% (e_K/γ = 0.3)
103 keV 22% (e_K/γ = 1.5)
70 keV 2.4% (e_K/γ = 5)

Fig. A.3.15.

Thulium – 170
[β^-, EC 0.2%; Q_{β^-} = 967 keV,
Q_{EC} = 500 keV]

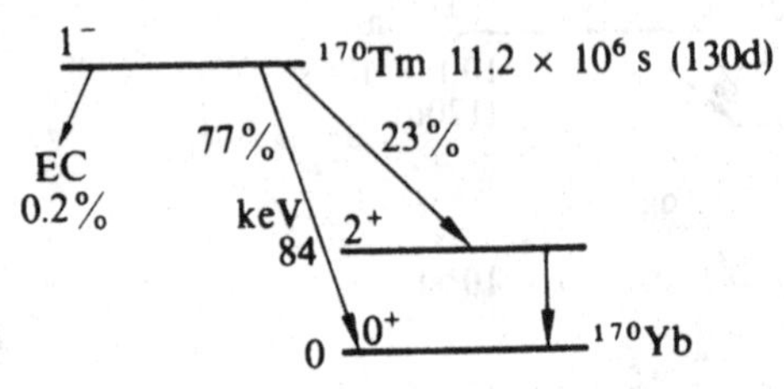

84 keV 3.3% (e_K/γ = 1.5, e_L/γ = 3.9)

Fig. A.3.16.

Plutonium – 238
[α-decay; Q_α = 5592 keV]

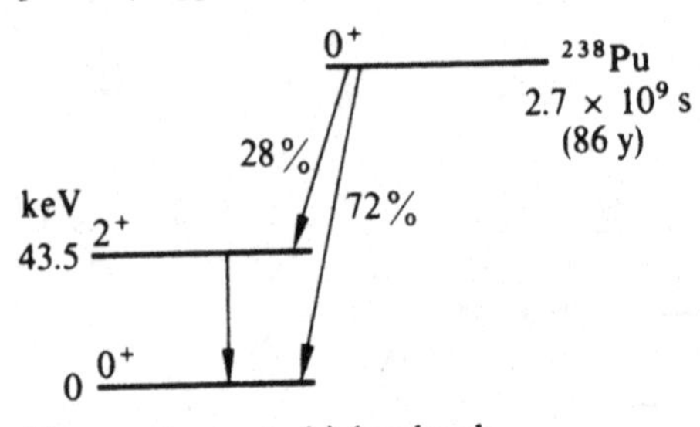

Also weaker αs to higher levels
43.5 keV ~ 0.04% (e_L/γ = 740)

Fig. A.3.17.

Americium – 241
[α-decay only]

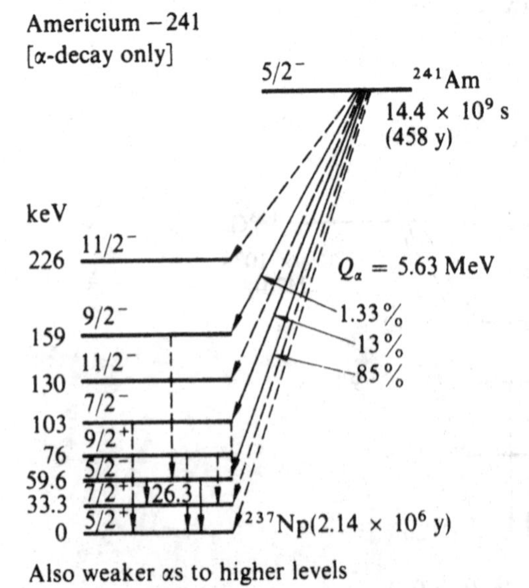

Also weaker αs to higher levels
26.3 keV 2.5% (e_L/γ ~ 3)
59.6 keV 36% (e_L/γ ~ 0.9)

Fig. A.3.18.

Curium – 242
[α-decay only; Q_α = 6217 keV]

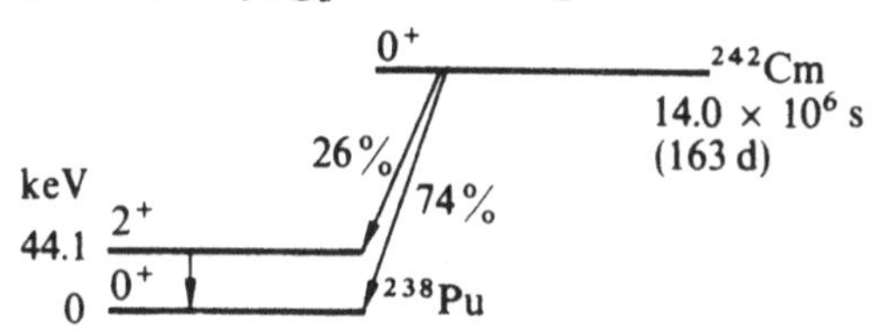

Also weaker αs to higher levels
44.1 keV ~ 0.05 % (e_L/γ ~ 520)

Fig. A.3.19.

Zinc – 65
[EC and β^+; Q_{EC} = 1349 keV]

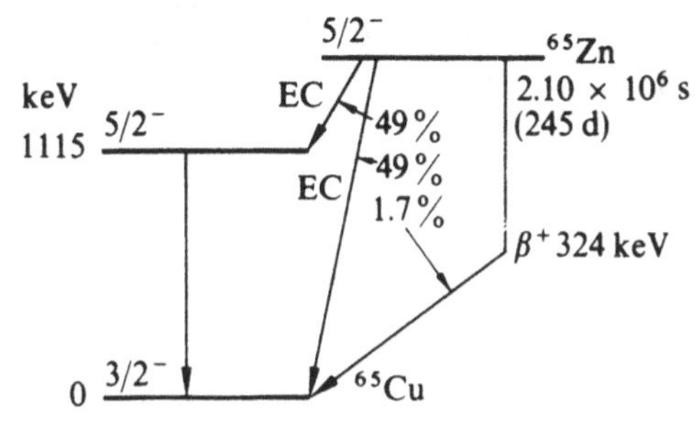

1115 keV 49 %
(511 keV $\gamma^\pm$ 3.4 %)

Fig. A.3.20.

Gallium – 67
[EC only; Q_{EC} = 1000 keV]

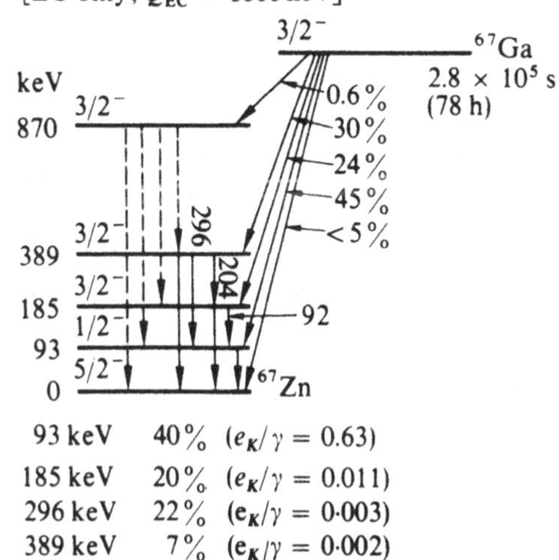

93 keV 40 % (e_K/γ = 0.63)
185 keV 20 % (e_K/γ = 0.011)
296 keV 22 % (e_K/γ = 0·003)
389 keV 7 % (e_K/γ = 0·002)
204 keV 1.5 % (e_K/γ ~ 0·02)
92 keV 2.3 % (e_K/γ = 0.074)

Fig. A.3.21.

Arsenic – 74
[EC, β^+ and β^-; Q_{EC} = 2564 keV, Q_β = 1360 keV]

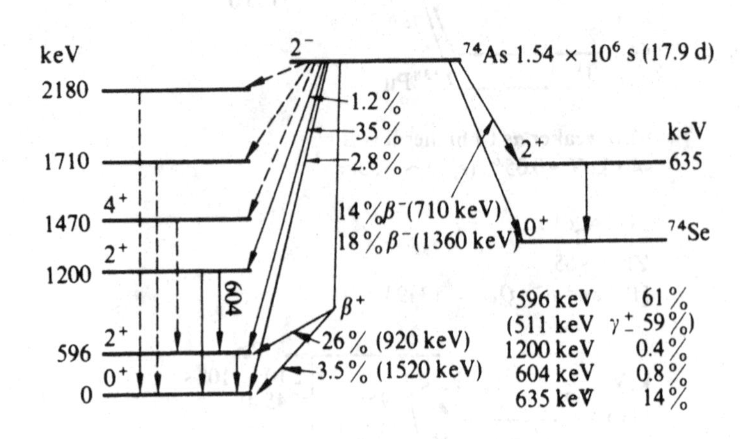

Fig. A.3.22.

Iodine – 126

[EC, β^+ and β^-; Q_{EC} = 2150 keV,
$Q_{\beta-}$ = 1251 keV]

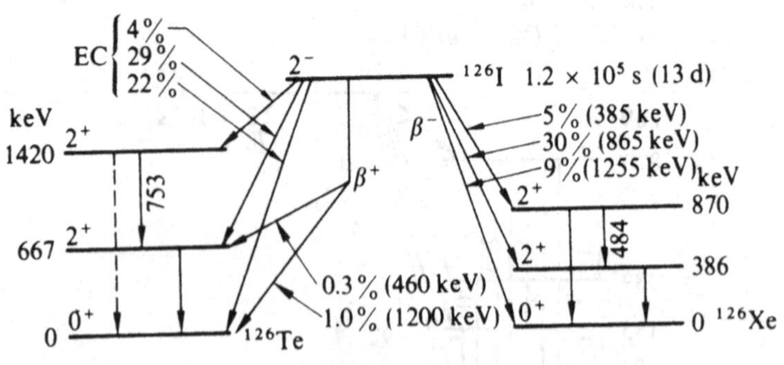

386 keV 34% (e_K/γ = 0.017, $K/L + M + \cdots = 7$)
667 keV 33%
484 keV 4.2%
753 keV 3.6%
870 keV 0.8%

Fig. A.3.23.

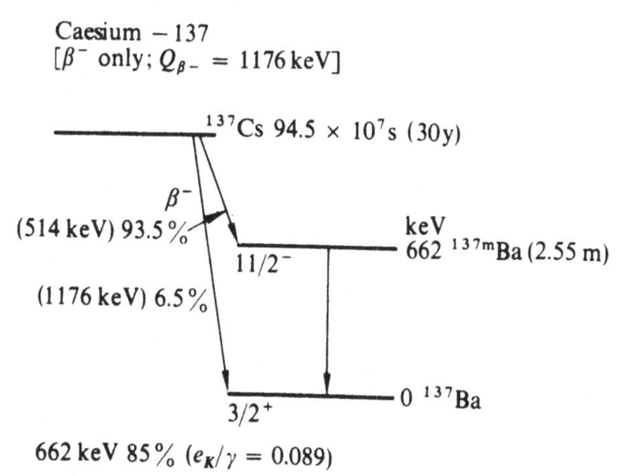

Caesium – 137
[β^- only; Q_{β^-} = 1176 keV]
^{137}Cs 94.5 × 10^{7}s (30y)
β^-
(514 keV) 93.5 %
(1176 keV) 6.5 %
keV
662 ^{137m}Ba (2.55 m)
11/2$^-$
0 ^{137}Ba
3/2$^+$
662 keV 85 % (e_K/γ = 0.089)

APPENDIX 4

Absorption edges and characteristic emission energies in keV

Atomic no.	Element	K series					L series							
		K_{ab}	$K_{\beta 2}$	$K_{\beta 1}$	$K_{\alpha 1}$	$K_{\alpha 2}$	L_{1ab}	L_{2ab}	L_{3ab}	$L_{\gamma 1}$	$L_{\beta 2}$	$L_{\beta 1}$	$L_{\alpha 1}$	$L_{\alpha 2}$
1	H	0.0136												
2	He	0.0246												
3	Li	0.055			0.052									
4	Be	0.116			0.110									
5	B	0.192			0.185									
6	C	0.283			0.282									
7	N	0.399			0.392									
8	O	0.531			0.523									
9	F	0.687			0.677									
10	Ne	0.874			0.851		0.048	0.022	0.022					
11	Na	1.08		1.067	1.041		0.055	0.034	0.034					
12	Mg	1.303		1.297	1.254		0.063	0.050	0.049					
13	Al	1.559		1.553	1.487	1.486	0.087	0.073	0.072					
14	Si	1.838		1.832	1.740	1.739	0.118	0.099	0.098					
15	P	2.142		2.136	2.015	2.014	0.153	0.129	0.128					
16	S	2.470		2.464	2.308	2.306	0.193	0.164	0.163					
17	Cl	2.819		2.815	2.622	2.621	0.238	0.203	0.202					
18	Ar	3.203		3.192	2.957	2.955	0.287	0.247	0.245					
19	K	3.607		3.589	3.313	3.310	0.341	0.297	0.294					

21	Sc	4.496		4.460	4.090	4.085	0.402	0.411	0.406			0.399		0.395
22	Ti	4.964		4.931	4.510	4.504	0.530	0.460	0.454			0.458		0.452
23	V	5.463		5.427	4.952	4.944	0.604	0.519	0.512			0.519		0.510
24	Cr	5.988		5.946	5.414	5.405	0.679	0.583	0.574			0.581		0.571
25	Mn	6.537		6.490	5.898	5.887	0.762	0.650	0.639			0.647		0.636
26	Fe	7.111		7.057	6.403	6.390	0.849	0.721	0.708			0.717		0.704
27	Co	7.709		7.649	6.930	6.915	0.929	0.794	0.779			0.790		0.775
28	Ni	8.331	8.328	8.264	7.477	7.460	1.015	0.871	0.853			0.866		0.849
29	Cu	8.980	8.976	8.904	8.047	8.027	1.100	0.953	0.933			0.948		0.928
30	Zn	9.660	9.657	9.571	8.638	8.615	1.200	1.045	1.022			1.302		1.009
31	Ga	10.368	10.365	10.263	9.251	9.234	1.30	1.134	1.117			1.122		1.096
32	Ge	11.103	11.100	10.981	9.885	9.854	1.42	1.248	1.217			1.216		1.186
33	As	11.863	11.863	11.725	10.543	10.507	1.529	1.359	1.323			1.317		1.282
34	Se	12.652	12.651	12.495	11.221	11.181	1.652	1.473	1.434			1.419		1.379
35	Br	13.475	12.465	13.290	11.923	11.877	1.794	1.599	1.552			1.526		1.480
36	Kr	14.323	14.313	14.112	12.648	12.597	1.931	1.727	1.675			1.638		1.587
37	Rb	15.201	15.184	14.960	13.394	13.335	2.067	1.866	1.806			1.752	1.694	1.692
38	Sr	16.106	16.083	15.834	14.164	14.097	2.221	2.008	1.941			1.872	1.806	1.805
39	Y	17.037	17.011	16.736	14.957	14.882	2.369	2.154	2.079			1.996	1.922	1.920
40	Zr	17.998	17.969	17.666	15.774	15.690	2.547	2.305	2.220	2.302	2.219	2.124	2.042	2.040
41	Nb	18.987	18.951	18.621	16.614	16.520	2.706	2.467	2.374	2.462	2.367	2.257	2.166	2.163
42	Mo	20.002	19.964	19.607	17.478	17.373	2.884	2.627	2.523	2.623	2.518	2.395	2.293	2.290
43	Tc	21.054	21.012	20.585	18.410	18.328	3.054	2.795	2.677	2.792	2.674	2.538	2.424	2.420
44	Ru	22.118	22.072	21.655	19.278	19.149	3.236	2.966	2.837	2.964	2.836	2.683	2.558	2.554
45	Rh	23.224	23.169	22.721	20.214	20.072	3.419	3.145	3.002	3.144	3.001	2.834	2.696	2.692
46	Pd	24.347	24.297	23.816	21.175	21.018	3.617	3.329	3.172	3.328	3.172	2.990	2.838	2.833
47	Ag	25.517	25.454	24.942	22.162	21.988	3.810	3.528	3.352	3.519	3.348	3.151	2.984	2.978
48	Cd	26.712	26.641	26.093	23.172	22.982	4.019	3.727	3.538	3.716	3.528	3.316	3.133	3.127
49	In	27.928	27.859	27.274	24.207	24.000	4.237	3.939	3.729	3.920	3.713	3.487	3.287	3.279
50	Sn	29.190	29.106	28.483	25.270	25.042	4.464	4.157	3.928	4.131	3.904	3.662	3.444	3.435

Atomic no.	Element	K series					L series							
		K_{ab}	$K_{\beta 2}$	$K_{\beta 1}$	$K_{\alpha 1}$	$K_{\alpha 2}$	L_{1ab}	L_{2ab}	L_{3ab}	$L_{\gamma 1}$	$L_{\beta 2}$	$L_{\beta 1}$	$L_{\alpha 1}$	$L_{\alpha 2}$
51	Sb	30.486	30.387	29.723	26.357	26.109	4.697	4.381	4.132	4.347	4.100	3.843	3.605	3.595
52	Te	31.809	31.698	30.993	27.471	27.200	4.938	4.613	4.341	4.570	4.301	4.029	3.769	3.758
53	I	33.164	33.016	32.292	28.610	28.315	5.190	4.856	4.559	4.800	4.507	4.220	3.937	3.926
54	Xe	34.579	34.446	33.644	29.802	29.485	5.452	5.104	4.782	5.036	4.720	4.422	4.111	4.098
55	Cs	35.959	35.819	34.984	30.970	30.623	5.720	5.358	5.011	5.280	4.936	4.620	4.286	4.272
56	Ba	37.410	37.255	36.376	32.191	31.815	5.995	5.623	5.247	5.531	5.156	4.828	4.467	4.451
57	La	38.931	38.728	37.799	33.440	33.033	6.283	5.894	5.489	5.789	5.384	5.043	4.651	4.635
58	Ce	40.449	40.231	39.255	34.717	34.276	6.561	6.165	5.729	6.052	5.613	5.262	4.840	4.823
59	Pr	41.998	41.772	40.746	36.023	35.548	6.846	6.443	5.968	6.322	5.850	5.489	5.034	5.014
60	Nd	43.571	43.928	42.269	37.359	36.845	7.144	6.727	6.215	6.602	6.090	5.722	5.230	3.208
61	Pm	45.207	44.955	43.945	38.649	38.160	7.448	7.018	6.466	6.891	6.336	5.956	5.431	5.408
62	Sm	46.846	46.553	45.400	40.124	39.523	7.754	7.281	6.721	7.180	6.587	6.206	5.636	5.609
63	Eu	48.515	48.241	47.027	41.529	40.877	8.069	7.624	6.983	7.478	6.842	6.456	5.846	5.816
64	Gd	50.229	49.961	48.718	42.983	42.280	8.393	7.940	7.252	7.788	7.102	6.714	6.059	6.027
65	Tb	51.998	51.737	50.391	44.470	43.737	8.247	8.258	7.519	8.104	7.368	6.979	6.275	6.241
66	Dy	53.789	53.491	52.178	45.985	45.193	9.083	8.621	7.850	8.418	7.638	7.249	6.495	6.457
67	Ho	55.615	55.292	53.934	47.528	46.686	9.411	8.920	8.074	8.748	7.912	7.528	6.720	6.680
68	Er	57.483	57.088	55.690	49.099	48.205	9.776	9.263	8.364	9.089	8.188	7.810	6.948	6.904
69	Tm	59.335	58.969	57.576	50.730	49.762	10.144	9.628	8.652	9.424	8.472	8.103	7.181	7.135
70	Yb	61.303	60.959	59.352	52.360	51.326	10.486	9.977	8.943	9.779	8.758	8.401	7.414	7.367
71	Lu	63.304	62.946	61.282	54.063	52.959	10.867	10.345	9.241	10.142	9.048	8.708	7.654	7.604
72	Hf	65.313	64.936	63.209	55.757	54.579	11.264	10.734	9.556	10.514	9.346	9.021	7.898	7.843
73	Ta	67.400	66.999	65.210	57.524	56.270	11.676	11.130	9.876	10.892	9.649	9.341	8.145	8.087
74	W	69.508	69.090	67.233	59.310	57.973	12.090	11.535	10.198	11.283	9.959	9.670	8.396	8.333
75	Re	71.662	71.220	69.298	61.131	59.707	12.522	12.955	10.531	11.684	10.273	10.008	8.651	8.584

76	Os	73.860	73.393	71.404	62.991	61.477	12.965	12.383	10.869	12.094	10.596	10.354	8.910	8.840
77	Ir	76.097	75.605	73.549	64.886	63.278	13.413	12.819	11.211	12.509	10.918	10.706	9.173	9.098
78	Pt	78.379	77.866	75.736	66.820	65.111	13.873	13.268	11.559	12.939	11.249	11.069	9.441	9.360
79	Au	80.713	80.165	77.968	68.794	66.980	14.353	13.733	11.919	13.379	11.582	11.439	9.711	9.625
80	Hg	83.106	82.526	80.258	70.821	68.894	14.841	14.212	12.285	13.828	11.923	11.823	9.987	9.896
81	Tl	85.517	84.904	82.558	72.860	70.820	15.346	14.697	12.657	14.288	12.268	12.210	10.266	10.170
82	Pb	88.001	87.343	84.922	74.957	72.794	15.870	15.207	13.044	14.762	12.620	12.611	10.549	10.448
83	Bi	90.521	89.833	87.335	77.097	74.805	16.393	15.716	13.424	15.244	12.977	13.021	10.836	10.729
84	Po	93.112	92.386	89.809	79.296	76.868	16.935	16.244	13.817	15.740	13.338	13.441	11.128	11.014
85	At	95.740	94.976	92.319	81.525	78.956	17.490	16.784	14.215	16.248	13.705	13.873	11.424	11.304
86	Rn	98.418	97.616	94.877	83.800	81.080	18.058	17.337	14.618	16.768	14.077	14.316	11.724	11.597
87	Fr	101.147	100.305	97.483	86.119	83.243	18.638	17.904	15.028	17.301	14.459	14.770	12.029	11.894
88	Ra	103.927	103.048	100.136	88.485	85.446	19.233	18.481	15.442	17.845	14.839	15.233	12.338	12.194
89	Ac	106.759	105.838	102.846	90.894	87.681	19.842	19.078	15.865	18.405	15.227	15.712	12.650	12.499
90	Th	109.630	108.671	105.592	93.334	89.942	20.460	19.688	16.296	18.977	15.620	16.200	12.966	12.808
91	Pa	112.581	111.575	108.408	95.851	92.271	21.102	20.311	16.731	19.559	16.022	16.700	13.291	13.120
92	U	115.591	114.549	111.289	98.428	94.648	21.753	20.943	17.163	20.163	16.425	17.218	13.613	13.438
93	Np	118.619	117.533	114.181	101.005	97.023	22.417	21.596	17.614	20.774	16.837	17.740	13.945	13.758
94	Pu	121.720	120.592	117.146	103.653	99.457	23.097	22.262	18.066	21.401	17.254	18.278	14.279	14.082
95	Am	124.876	123.706	120.163	106.351	101.932	23.793	22.944	18.525	22.042	17.677	18.829	14.618	14.411
96	Cm	128.088	126.875	123.235	109.098	104.448	24.503	23.640	18.990	22.699	18.106	19.393	14.9061	14.743
97	Bk	131.357	130.101	126.362	111.896	107.023	25.230	24.352	19.461	23.370	18.540	19.971	15.309	15.079
98	Cf	134.683	133.383	129.544	114.745	109.603	25.971	25.080	19.938	24.056	18.562	20.562	15.661	15.420
99	Es	138.067	136.724	132.781	117.646	112.244	26.729	25.824	20.422	24.758	19.426	21.166	16.018	15.764
100	Fm	141.510	140.122	136.075	120.598	114.926	27.503	26.584	20.912	25.475	19.879	21.785	16.379	16.113

With acknowledgements to Twentieth Century Electronics Ltd. Additional relevant information is given by Fine and Hendee (1955).

APPENDIX 5

K-shell fluorescence yields

Element	*Atomic number*	ω_K
C	6	0.0035
Al	13	0.0380
Si	14	0.043
P	15	0.060
S	16	0.082
Cl	17	0.0955
Ar	18	0.122
Sc	21	0.190
Ti	22	0.221
V	23	0.253
Cr	24	0.283
Mn	25	0.313
Fe	26	0.342
Co	27	0.366
Cu	29	0.443
Ga	31	0.528
Ge	32	0.554
As	33	0.588
Kr	36	0.660
Rb	37	0.669
Sr	38	0.702
Ag	47	0.834
Te	52	0.857
Xe	54	0.894
Cs	55	0.889
Eu	63	0.925
Dy	66	0.943
Pt	78	0.967
Hg	80	0.958
Pb	82	0.972
U	92	0.970

Sources of data

Carbon: determination by Hink and Paschke (1971)

All other values are from the table of experimental values compiled by Bambynek *et al.* (1972).

Bibliography

General works of reference

Blatt, J.M. and Weisskopf, V.F. (1952). *Theoretical Nuclear Physics.* Wiley.

Blokhin, M. A. (1957). Second revised edition, Moscow, English Translation 1961, US AEC Office of Technical Information. *The Physics of X-rays.*

Bransden, B.H. and Joachain, C.J. (1983). *Physics of Atoms and Molecules.* Longman.

Compton, A.H. and Allison, S.K. (1935). *X-rays in Theory and Experiment.* Van Nostrand: New York.

Evans, R.D. (1955). *The Atomic Nucleus.* McGraw-Hill.

Flügge, S. (Ed.). (1957). *Handbuch der Physik* (Volumes 30 and 34). Springer.

Heitler, W. (1954). *The Quantum Theory of Radiation.* Oxford.

Johns, H.E. (1964). *The Physics of Radiology.* Thomas.

Mott, N.F. and Massey, H.S.W. (1949). *The Theory of Atomic Collisions.* Oxford.

Panofsky, W. K. H. and Philips, M. (1962). *Classical Electricity and Magnetism*, 2nd Ed. Addison-Wesley.

Segrè, E. (1977). *Nuclei and Particles.* Benjamin.

Siegbahn, K. (Ed.). (1965). *α-, β-, γ-ray Spectroscopy.* Amsterdam.

White, H.W. (1934). *Introduction to Atomic Spectra.* McGraw-Hill.

Chapter 1

Barkla, C.G. (1909). *Proc. Camb. Phil. Soc.*, **15**, 257.

Bragg, W.H. and Bragg, W. L. (1913). *Proc. Roy. Soc.* **88A**, 428.

Cajori, Florian (1896). *Phil. Mag.* **42**, 45.

Cooper (1971). *See bibliography of chapter* 8, *section 2.*

Greening, J.R. (1950). *Proc. Phys. Soc.*, **A63**, 1227.

Jahoda, *et al.* (1960). *See bibliography for chapter 8, section 5.*

Jones, D.E.A. (1940). *Brit. J. Radiol.*, **13**, 95.

Kaye (1936). *Rep. Prog. Phys.* **3**, 338.

Lea, M. Carey (1896). *Phil. Mag.*, **41**, 382.

Nakel (1966). *See bibliography for chapter 2.*

Quimby, E.H. (1945). *Amer. J. Roentgenol. and Rad. Ther.*, **54**, 688.

Röntgen, W.C. (1896). (Trans. A. Stanton.) *Nature* **53**, 274.

Swinton, A.A.C. (1896). *Nature*, **53**, 276.

Thomson, J.J. (1896). *Proc. Roy. Soc.*, *59*, 274.

Wilson, C.T.R. (1896). *Proc. Roy. Soc.*, **59**, 338.

Chapter 2

Amrehn, H. and Kuhlenkampff, H. (1955). *Z. Phys.* **140**, 452.

Amrehn, H. (1956). *Z. Phys.*, **144**, 529.

Bem, J., Grishin, V.G. and Ryabtsov, V.D. (1966). *JETP Letters*, **4**, 72.
Bergwall, S. and Elango, M. (1967). *Phys. Lett.*, **A24**, 230.
Bethe, H. (1934). *Proc. Camb. Phil. Soc.* **30**, 524.
Bethe, H. and Heitler, W. (1934). *Proc. Roy. Soc.*, **146**, 83.
Bisi, A. and Zappa, L. (1959). *Nucl. Phys.*, **10**, 331.
Bisi, A., Fasana, A. and Zappa, L. (1963). *Nucl. Phys.*, **45**, 405.
Bogdankevich, O.V. and Nikolaev, F.A. (1966). *Methods in Bremsstrahlung Research.* Eng. trans. Academic Press.
Bohm, K. (1937). *Z. Phys.*, **38**, 334.
Bohm, K. (1938). *Ann. der Phys.*, **33** 315.
Bohr (1913). *See bibliography for appendix 1.*
Botden, P.J.M., Combee, B., and Houtman, J. (1952). *Philips Tech. Rev.*, **14**, 165.
Bourgoignie, R.J., Vanhuyse, V.J. and Creten, W.L. (1965). *Z. Phys.* **188**, 303.
Bouwers, A. and Diepenhorst, P. (1933). *X-ray Research and Development.* Philips, Eindhoven.
Buechner, W.W., Van de Graaff, R.J., Burrill, E.A. and Sperduto, A. (1948). *Phys. Rev.*, **74**, 1348.
Clark, J.C. and Kelly, H.R. (1941). *Phys. Rev.*, **59**, 220.
Compton and Allison (1935). *See General works of reference.*
Cosslett, V.E. and Dyson, N.A. (1957). *See* Cosslett, V.E. Engstrom, A. and Pattee, H. *in bibliography for chapter 8.*, Section 8.1.
Cosslett, V.E. and Thomas, R.N. (1964). *See bibliography for appendix 1.*
Determann, H. (1937). *Ann. der Phys.*, **30**, 481.
Diambrini, G., Figuera, A.S., Rispoli, B., Serra, A. (1961). *Nuovo Cim.*, **19**, 250.
Doffin, H. and Kuhlenkampff, H. (1957). *Z. Phys.*, **148**, 496.
Duane, W.D. and Hunt, F.L. (1915). *Phys. Rev.*, **6**, 166.
Duane, W. (1927). *Proc. Nat. Acad. Sci.*, **13**, 662.
Duane, W. (1928). *Proc. Nat. Acad. Sci.*, **14**, 450.
Duane, W. (1929). *Proc. Nat. Acad. Sci.*, **15**, 805.
DuMond, J. and Bollman, V. (1937). *Phys. Rev.*, **51**, 400.
Dyson, N.A. (1959). *Proc. Phys. Soc.*, **73**, 924.
Edelsack, E.A., Kreger, W.E., Mallet, W. and Scofield, N.E. (1960). *Health Physics*, **4**, 1.
Fano, U., McVoy, K.W. and Albers, J. (1959). *Phys. Rev.*, **116**, 1147.
Felbinger, K., Häufglöckner, H., Niemann, J. and Scheer, M. (1960). *Naturwiss*, **47**, 55.
Fronsdal, C. and Uberall, H. (1958). *Phys. Rev.*, **111**, 580.
Galster, S. (1964). *Nucl. Phys.* **58**, 72.
Gluckstern, R.L. and Hull, M.H. (1953). *Phys. Rev.* **90**, 1030.
Goldhaber, M., Grodzins, L. and Sunyar, A.W. (1957). *Phys. Rev.*, **106**, 826.
Green, M. and Cosslett, V.E. (1968). *J. Phys. D., Appl. Phys.* **1**, 425.
Hall, H.E., Hanson, A.O. and Jamnik, D. (1963). *Phys. Rev.*, **129**, 2207.
Harworth, K. and Kirkpatrick, P. (1942). *Phys. Rev.*, **62**, 334.
Hinz, H. (1941). *Ann. der Phys.*, **39**, 573.
Honerjäger, R. (1940). *Ann. der Phys.*, **38**, 33.
Jamnik, D. and Axel, P. (1960). *Phys. Rev.*, **117**, 194.
Johansson, P. (1960). *Ark. Fys.*, **18**, 329.
Johnson, W.R. and Rozics, J.D. (1962). *Phys. Rev.*, **128**, 192.
Kanter (1961). *See bibliography for appendix 1.*
Katz and Penfold (1952). *See bibliography for appendix 1.*
Kerscher, R. and Kuhlenkampff, H. (1955). *Z. Phys.* **140**, 632.
Kimura, M., Mutsuro, N., Ohnuki, Y., Shoda, K., Sugawara, M., Tohei, T. and Yuta, H. (1959). *J. Phys. Soc. Japan*, **14**, 387.
Kirkpatrick, P. (1923). *Phys. Rev.*, **22**, 226.
Kirkpatrick, P. and Wiedmann, L. (1945). *Phys. Rev.*, **67**, 321.

Koch, H. W. and Carter, R.E. (1950). *Phys. Rev.*, **77**, 165.
Koch, H.W. and Motz, J.W. (1959). *Rev. Mod. Phys.*, **31**, 920.
Kramers, H.A. (1923). *Phil. Mag.*, **46**, 836.
Kuckuck, R.W. and Ebert, P.J. (1973). *Phys. Rev.*, **A7**, 456.
Kuhlenkampff, H. (1922). *Ann. der Phys.*, **69**, 548.
Kuhlenkampff, H. (1928). *Ann. der Phys.*, **87**, 597.
Kuhlenkampff, H. (1938). *Ann. der Phys.* **33**, 600.
Kuhlenkampff, H. (1959). *Z. Phys.*, **157**, 282.
Kuhlenkampff, H. and Ross, D. (1961). *Z. Phys.*, **161**, 424.
Kuhlenkampff, H., Scheer, M. and Zeitler, E. (1959). *Z. Phys.*, **157**. 275.
Kuhlenkampff, H. and Schmidt, L. (1943). *Ann. der Phys.*, **43**, 494.
Kuhlenkampff, H. and Zinn, W. (1961). *Z. Phys.*, **161**, 428.
Lane and Zaffarano (1954). *See bibliography for appendix 1.*
Lanzl, L.H. and Hanson, A.O. (1951). *Phys. Rev.*, **83**, 959.
Lawson, J.D. (1950). *Proc. Phys. Soc.*, **63A**, 653.
Lenz (1954). *See bibliography for appendix 1.*
Liden, B. and Auleytner, J. (1962). *Ark. Fys.*, **22**, 549.
Lohrmann, E. (1961). *Phys. Rev.*, **122**, 1908.
McVoy, K.W. (1958). *Phys. Rev.*, **111**, 1333.
Massey, H.S.W. and Burhop, E.H.S. (1952). *Electronic and Ionic Impact Phenomena.* Oxford.
Motz, J.W. and Placious, R.C. (1958). *Phys. Rev.*, **109**, 235.
Motz, J.W. and Placious, R.C. (1960). *Nuovo Cim.*, **15**, 571.
Muirhead, E.G., Spicer, B.M. and Lichtblau, H. (1952). *Proc. Phys. Soc.*, **A65**, 59.
Nakel, W. (1966). *Phys. Lett.*, **22**, 614.
Nakel, W. (1967). *Phys. Lett.*, **25A**, 569.
Neff, H. (1951). *Zeit. Phys.*, **131**, 1.
Neumcke, B. (1966). *Phys. Lett.*, **23**, 382.
Nicholas, W.W. (1927). *Phys. Rev.*, **29**, 619.
Nicholas, W.W. (1929). *Bur. of Stand. J. of Res.* **2**, 837.
Nijboer, B.R.A. (1946). *Physica*, **12**, 461.
Ogier, W.T., Carlson, R.D. and Knoche, J. (1966). *Phys. Rev.*, **142**, 50.
Olsen, H. and Maximon, L.C. (1958). *Phys. Rev.*, **110**, 589.
O'Dell, A.A., Sandifer, C.W., Knowlen, R.D. and George, W.D. (1968). *Nucl. Instrum. Meth.* **61**, 340.
Panofsky and Philips (1962). *See General works of reference.*
Peterson, T.J. and Tomboulian, D.H. (1962). *Phys. Rev.* **125**, 235.
Piston, D.S. (1936). *Phys. Rev.* **49**, 275.
Rester, D.H., Edmonson, N. and Peasley, Q. (1970). *Phys. Rev.* **A2**, 2190.
Sandström, R. (1960). *Ark. Fys.*, **18**, 305.
Sarkar, S. (1963). *Nuovo Cimento*, **28**, 1.
Scheer, M. and Zeitler, E. (1955). *Z. Phys.*, **140**, 642.
Scheer, M., Trott, E. and Zahs, G. (1968). *Z. Phys.*, **209**, 68.
Scherzer, O. (1932). *Ann. der Phys.*, **13**, 137.
Sesemann, G. (1941). *Ann. der Phys.*, **40**, 66.
Smick, A.E. and Kirkpatrick, P. (1941). *Phys. Rev.*, **60**, 162.
Sokolov, A.A. and Kerimov, K.B. (1966). *Proc. of 12th Inter. Conf. on High Energy Physics.* Dubna, 1964, Vol. 1, pp. 973–4.
Sommerfeld, A. (1929). *Frank. Inst. J.*, **208**, 571.
Sommerfeld, A. (1931). *Ann. der Phys.*, **11**, 257.
Soole, B.W. (1972). *J. Phys.* **B**, **5**, 1583.
Soole, B.W. (1977). *Phys. Med. Biol.*, **22**, 187.
Spijkerman, J.J. and Bearden, J.A. (1964). *Phys. Rev.*, **A134**, 871.

Stabler, R.C. (1965). *Nature*, **206**, 922.
Stephenson, S.T. and Mason, F.D. (1949). *Phys. Rev.*, **75**, 1711.
Stephenson, S.T. (1957). *See* Flügge (Ed.) *in General works of reference.*
Taylor, B.N., Parker, W.H. and Langenberg, D.N. (1969). *Revs. Mod. Phys.*, **41**, 375.
Thomsen, J.S. and Burr, A.F. (1968). *Am. J. Phys.*, **36**, 803.
Thordarson, S. (1939). *Ann. der Phys.*, **35**, 135.
Ulrey, C.T. (1918). *Phys. Rev.*, **11**, 401.
Wagner, E. (1918). *Ann. der Phys.*, **57**, 401.
Webster, D. L. and Hennings, A.E. (1923). *Phys. Rev.*, **21**, 312.
Weinstock, R. (1942). *Phys. Rev.*, **61**, 584.
Weinstock, R. (1943). *Phys. Rev.*, **64**, 276.
Whiddington (1912). *See bibliography for appendix 1.*
Whiddington (1914). *See bibliography for appendix 1.*
Williams (1931). *See bibliography for appendix 1.*
Zdarko, R., Drickey, D. and Mozley, R. (1964). *Phys. Rev.*, **B136**, 1674.

Chapter 3

Archard, G.D. (1960). *See* Engstrom, A., Cosslett, V.E. and Pattee, H. (1960) *in bibliography for chapter 8.*
Arthurs, A.M. and Moiseiwitsch, B.L. (1958). *Proc. Roy. Soc.*, **A247**, 550.
Bambynek, W., Crasemann, B., Fink, R.W., Freund, H.-U., Mark, H., Swift, C.D., Price, P.E. and Venogopala Rao, P. (1972). *Rev. Mod. Phys.* **44**, 716.
Bearden, J.A. (1967). *Rev. Mod. Phys.*, **39**, 78.
Beckman, O. (1955). *Ark Fys.*, **9**, 495.
Bekk, K.-J. (1974). Diplom – Thesis, University of Freiburg.
Bergström, I. and Nordling, C. (1965). *See* Siegbahn, K. (Ed.) *In General works of reference.*
Bethe, A. (1930). *Ann. Phys.* **5**, 325.
Bishop, H.E. (1965). *Proc. Phys. Soc.* **85**, 855.
Blokhin (1957). *See General works of reference.*
Bohr, N. and Coster, D. (1923). *Z. Phys.*, **12**, 342.
Bransden, B.H. and Joachain, C.J. (1983). *See General works of reference.*
Brogren, C. (1963). *Ark. Fys.*, **23**, 219.
Burbank, B.G. (1944). *Phys. Rev.*, **66**, 160.
Burhop, E.H.S. (1940). *Camb. Phil. Soc. Proc.*, **36**, 43.
Burhop, E.H.S. (1952). *The Auger Effect and Other Radiationless Transitions.* Cambridge.
Campbell, A.J. (1963). *Proc. Roy. Soc.*, **274**, 319.
Candlin, D.J. (1955). *Proc. Phys. Soc.*, **A68**, 322.
Castaing, R. and Descamps, J. (1955). *J. Phys. Rad.*, **16**, 304.
Chattarji, D. (1976). *The Theory of Auger Transitions.* Academic Press.
Christofzik, H.-J. (1970). Diplom-Thesis, University of Münster.
Clark, J.C. (1935). *Phys. Rev.*, **48**, 40.
Compton and Allison (1935). *See General works of reference.*
Dangerfield, G.R. and Spicer, B.M. (1975). *J. Phys. B.*, **8**, 1744.
Dolby, R. (1960). *Brit. J. App. Phys.*, **11**, 64.
Druyvesteyn, M.J. (1927). *Z. Phys.*, **43**, 707.
Duncumb, P. and Reed, S.J.B. (1968), ed. K.F.J. Heinrich, N.B.S. Special Publication 298, *Quantitative Electron Probe Microanalysis.*
Dyson, N.A. (1975). *Phys. Med. Biol.*, **20**, 1.
Edamoto, I. (1950). *Sci. Rep. Res. Inst. Tohoku Univ.*, **A2**, 561.
Evans, R.D. (1955). See *General works of reference.*
Fink, R.W., Jopson, R.C., Mark, H. and Swift, C.D. (1966). *Rev. Mod. Phys.*, **38**, 513.

Glupe, G. (1972). Ph.D. thesis, University of Münster.
Glupe, G. and Melhorn, W. (1967). *Phys. Lett.*, **25A**, 274.
Glupe, G. and Melhorn, W. (1971). *J. Phys.* (*Paris*) **C4**, 40.
Goldberg, M. (1961). *J. Phys. Radium*, **22**, 743.
Green, M. (1963a). *Proc. Phys. Soc.*, **92**, 204.
Green, M. (1963b). *See* Pattee, H., Cosslett, V.E. and Engstrom, A., Eds. *in bibliography for chapter 8, section 8.1.*
Green, M. (1964). *Proc. Phys. Soc.*, **83**, 435.
Green, M. and Cosslett, V.E. (1961). *Proc. Phys. Soc.*, **78**, 1206.
Green, M. and Cosslett, V.E. (1968). *Brit. J. Appl. Phys.* (*J. Phys. D.*) **1**, 425.
Hansen, H. and Flammersfeld, H. (1966). *Nucl. Phys.* **79**, 135.
Hansen, W.W. and Stoddard, K.B. (1933). *Phys. Rev.*, **43**, 701.
Hansen, H., Weigmann, H. and Flammersfeld, H. (1964). *Nucl. Phys.* **58**, 241.
Hink, W. and Paschke, H. (1971a). *Z. Phys.* **244**, 140.
Hink, W. and Paschke, H. (1971b). *Phys. Rev.* **A4**, 507.
Hink, W. and Ziegler, A. (1969). *Z. Phys.*, **226**, 222.
Hirst, H. and Alexander, E. (1935). *Phil. Mag.*, **19**, 918.
Kirkpatrick, P. and Hare, D.G. (1934). *Phys. Rev.*, **46**, 831.
Kirkpatrick, P. and Baez, A.V. (1947). *Phys. Rev.*, **71**, 521.
Metchnik, V. and Tomlin, S.G. (1963). *Proc. Phys. Soc.*, **81**, 956.
Moseley, H.G.J. (1913). *Phil. Mag.* **26**, 1024.
Moseley, H.G.J. (1914). *Phil. Mag.*, **27**, 703.
Mott and Massey (1949). *See General works of reference.*
Motz, J.W. and Placious, R.C. (1964). *Phys. Rev.*, **136**, A662.
Ogurtsov, G.N. (1973). *Sov. Phys.* - JETP **37**, 584.
Parratt, L.G. (1936a). *Phys. Rev.*, **49**, 132.
Parratt, L.G. (1936b). *Phys. Rev.*, **49**, 502.
Parratt, L.G. (1959). *Rev. Mod. Phys.*, **31**, 616.
Pauling, L. (1927). *Z. Phys.*, **40**, 344.
Perlman, H.S. (1960). *Proc. Phys. Soc.* **76**, 623.
Pockman, L.T., Webster, D.L., Kirkpatrick, P. and Harworth, K. (1947). *Phys. Rev.* **71**, 330.
Powell, L.J. (1976). *Rev. Mod. Phys.*, **48**, 33.
Ramberg, E. and Richtmeyer, F.K. (1937). *Phys. Rev.* **51**, 913.
Rao, P.V., Chen, M.H. and Crasemann, B. (1972). *Phys. Rev.* **A5**, 997.
Reed, S.J.B. (1975). *Electron Microprobe Analysis.* Cambridge.
Rester, D.H. and Dance, W.E. (1966). *Phys. Rev.* **152**, 1.
Sandstrom, A.L. (1957). *See* Flügge (Ed.) *in General works of reference.*
Schörling, P.O. (1961). *Ark. Fys.*, **19**, 47.
Schörling, P.O. (1962). *Ark. Fys.*, **21**, 371.
Schörling, P.O. (1964). *Ark. Fys.*, **27**, 143.
Schörling, P.O. (1965). *Ark. Fys.*, **29**, 375.
Segrè (1977). *See General works of reference.*
Smick, A.E. and Kirkpatrick, P. (1945). *Phys. Rev.*, **67**, 153.
Sommerfeld, A. (1934). *Atomic Structure and Spectral Lines*, (translation from the 5th German edition). Methuen.
Stoddard, K.B. (1934). *Phys. Rev.*, **46**, 837.
Stoddard, K.B. (1935). *Phys. Rev.*, **48**, 43.
Tothill, P. (1968). *Brit. J. Appl. Phys.*, (*J. Phys. D.*) Ser. 2, **1**, 1093.
Unsworth, M.H. and Greening, J.R. (1970a). *Phys. Med., Biol.*, **15**, 621.
Unsworth, M.H. and Greening, J.R. (1970b). *Phys. Med. Biol.*, **15**, 631.
Victor, C. (1961). *Ann. Phys.*, **6**, 183.
Vignes, A. and Dez, G. (1968). *Brit. J. Appl. Phys.* (*J. Phys. D*), Ser. 2, **1**, 1309.

Vrakking, J.J. and Meyer, F. (1974). *Phys. Rev.* **A9**, 1932.
Webster, D.L. (1928a). *Proc. Nat. Acad. Sci.*, **14**, 330.
Webster, D.L. (1928b). *Proc. Nat. Acad. Sci.*, **14**, 339.
Webster, D.L., Hansen, W.W. and Duveneck, F.B. (1933a). *Phys. Rev.*, **43**, 839.
Webster, D.L., Hansen, W.W. and Duveneck, F.B. (1933b). *Phys. Rev.*, **44**, 258.
Wentzel, G. (1923). *Z. Phys.*, **16**, 46.
White (1934). *See General works of reference.*
Worthington, C.R. and Tomlin, S.G. (1956). *Proc. Phys. Soc.*, **A69**, 401.

Chapter 4

Alkhazov, G.D., Komar, A.P. and Vorab'ev, A.A. (1967). *Nucl. Instr. Meth.*, **48**, 1.
Arndt, U.W., Coates, W.A. and Crathorn, A.R. (1954). *Proc. Phys. Soc.*, **67B**, 357.
Attix, F.H., Roesch, W.C. and Tochilin, E. (1966–9), Eds. *Radiation Dosimetry* (3 Vols.). Academic Press.
Bailey, L.E. and Swedlund, J.B. (1967). *Phys. Rev.*, **158**, 6.
Bede, D.E. and Tomboulian, D.H. (1961). *Rev. Sci. Instr.*, **32**, 184.
Bearden, J.A., Huffman, F.N. and Spijkermann, J.J. (1964). *Rev. Sci. Instrum.*, **35**, 1681.
Birks, J.B. (1964). *The Theory and Practice of Scintillation Counting.* Pergamon Press.
Bishop, H.E. (1966). *See* Castaing, R., Descamps, P. and Philibert, J. Eds. *in bibliography for chapter 8, Section 1.*
Bisi, A. and Zappa, L. (1955). *Nuovo Cim.*, **2**, 988.
Boag, J.W. (1966). *In* Attix, Roesch & Tochilin, Vol. 2. *See above.*
Botden, *et al.* (1952). *See bibliography for chapter 2.*
Brown, J.G. (1966). *X-rays and Their Applications.* Iliffe.
Bruining, H. and De Boer, J.H. (1938). *Physica* **5**, 17.
Burhop (1952). *See bibliography for chapter 3.*
Burlin, T.E. (1968). *In* Attix, Roesch & Tochilin, Vol. 1, p. 331. *See above.*
Campbell (1963). *See bibliography for chapter 3.*
Campbell, J.L. and Ledingham, K.N.D. (1966). *Brit. J. Appl. Phys.* **17**, 769.
Cauchois, Y. (1932). *Journ. de Phys.* **3**, 320.
Cosslett, V.E. and Nixon, W.C. (1952). *See bibliography for chapter 8, section 1.*
Culhane, J.L., Herring, J., Sanford, P.W., O'Shea, G. and Philips, R.D. (1966). *J. Sci. Instrum.*, **43**, 908.
Culhane, J.L., Willmore, A.R., Pounds, K.A. and Sanford, P.W. (1964). *Space Res.* **4**, 741.
Curran, S.C., Cockcroft, A.L. and Angus, J. (1949). *Phil. Mag.*, **40**, 929.
Curran, S.C. and Wilson, H.W. (1965). *See* Siegbahn, K. (Ed.) *in General work of reference.*
Dolby, R.M. (1959). *Proc. Phys. Soc.*, **73**, 81.
Dumond, J.W.M. and Kirkpatrick, H.A. (1930). *Rev. Sci. Inst.*, **1**, 88.
Duncumb (1960). *See p. 365* of Engstrom, A., Cosslett, V.E. and Pattee, H. *in bibliography for chapter 8, section 1.*
Ehrenberg and Franks (1953). *See bibliography for appendix 1.*
Ehrenberg and King (1963). *See bibliography for appendix 1.*
England, J.B.A. (1974). *Techniques in Nuclear Structure Physics.* (2 Vols). Macmillan.
Fano, U. (1947). *Phys. Rev.* **72**, 26.
Feldman (1960). *See bibliography for appendix 1.*
Franks, A. (1964). *Nature*, **201**, 913.
Goldhaber, M., Grodzins, L. and Sunyar, A.W. (1957). *Phys. Rev.*, **106**, 826.
Gibbons, P.E. and Northropp, D.C. (1962). *Proc. Phys. Soc.*, **80**, 276.
Green (1964). *See bibliography for chapter 3.*
Hanna, G.C., Kirkwood, D.H.W. and Pontecorvo, B. (1949). *Phys. Rev.* **75**, 985.
Hanson, *et al.* (1952). *See bibliography for appendix 1.*
Henke (1957). *See p. 72 of* Cosslett, V.E., Engstrom, A. and Pattee, H. Eds. *in bibliography for chapter 8, section 1.*

Henke (1960). *See p.* 10 *of* Engstrom, A., Cosslett, V.E. and Pattee, H. Eds. *in bibliography for chapter 8, section 1.*
Henke (1963). *See p. 157 of* Pattee, H., Cosslett, V.E. and Engstrom, A. Eds. *in bibliography for chapter 8, section 1.*
Henke (1966). *See p. 440 of* Castaing, R., Descamps, P. and Philibert, J. Eds. *in bibliography for chapter 8, section 1.*
Hereford and Swann (1950). *See bibliography for appendix 1.*
Jacob, L., Noble, R. and Yee, H. (1960). *J. Sci. Instrum.*, **37**, 460.
Jenkins, R. (1972). *X-ray spectrometry*, **1**, 23.
Jenkins, R. and de Vries, J.L. (1970). *Practical X-ray Spectrometry.* Macmillan.
Johann, H.H. (1931). *Z. Phys.* **69**, 185.
Johansson, T. (1933). *Z. Phys.*, **82**, 507.
Johns, H.E. (1964). *See General works of reference.*
Kerst, D.W., Adams, G.D., Koch, H.W. and Robinson, C.S. (1950). *Phys. Rev.* **78**, 297.
Kirkpatrick, P. (1939). *Rev. Sci. Inst.*, **10**, 186.
Kirkpatrick, P. (1944). *Rev. Sci. Inst.*, **15**, 223.
Koller and Alden (1951). *See bibliography for appendix 1.*
Lane and Zaffarano (1954). *See bibliography for appendix 1.*
Large, L.N. and Whitlock, W.S. (1962). *Proc. Phys. Soc.*, **72**, 148.
Marshall and Ward (1937). *See bibliography for appendix 1.*
McCrary, J.H., Plassmann, E.H., Puckett, J.M., Conner, A.L. and Zimmermann, G.W. (1967). *Phys. Rev.*, **153**, 307.
McKay, K.G. (1948). *Adv. in Electron.* **1**, 65.
Medved, D.B. and Strausser, Y.E. (1965). *Adv. in Electron. and Electron Phys.*, **21**, 101.
Metzger, F. and Deutsch, M. (1950). *Phys. Rev.* **78**, 551.
Miller, C.W. (1954). *Proc. I.E.E.* **101**, 207.
Mulvey, T. and Campbell, A.J. (1958). *Brit. J. Appl. Phys.*, **9**, 406.
Neiler and Bell (1965). *See* Siegbahn, K. Ed. *in General works of reference.*
Nicholas (1929). *See bibliography for chapter 2.*
Nordfors, B. (1956). *Ark. Fys.* **10**, 279.
O'Neill and Scott (1950). *See bibliography for appendix 1.*
Orphan, W.J. and Rasmussen, N.C. (1967). *Nucl. Instr. and Meth.* **48**, 282.
Rudberg, E. (1936). *Phys. Rev.* **50**, 138.
Sandström (1957). *See* Flügge. Ed. *in General works of reference.*
Sharpe, J. (1961). E.M.I. Electronics Ltd. Rpt. CP5306.
Sood, B.S. (1958). *Proc. Roy. Soc.*, **A247**, 375.
Thordarson (1939). *See bibliography for chapter 2.*
Tomboulian (1957). *See* Flügge, Ed. *in General works of reference.*
Trump, J.G. and Van de Graaff, R.J. (1947). *J. Appl. Phys.*, **18**, 327.
Underwood (1964). Ph.D. thesis, Leicester.
Valentine, J.M. (1952). *Proc. Roy. Soc.* **211A**, 75.
Van de Graaff, R.J., Trump, J.G. and Buechner, W.W. (1946). *Rep. Prog. Phys.*, **11**, 1.
Weiss, J. and Bernstein, W. (1955). *Phys. Rev.*, **98**, 1828.
Williams (1931). *See bibliography for appendix 1.*
Young (1956a,b, 1957). *See bibliography for appendix 1.*

Chapter 5

Anderson, C.D. (1930). *Phys. Rev.*, **25**, 1139.
Black, P.J., Evans, D.E. and O'Connor, D.A. (1962). *Proc. Roy. Soc.*, **A270**, 168.
Bosch, R., Lang, T., Muller, R. and Wolfi, W. (1962). *Phys. Lett.*, **2**, 16.
Champeney, D.C. (1979). *Rep. Prog. Phys.* **42**, 1017.
Champeney, D.C. and Woodhams, F.W.D. (1966). *Phys. Lett.* **20**, 275.
Chipman, D.R. (1955). *J. App. Phys.* **26**, 1387.

Compton and Allison (1935). *See General works of reference.*
Davey, W.G. (1953). *Proc. Phys. Soc.*, **66**, 1059.
Davey, W.G. and Moon, P.B. (1953). *Proc. Phys. Soc.*, **A66**, 956.
Davisson, C.M. (1965). *See* Siegbahn, K. Ed. *in General works of reference.*
Davisson, C.M. and Evans, R.D. (1952). *Rev. Mod. Phys.*, **24**, 79.
Eliseenko, L.G., Shchemelev, V.N. and Rumsh, M.A. (1967). *Sov. Phys. JETP* (English trans.) **25**, 211.
Evans (1955). *See General works of reference.*
Evans, R.D. and Evans, R.O. (1948). *Rev. Mod. Phys.*, **20**, 305.
Franz, W. (1935). *Z. Phys.*, **98**, 314.
Gibb, T.C. (1976). *Principles of Mössbauer Spectroscopy.* Chapman and Hall.
Goldstein, H. (1959). *Fundamental Aspects of Reactor Shielding.* Pergamon.
Grodstein, G., (White) (1957). NBS Circular 583. National Bureau of Standards.
Grodzins, L. (1958). *Phys. Rev.*, **109**, 1014.
Hall, H. (1936). *Revs. Mod. Phys.* **8**, 358.
Heitler (1954). *See General works of reference.*
Hubbell, J.H. (1969). NBS Circular NBS29 National Bureau of Standards.
Ilakovac, K. (1954). *Proc. Phys. Soc.* **A67**, 601.
James, R.W. (1967). *The Optical Principles of the Diffraction of X-rays.* Bell.
Johns (1964). *See General works of reference.*
Lutze, E. (1931). *Ann. Phys.*, **9**, 853.
Malmfors, K.G. (1953). *Ark. Fys.*, **6**, 49.
Manninen, S., Pitkanen, T., Koikkalainen, S. and Paakkari, T. (1984). *Int. J. Appl. Radiat. Isotopes*, **35**, 93.
May, L. (1971). *An Introduction to Mössbauer Spectroscopy.* Adam Hilger.
Metzger, F.R. (1959). *Prog. Nucl. Phys.*, **7**, 54.
Moffat, J. and Stringfellow, M.W. (1958). *Phil. Mag.* **3**, 540.
Moffat, J. and Stringfellow, M.W. (1960). *Proc. Roy. Soc.*, **A254**, 242.
Moon, P.B. (1950). *Proc. Phys. Soc.*, **A63**, 1189.
Moon, P.B (1951). *Proc. Phys. Soc.*, **A64**, 76.
Moon, P.B. (1961). *Proc. Roy. Soc.*, **A263**, 309.
Moreh, R. and Kahana, S. (1973). *Phys. Lett.* **47B**, 351.
Motz, J.W. and Missoni, G. (1961). *Phys. Rev.* **124**, 1458.
Rasmussen, V.K., Metzger, F.R. and Swann, C.P. (1958). *Phys. Rev.*, **110**, 154.
Reibel, K. and Mann, A.K. (1960). *Phys. Rev.*, **118**, 701.

Chapter 6

Avaldi, L., Mitchell, I.V. and Milazzo, M. (1983). *J. Phys. B.* (*At. Mol. Phys.*) **16**, 4555.
Bang, J. and Hansteen, J.M. (1959). *K. Danske Vidensk. Selsk. Mat. Fys. Medd.*, **31**, No. 13Q.
Basbas, G., Brant, W. and Laubert, R. (1973). *Phys. Rev.*, **7A**, 983.
Bissinger, G.A., Joyce, J.M., Ludwig, E.J., McEver, W.S. and Shaforth, S.M. (1970). *Phys. Rev.*, **A1**, 841.
Bothe, W. and Franz, S. (1928). *Z. Phys.*, **52**, 466.
Briand (1971). *See bibliography for chapter* 7.
Cahill, T.A. (1980). *Ann. Rev. Nuc. Sci.*, **30**, 211.
Chen, J.R., Reber, J.D., Ellis, D.J. and Miller, J.E. (1976). *Phys. Rev.* **A13**, 941.
Chen, J.R. (1977). *Nucl. Inst. Meth.*, **142**, 9.
Cork, J.M. (1941). *Phys. Rev.*, **59**, 957.
Drell, S.D. and Huang, K. (1955). *Phys. Rev.*, **99**, 686.
Fano, U. and Lichten, W. (1965). *Phys. Rev. Lett.* **14**, 627.
Folkmann, F., Gaarde, C., Huus, T. and Kemp, K. (1974). *Nucl. Inst. Meth.*, **116**, 487.
Garcia, J.D. (1970a). *Phys. Rev.*, **A1**, 280.

Garcia, J.D. (1970b). *Phys. Rev.*, **A1**, 1402.
Garcia, J.D., Fortner, R.J. and Kavanagh, T.M. (1973). *Rev. Mod. Phys.* **45**, 111.
Hansteen, J.M. and Mosebekk, O.P. (1973). *Nucl. Phys.* **A201**, 541.
Hopkins, F., Elliott, D.O., Bhalla, C.P. and Richard, P. (1973). *Phys. Rev.*, **A8**, 2952.
Huus, T., Bjerregaard, J.H. and Elbek, B. (1956). *K. Danske Vidensk. Selsk., Mat. Fys. Medd.* **30**, 1.
Ishii, K., Morita, S. and Tawara, H. (1976). *Phys. Rev.* **A13**, 131.
Kaufmann, R., McGuire, J.H., Richard, P. and Moore, C.F. (1973). *Phys. Rev.* **A8**, 1233.
Kavanagh, T.M., Cunningham, M.E., Der, R.C., Fortner, R.J., Khan, J.M., Zaharis, E.J. and Garcia, J.D. (1970). *Phys. Rev. Lett.*, **25**, 1473.
Khan, J.M. and Potter, D.L. (1964). *Phys. Rev.* **A133**, 890.
Khan, M.R., Crumpton, D. and Francois, P.E. (1976). *J. Phys. B. (At. Mol. Phys.)* **9**, 455.
Khan, M.R., Hopkins, A.G., Crumpton, D. and Francois, P.E. (1977). *X-ray Spec.* **6**, 140.
Krause, M.O. (1979). *J. Phys. Chem. Ref. Data* **8**, 307.
Laegsgaard, E., Andersen, J.U. and Feldman, L.C. (1972). *Phys. Rev. Lett.*, **29**, 1206.
McWherter, J., Bolger, J., Moore, C.F. and Richard, P. (1973). *Z. Phys.*, **263**, 283.
Merzbacher, E. and Lewis, H.W. (1958). *Encyclop. of Phys.* **34**, 166, Ed. S. Flügge.
Messelt, S. (1958). *Nucl. Phys.* **5**, 435.
Moore, C.F., Senglaub, M., Johynson, B. and Richard, P. (1972). *Phys. Lett.*, **40A**, 107.
Parratt, (1936a). *See bibliography for chapter 3.*
Saris, F.W. and Bierman, D.J. (1971). *Phys. Lett.*, **35A**, 199.
Sokhi, R.S. and Crumpton, D. (1984). *At. Data and Nucl. Data Tables.*, **30**, 49.
Sokhi, R.S. and Crumption, D. (1986). *J. Phys. B. (At. Mol. Phys.)* **19**, 4193.
Walske, M.C. (1956). *Phys. Rev.* **101**, 940.
Ward, T.R. and Dyson, N.A. (1978). *J. Phys. B. (At Mol. Phys.)* **11**, 2705.

Chapter 7

Abelson, P.H. (1939). *Phys. Rev.*, **56**, 753.
Allison, S.K. (1927). *Phys. Rev.*, **30**, 245.
Allison, S.K. (1928). *Phys. Rev.*, **32**, 1.
Avignone, III, F.T., Khalil, A.E. and Grabowski, Z.W. (1981). *Phys. Rev.*, **A24**, 1198.
Babu, R.P., Murty, K.N. and Murty, V.A.N. (1976). *Phys. G., Nucl. Phys.*, **2**, 331.
Bambynek, W., Behrens, H., Chen, M.H., Crasenah, B., Fitzpatrick, M.L., Ledingham, K.W.D., Genz, H., Mutterer, M. and Intermann, R.L. (1977). *Rev. Mod. Phys.*, **49**, 77.
Barton, G.W. (Jr), Robinson, H.P., Perlman, I. (1951), *Phys. Rev.*, **81**, 208.
Bell, P.R., Rauch, J. and Cassidy, J.M. (1952). *Science*, **115**, 12.
Berenyi, D. and Varga, D. (1969). *Nucl. Phys.* **A138**, 685.
Berenyi, D., Scharbert, T. and Vatai, E. (1969). *Nucl. Phys.* **A124**, 464.
Biavati, M.H., Nassiff, S.J. and Wu, C.S. (1962). *Phys. Rev.* **125**, 1364.
Blatt and Weisskopf (1952). *See General works of reference.*
Boehm, F. and Wu, C.S. (1954). *Phys. Rev.*, **93**, 518.
Bolgiano, P., Madansky, L. and Rasetti, F. (1953). *Phys. Rev.*, **89**, 679.
Bosch, H.E., Farinolli, M.A., Martin, N., Simon, M.C. (1969). *Nucl. Instr. Meth.*, **73**, 323.
Bothe, W. and Franz, S. (1928). *Z. Phys.*, **52**, 466.
Briand, J.P., Chevallier, P., Tavernier, M., Rozet, J.P. (1971). *Phys. Rev. Lett.*, **27**, 777.
Briand, J.P., Chevallier, P., Johnson, A., Rozet, J.P., Tavernier, M. and Tounti, A. (1974). *Phys. Lett.* **49A**, 51.
Buechner, W.W., Van de Graaff, R.J., Burrill, E.A. and Sperduto, A. (1948). *Phys. Rev.*, **74**, 1348.
Buhring, W. and Haxel, O. (1957). *Z. Phys.* **148**, 653.
Bustard, T.S. and Silverman, J. (1967). *Nucl. Sci. Eng.*, **27**, 586.
Cameron, J.F. and Rhodes, J.R. (1962). In *Radioisotopes in the Physical Sciences and Industry*, Vol. 2, p. 23. International Atomic Energy Agency, Vienna, 1962.

Cameron, J.F., Rhodes, J.R. and Berry, P.F. (1963). UKAEA, AERE Report 3086.
Coleman, E.W., Brownell, L.E. and Fox, C.J. (1958). *Second United Nations International Conference on the Peaceful Uses of Atomic Energy*, **26**, 272.
Compton and Allison (1935). *See General works of reference.*
Cook, G.B., Mellish, C.E. and Paynes, J.A. (1958). *Second United Nations International Conference on the Peaceful Uses of Atomic Energy*, **19**, 127.
Daggs, R.G. (1956). *First United Nations International Conference on the Peaceful Uses of Atomic Energy*, **15**, 174.
Dyson, N.A. (1975). *Phys. Med. Biol.*, **20**, 1.
Edwards, J.E. and Pool, M.L. (1946). *Phys. Rev.*, **69**, 549.
Evans (1955). *See General works of reference.*
Feenberg, E. and Trigg, G. (1950). *Rev. Mod. Phys.* **22**, 399.
Filosofo, I., Reiffel, L., Stone, C.A. and Voyvodic, L. (1962). *Radioisotopes in the Physical Sciences and Industry*, Vol. 2, p. 3. International Atomic Energy Agency, Vienna.
Freedman, M.S. (1974). *Ann. Nucl. Sci.* **24**, 209.
Gelberg, A. and Piticu, I. (1971). *Rev. Roum. Phys. (Rumania)* **16**, 153.
Gellman, H., Griffith, B.A., Stanley, I.P.S. (1950). *Phys. Rev.*, **85**, 944.
Goodrich, M., Levinger, J.S. and Payne, W. (1953). *Phys. Rev.* **91**, 1225.
Goodrich, M. and Payne, W.B. (1954). *Phys. Rev.* **94**, 405.
Green, F.L. and Cheek, W.D. (1958). *Second United Nations International Conference on the Peaceful Uses of Atomic Energy*, **19**, 169.
Hartmann, F.X. and Naumann, R.A. (1985). *Phys. Rev.* **C31**, 1594.
Isozumi, Y. and Shimizu, S. (1971). *Phys. Rev.* **C4**, 522.
Kadar, I., Berenyi, D. and Myslck, B. (1970). *Nucl. Phys.* **A153**, 383.
Kereiakes, J.G., Kraft, G.R., Weir, O.E. and Krebs, A.T. (1958). *Nucleonics*, **16** (1), 80.
Kereiakes, J.G. and Krebs, A.T. (1958). *Second United Nations International Conference on the Peaceful Uses of Atomic Energy*, **20**, 234.
Khalil, A.E. (1983). *Phys. Rev.* **28A**, 2414.
Knipp, J.K. and Uhlenbeck, G.E. (1936). *Physica* **3**, 425.
Liden, K. and Starfelt, N. (1954). *Ark. Fys.* **7**, 193.
Madansky, L. and Rasetti, F. (1954). *Phys. Rev.* **94**, 407.
Mandal, A.M. and Patro, A.P. (1985). *J. Phys. G., Nucl. Phys.*, **1**, 1025.
Morrison, P. and Schiff, L.I. (1940). *Phys. Rev.*, **58**, 24.
Mukoyama, T. and Shimizu, S. (1978). *J. Phys. G., Nucl. Phys.* **4**, 1509.
Nagy, H.J. and Schupp, G. (1984). *Phys. Rev.* **C30**, 2031.
Narasimhamarty, K. and Jnanananda, S. (1967). *Proc. Phys. Soc.* **90**, 109.
Parratt (1959). *See bibliography for chapter 3.*
Rudraswamy, B., Gopala, K., Venkatarmaiah, P. and Sanjeeviah, H. (1984). *J. Phys. G.*, **10**, 1579.
Rester, D.H., Dance, W.E. and Derrickson, J.H. (1970). *J. Appl. Phys.* **41**, 2682.
Robinson, B.L. and Fink, R.W. (1960). *Rev. Mod. Phys.* **32**, 117.
Rose, M.E. (1955). *J. Phys. Rad.*, **16**, 520.
Rose, M.E., Goertzel, G.H., Spinrad, B.I., Harr, J. and Strong, P. (1951). *Phys. Rev.* **83**, 79.
Rothwell, P. and West, D. (1950). *Proc. Roy. Soc.*, **A63**, 541.
Sama, K.V.N. and Morty, K.N. (1976). *J. Phys. G., Nucl. Phys.*, **2**, 387.
Schupp, G. and Nagy, H.J. (1984). *Phys. Rev.*, **C30**, 2031.
Scott, R.D. (1980). *J. Phys. G., Nucl. Phys.*, **6**, 1427.
Scott, R.D. (1984). *J. Phys. G., Nucl. Phys.*, **10**, 1559.
Sen, S.K., Salic, D.L. and Tomchuk, E. (1972). *Phys. Rev. Lett.*, **28**, 1295.
Siegbahn (1965). *See General works of reference.*
Starfelt, N., Cederlund, J. and Liden, K. (1957). *Int. J. App. Radiation and Isotopes*, **2**, 165.

Victor (1961). *See bibliography for chapter 3.*
Wu, C.S. (1941). *Phys. Rev.*, **59**, 481.
Wyard, S.J. (1952). *Proc. Phys. Soc.* **A65**, 377.

Chapter 8

Section 8.1 X-ray microscopy and microanalysis

The following volumes of conference proceedings are listed together chronologically here:

Cosslett, V.E., Engstrom, A. and Pattee, H. Eds. (1957). *X-ray Microscopy and Microradiography*. Academic Press.
Engstrom, A., Cosslett, V.E. and Pattee, H. Eds. (1960). *X-ray Microscopy and Microanalysis.* Elsevier.
Pattee, H.H., Cosslett, V.E. and Engstrom, A. Eds. (1963). *X-ray Optics and X-ray Microanalysis.* Academic Press.
Castaing, R., Descamps, P. and Philibert, J. Eds. (1966). *X-ray Optics and Microanalysis.* Hermann.
Möllenstedt, G. and Gaukler, K.H. Eds. (1969). *X-ray Optics and Microanalysis.* Springer, Berlin.
Shinoda, G., Kohra, K. and Ichinokawa, T. Eds. (1972). *X-ray Optics and Microanalysis.* University of Tokyo Press.
Borovsky, I. and Komyak, N. Eds. (1976). *X-ray Optics and Microanalysis.* Leningrad Maschinostroennie.
Beaman, D.R., Ogilvie, R.E. and Wittry, D.B. Eds. (1980). *X-ray Optics and Microanalysis.* Pendell Publishing Co.

Anderton, H. (1967). *Sci. Prog. Oxf.*, **55**, 337.
Ardenne, M. von (1939). *Naturwiss*, **26**, 485.
Asunmaa, S.K. (1960), p. 66 of Engstrom, Cosslett and Pattee, Eds. *See above.*
Castaing, R. (1951). Thesis, Paris.
Cosslett, V.E. (1952). *Proc. Phys. Soc.*, **B65**, 782.
Cosslett, V.E. (1965). *Rep. Prog. Phys.* **28**, 381.
Cosslett, V.E. and Nixon, W.C. (1952). *Proc. Roy. Soc. B.*, **140**, 422.
Dolby (1959). *See bibliography for chapter 4.*
Duncumb, P. (1957), p. 617 of Cosslett, Engstrom and Pattee, Eds. *See above.*
Duncumb, P. (1967). *Sci. Prog. Oxf.* **55**, 511.
Dyson, N.A. (1957), p. 310 of Cosslett, Engstrom and Pattee, Eds. *See above.*
Engstrom, A. (1963), p. 23 Cosslett, Engstrom and Pattee, Eds. *See above.*
Feder, R., Spiller, E., Topalian, J., Broers, A.N., Gudat, W., Panessa, B.J. and Zadunaisky, Z.A. (1977). *Science* **197**, 259.
Goldstein, J.I. (1969). *In* Tousimis, A.J. and Marton, L. (Eds.). Electron probe microanalysis, *Adv. in Electronics and Electron Phys. Suppl. 6.* Academic Press.
Heinrich, K.F.J. (1981). *Electron beam X-ray microanalysis.* Van Nostrand.
Keil, K. (1973). *In* Andersen, C.A. (Ed.), *Microprobe analysis.* Wiley.
Liebmann, G. and Grad, E.M. (1951). *Proc. Phys. Soc.*, **64B**, 956.
Lindstrom, B. (1963), p. 23 of Pattee, Cosslett and Engstrom, Eds. *See above.*
Long, J.V.P. (1963), p. 279 of Pattee, Cosslett and Engstrom, Eds. *See above.*
Kirkpatrick, P. and Pattee, H. (1957). *See* Flügge, Ed. *In General works of reference.*
Kirkpatrick, P. and Baez, A.V. (1948). *J. Opt. Soc. Amer.* **38**, 766.
Polack, F., Lowenthal, S., Petroff, Y. and Farge, Y. (1977). *Appl. Phys. Lett.* **31**, 785.
Reed, S.J.B. (1975). *Electron Microprobe Analysis.* Cambridge University Press.
Rieser, (Jr.), L.M. (1957), p. 195 of Cosslett, Engstrom and Pattee, Eds. *See above.*
Robertson, A.J. (1968). *Phys. Med. Biol.*, **13**, 505.
Shinoda, G. (1963), p. 297 of Pattee, Cosslett and Engstrom, Eds. *See above.*
Sievert, R. (1936). *Acta. Radiol.*, **17**, 299.

Section 8.2 Chemical influences; the isotope effect

Appleton, A. (1964). *Contemp. Phys.*, **6**, 50.
Baun, W.L. (1969). Electron Probe Microanalysis, p. 155. *Advances in Electron Physics*, Supplement 6.
Bhattacherjee, S.K., Boehm, F. and Lee, P.L. (1969). *Phys. Rev.* **188**, 1919.
Boehm, F. and Chesler, R.B. (1968). *Phys. Rev.* **166**, 1206.
Brockmeier, R.T., Boehm, F. and Hatch, E.N. (1965). *Phys. Rev. Lett.* **15**, 132.
Chesler, R.B., Boehm, F., Brockmeier, R.T. (1967). *Phys. Rev. Lett.* **18**, 953.
Chivate, P., Damle, P.S., Joshi, N.V. and Mande, C. (1968). *Proc. Phys. Soc.* (*J. Phys. C.*), ser. 2, **1**, 1171.
Cooper, M. (1971). *Adv. Phys.*, **20**, 453.
Cooper, M. (1977). *Contemp. Phys.*, **18**, 489.
Cooper, M. (1985). *Rep. Prog. Phys.*, **48**, 415.
Fabian, D.J., Watson, L.M. and Marshal, C.A.W. (1971). *Rep. Prog. Phys.*, **34**, 601.
Fischer, D.W. (1965). *J. Chem. Phys.*, **42**, 3814.
Fujimoto, H. (1965a). *Sci. Rep. Tohoku Univ.*, 1st ser., **49**, 32.
Fujimoto, H. (1965b). *Sci. Rep. Tohoku Univ.*, 1st ser., **49**, 36.
Houston (1931). *Phys. Rev.*, **38**, 1791.
Kawata, S. and Maeda, K. (1977). *J. Phys. F., Metal. Phys.*, **7**, 2243.
Khasbardar, B.V., Vaingankar, A.S. and Patil, R.N. (1980). *J. Phys. F., Metal. Phys.* **10**, 1879.
Kushwaha, M.S., Shrivastava, B.D. and Dubey, V.S. (1975). *J. Phys. F., Metal. Phys.* **5**, 597.
Lee, P.L. and Boehm, F. (1971). *Phys. Lett.* **35B**, 33.
Lindh, A.E. and Lundquist, O. (1924). *Ark. Mat. Fys.*, **18**, 3.
Nordling, C. (1959). *Ark. Fys.*, **15**, 241.
Parratt (1959). *See bibliography for chapter 3.*
Sarode, P.R. and Chetal, A.R. (1977). *J. Phys. F. Metal. Phys.*, **7**, 745.
Seltzer, E.C. (1969). *Phys. Rev.*, **188**, 1916.
Skinner, H.W.B. (1938). *Rep. Prog. Phys.*, **5**, 257.
Skinner, H.W.B. (1940). *Phil. Trans. Roy. Soc.*, **A239**, 95.
Sugawara, H., Fujimoto, H. and Hayasi, T. (1967). *Sci. Rep. Tohoku Univ.*, 1st ser., **50**, 152.
Sumbaev, O.I. (1970). *Sov. Phys. JETP* (Eng. Trans.) **20**, 927.
Sumbaev, O.I. and Mezentsev, A.F. (1965). *Sov. Phys. JETP* (Eng. Trans.) **21**, 295.
Sumbaev, O.I. and Mezentsev, A.F. (1966). *Sov. Phys. JETP* (Eng. Trans.) **22**, 323.
Sumbaev, O.I. and Mezentsev, A.F., Marushenko, V.I., Petrovich, E.V. and Ryl'nikov, A.S. (1966). *Sov. Phys. JETP* **23**, 572.
Sumbaev, O.I., Petrovich, E.V., Smirnov, Yu. P., Egorov, A.I., Zykov, V.S. and Grushko, A.I. (1968). *Sov. Phys. JETP* (Eng. Trans.) **26**, 891.
Tomboulian (1957). *See* Flugge, Ed. *In General works of reference.*
Tsustumi, K., Obashi, M. and Sawada, M. (1958). *J. Phys. Soc. Japan*, **13**, 43.
Vaingankar, A.S., Khasbardar, B.V. and Patil, R.N. (1979). *J. Phys. F., Metal. Phys.*, **9**, 2301.
Valasek, J. (1933). *Phys. Rev.*, **43**, 612.
Verma, L.P. and Agarwal, B.K. (1968). *J. Phys. C.* (*Proc. Phys. Soc.*), ser. 2, **1**, 1658.
Wells, A.F. (1962). *Structural Inorganic Chemistry*, Oxford.

Section 8.3 Mesonic X-rays

Backenstross, G. (1970). *Ann. Rev. Nucl. Sci.*, **20**, 467.
Burleson, G.R., Cohen, D., Lamb, R.C., Michael, D.N., Schlutter, R.A. and White, J.O. (1965). *Phys. Rev. Lett.* **15**, 70.
Chang, W.Y. (1949). *Rev. Mod. Phys.*, **21**, 166.

Dubler, T., Käses, K., Robert-Tissot, B., Schaller, L.A., Schellenberg, L. and Schnewley, H. (1976). *Phys. Lett.* **57A**, 325.
Fitch, V.L. and Rainwater, J. (1953). *Phys. Rev.*, **92**, 789.
Kessler, D., Anderson, H.L., Dixit, M.S., Evans, H.J., McKee, R.J., Hargrove, C.K., Barton, R.D., Hincks, E.P. and McAndrew, J.D. (1967). *Phys. Rev. Lett.*, **18**, 1179.
Lewis, C.A., O'Leary, K., Jackson, D.F. and Lam, G.K.Y. (1982). *Phys. Med. Biol.*, **27**, 683.
Macagno, E.R., Bernow, S., Devons, S., Duerdoth, I., Hitlin, D., Kast, J.W., Rainwater, J., Runge, K. and Wu, C.S. (1967). *Proc. Intern. Conf. Electromagnetic Sizes of Nuclei.* Ottawa, May 22–24, 1967.
Ponomarev, L.I. (1973). *Ann. Rev. Nucl. Sci.* **23**, 395.
Quitmann, D., Engfer, R., Hegel, U., Brix, P., Backenstross, G., Goebel, K. and Stadler, B. (1964). *Nucl. Phys.* **51**, 609.
Taylor, M.C., Coulson, L. and Phillips, G.C. (1973). *Radiat. Res.*, **54**, 335.
Wiegand, C.E. and Mack, D.A. (1967). *Phys. Rev. Lett.* **18**, 685.
Wu, C.S. and Wilets, L. (1969). *Ann. Rev. Nucl. Sci.* **19**, 527.

Section 8.4 Synchrotron radiation

Ansaldo, E.J. (1977). *Contemp. Phys.*, **18**, 527.
Schwinger, J. (1949). *Phys. Rev.* **75**, 1912.
Suller, V.P. and Thompson, D.J. (1978). *Nucl. Inst. Meth.* **152**, 1.
Tomboulian, D.H. and Hartmann, P.L. (1956). *Phys. Rev.* **102**, 1423.
Winick, H. and Bienenstock, A. (1978). *Ann. Rev. Nucl. Part. Sci.*, **28**, 33.

Section 8.5 Plasma physics and astrophysics

Adams, D.J. (1971). *Contemp. Phys.* **12**, 471.
Adams, D.J. (1980). *Cosmic X-ray Astronomy*. Adam Hilger.
Beckner, E.H. (1967). *Rev. Sci. Instrum.* **38**, 507.
Boyer, K., Little, E.M., Quinn, W.E., Sawyer, G.A. and Stratton, T.F. (1959). *Phys. Rev. Lett.*, **2**, 279.
Bradt, H., Rappaport, S., Mayer, W., Nather, R.E., Warner, B., McFarlane, M. and Kristian, J. (1969). *Nature* **222**, 728.
Fabian, A.C. (1985). *In Interacting Binary Stars.* Pringle, J.E. and Wade, R.A. Eds. Cambridge University Press.
Giacconi, R., Gursky, H., Kellogg, E., Schreier, E. and Tananbaum, H. (1971). *Astrophys. J.* **167**, L67.
Grader, R.J., Hill, R.W., Seward, F.D. and Hiltner, W.A. (1970). *Astrophys. J.* **159**, 201.
Griem, H.R., Kolb, A.C. and Faust, W.R. (1959). *Phys. Rev. Lett.* **2**, 281.
Jahoda, F.C., Little, E.M., Quim, W.E., Sawyer, G.A. and Stratton, J.F. (1960). *Phys. Rev.* **119**, 843.
Maxon, M.S. and Corman, E.G. (1967). *Phys. Rev.*, **163**, 156.
Rappaport, S., Bradt, H.V. and Mayer, W. (1971). *Nat. Phys. Sci.* **229**, 40.
White, N.E. (1983). *Advance in High Energy Astrophysics and Cosmology*. Meeting, AU/COSPAR, Bulgaria, July 1983.
Willmore, A.P. (1978). *Rep. Prog. Phys.* **41**, 511.

Appendix 1

Bohr, N. (1913). *Phil. Mag.*, **25**, 10.
Cosslett, V.E. (1964). *Brit. J. Appl. Phys.*, **15**, 107.
Cosslett, V.E. (1966). *In X-ray Optics and Microanalysis*, p. 85ff. Castaing, R., Descamps, P. and Philibert, J., Eds. Hermann.
Cosslett, V.E. and Thomas, R.N. (1964a). *Brit. J. Appl. Phys.*, **15**, 235.
Cosslett, V.E. and Thomas, R.N. (1964b). *Brit. J. Appl. Phys.*, **15**, 883.

Eddy, C.E. (1929). *Proc. Camb. Phil. Soc.*, **25**, 50.
Ehrenberg, W. and Franks, J. (1953). *Proc. Phys. Soc.*, **66B**, 1057.
Ehrenberg, W. and King, D.E.N. (1963). *Proc. Phys. Soc.*, **81**, 751.
Everhart, T.E. and Hoff, P.H. (1971). *J. Appl. Phys*, **42**, 5837.
Feldman, C. (1960). *Phys. Rev.*, **117**, 455.
Hanson, A.O., Goldwasser, E.L. and Mills (1952). *Phys. Rev.* **86**, 617.
Hereford, F.L. and Swann, C.P. (1950). *Phys. Rev.* **78**, 727.
Kanicheva, I.R. and Burtsev, V.A. (1960). *Soviet Physics Solid State*, **1**, 1146.
Kanter, H. (1961). *Phys. Rev.*, **121**, 461.
Katz, L. and Penfold, A.S. (1952). *Rev. Mod. Phys.*, **24**, 28.
Knop, G. and Paul, W. (1965). *See* Siegbahn, K. (Ed.) *in General works of reference.*
Koller, L.R. and Alden, E.D. (1951). *Phys. Rev.*, **83**, 684.
Lane, R.O. and Zaffarano, D.J. (1954). *Phys. Rev.* **94**, 960.
Lenz, F. (1954). *Z. Naturforsch.*, **9a**, 185.
Marshall, J.S. and Ward, A.G. (1937). *Can. J. Research*, **15**, 39.
Nelms, A.T. (1956). *Circ. Nat. Bur. Stand.*, No. 577.
O'Neill, G.F. and Scott, W.T. (1950). *Phys. Rev.* **80**, 473.
Whiddington, R. (1912). *Proc. Roy. Soc.*, **A86**, 360.
Whiddington, R. (1914). *Proc. Roy. Soc.*, **A89**, 554.
Williams, E.J. (1931). *Proc. Roy. Soc.* **A130**, 310.
Young, J.R. (1956a). *J. Appl. Phys.*, **27**, 1.
Young, J.R. (1956b). *Phys. Rev.*, **103**, 292.
Young, J.R. (1957). *J. Appl. Phys.*, **28**, 524.

Appendix 2

Cooke, B.A. and Stewardson, E.A. (1964). *Brit. J. Appl. Phys.*, **15**, 1315.
Davisson (1965). *See* Siegbahn, K. (Ed.) *in General works of reference.*
Grodstein (1957). *See bibliography for chapter 5.*
Heinrich, K.J.F. (1966). *In The Electron Microprobe.* Wiley.
Henke, B.L. and Elgin, R.L. (1970). *Advances in X-ray Analysis*, p. 639. Plenum Press.
Hopkins, J.I. (1959). *J. Appl. Phys.*, **30**, 185.
Hubbell, J.H. (1971). *Atomic Data*, **3**, 241. Academic Press.
Hughes, G.D. and Woodhouse, J.B. (1966). *See* Castaing, Descamps and Philibert (Eds.) *in bibliography for chapter 8, section 1.*
Hughes, G.D., Woodhouse, J.B. and Bucklow, I.A. (1968). *Brit. J. Appl. Phys.*, Ser. 2, **1**, 695.
Laubert, S. (1941). *Ann. der Phys.* **40**, 553.
McCrary *et al.* (1967). *See bibliography for chapter 4.*

Appendix 3

Lederer, C.M., Hollander, J.M. and Perlman, I. (1966). *Table of Isotopes*, sixth ed. Wiley.

Appendix 4

Fine, S. and Hendee, C.F. (1955). *Nucleonics*, **13**, No. 3, 36.

Appendix 5

Bambynek, W., Crasemann, B., Fink, R.W., Freund, H.-U., Mark, H., Swift, C.D., Price, R.E. and Rao, P.V. (1972). *Rev. Mod. Phys.*, **44**, 716.
Hink, W. and Paschke, H. (1971). *Phys. Rev.*, **A4**, 507.

Index